British Atlantic, American Frontier

This book was published in association with
the Center for American Places,
Santa Fe, New Mexico, and
Harrisonburg, Virginia
(www.americanplaces.org)

British Atlantic, American Frontier

SPACES OF POWER IN EARLY MODERN BRITISH AMERICA

Stephen J. Hornsby

With cartography by Michael J. Hermann

University Press of New England • HANOVER AND LONDON

Published by University Press of New England, One Court Street, Lebanon, NH 03766
www.upne.com

Printed in the United States of America
5 4 3 2 1

LIBRARY OF CONGRESS CATALOGING-IN-PUBLICATION DATA
Hornsby, Stephen, 1956–
British atlantic, American frontier : spaces of power in early modern British America / Stephen J. Hornsby ; with cartography by Michael J. Hermann.
p. cm.
Includes bibliographical references and index.
ISBN 1–58465–426–0 (alk. paper) — ISBN 1–58465–427–9 (pbk. : alk. paper)
1. North America — Geography. 2. West Indies, British — Geography. 3. North America — History — Colonial period, ca. 1600–1775. 4. Agriculture — North America — History. 5. Agriculture — West Indies, British — History. 6. Great Britain — Colonies — America — History. I. Hermann, Michael, cartographer. II. Title.
E40.5.H67 2004
970'.0971241 — dc22 2004016103

COVER: Detail from *A View of the Pierced Island, a Remarkable Rock in the Gulf of St. Lawrence*, engraving after Hervey Smyth, 1760.

For Thomas and Emma

Contents

Figures

Acknowledgments

It is a great pleasure to acknowledge and thank the numerous people who have provided intellectual stimulus, field-experience, and practical help during the writing of *British Atlantic, American Frontier.* On the intellectual front, I owe much to the work and ideas of Cole Harris. In the early 1980s, he proposed a framework for understanding European settlement in northeastern North America, which he later elaborated on in the first volume of the *Historical Atlas of Canada.* During the same years, I explored elements of that framework in a study examining trans-Atlantic migration, staple industries, and agricultural settlement on Cape Breton Island in the nineteenth century. It was a framework that provided an essential starting point for this broader, more comparative treatment of the Atlantic world. My other intellectual debt is to Elizabeth Mancke. Her writings on the early British empire, particularly those dealing with the relationship between early Canada and the thirteen colonies, have been enormously stimulating, and she has always been generous in sharing her ideas. I am grateful to both Cole Harris and Elizabeth Mancke for encouraging the project and commenting on a first draft.

Yet *British Atlantic, American Frontier* would not have taken shape if I had not taken a position at the Canadian-American Center at the University of Maine. An appointment in Canadian Studies at an American university opened up possibilities for comparative study of Canada and the United States, a long-standing tradition and particular strength of the program at Maine. The colonial period seemed to offer considerable comparative potential, especially as it allowed me to triangulate among Britain, Canada, and the United States. Colleagues in Canadian Studies, particularly Robert Babcock, provided encouragement, and the superb Canadian collection in the university library made research relatively easy. The geographic location of Maine also helped. Quebec, the Maritimes, and southern New England were in easy reach, as was a trans-Atlantic flight. I also now realize that Maine served as a fault line between the two geographic systems of the continent: on the eastern side of Maine, the Penobscot River served as the outermost frontier of the French fur trade in Acadia, while islands in the Gulf of Maine marked the farthest reach of the West Country migratory fishery to the New World; on the western side of Maine, a resident New England fishery and an agricultural empire eventually emerged. On the margins of Canada and the United States, eastern Maine is as good a vantage point as I know from which to look out over the Atlantic as well as the continent.

Nevertheless, I could not have contemplated, let alone written *British Atlantic, American Frontier* without first-hand field experience of the landscapes of eastern

North America. The Canadian Studies program at Maine has supported several field trips through Quebec and Atlantic Canada, where regional scholars, particularly John Mannion and Gordon Handcock in Newfoundland and Ed MacDonald in Prince Edward Island, have been generous in guiding me through their local landscapes. Martyn Bowden of the Department of Geography at Clark University also has been a generous mentor. The founder and inspiration behind the Eastern Historical Geographers Association (EHGA), a loose grouping of geographers and historians, Martyn has helped organize annual meetings at some of the most significant colonial sites along the eastern seaboard, including Halifax, Quebec City, Portsmouth, Newport, Lewes, Alexandria, Richmond, Williamsburg, Charleston, and, most evocatively of all, Codrington College in Barbados. Indeed, it was this conference and series of field trips in Barbados that first opened my eyes to the fundamental similarities between the northern and southern margins of North America. I am further indebted to Martyn Bowden for generously making available his work, largely unpublished, on the landscape of the mercantile city, which underlies a good part of my writing in chapter 5. The meetings of the Vernacular Architecture Forum and its New England chapter, although covering some of the same ground as the EHGA, have also given me a detailed introduction to the colonial architecture of towns such as Portsmouth, Marblehead, Newport, Annapolis, and Charleston.

Numerous people have provided practical assistance. First and foremost, Michael Hermann has turned my sketch maps into a substantial and compelling cartography of the Atlantic world. During the several months that it took to prepare the maps, he was remarkably patient, immensely professional, and technically superb. He has been a great pleasure to work with. George F. Thompson and Randall B. Jones at the Center for American Places and Phyllis Deutsch and her staff at the University Press of New England also have been excellent to work with during the process of bringing the manuscript to publication. Betsy Hedler was a great help in getting the many permissions needed to reproduce the illustrations.

For the many historical maps and pictures, I would like to thank the staff of the British Library; the British Museum; the Victoria & Albert Museum; the National Maritime Museum, London; English Heritage, National Monuments Record; Bristol Museums and Art Gallery; the Dorset Natural History and Archaeological Society, Dorset County Museum; the Maine State Archives; Special Collections, Fogler Library, University of Maine; the Burton Historical Collection, Detroit Public Library; the Clements Library, University of Michigan; the Maryland Historical Society; Historic St. Mary's City, Maryland; the Museum of Southern Decorative Arts, Winston-Salem, North Carolina; the Library of Congress, Prints and Photographs Division; the National Gallery of Art, Washington, D.C.; the Provincial Archives of Newfoundland and Labrador; the Toronto Public Library; the National Archives of Canada, Ottawa; and the National Gallery of Canada, Ottawa.

I am grateful to Sarah Hornsby for supplying the images of Hearne pictures that have passed through the salerooms of Christie's in London, and to Steve Bicknell for photographing DesBarres's chart of Narragansett Bay. I would also like to thank the Canadian Embassy in Washington, D.C., for providing a Canadian Studies research grant, and to acknowledge the support received from the Canadian-American Center's endowment fund, which defrayed the cost of the illustrations.

Finally, it is a delight to dedicate this book to my two children, Thomas and Emma. Although they presently and understandably find their father's work less than interesting, I hope they may find, later in life, that this book helps them make sense of their American, British, and Canadian world.

British Atlantic, American Frontier

Introduction

> Is there anything more dangerous to scientific inquiry than the temptation to regard all things as natural?
>
> —Marc Bloch[1]

The shift toward an Atlantic paradigm in colonial American history offers geographers and historians exciting opportunities to reconceptualize early modern North America. Building on a growing interest in transnational approaches to the past, an Atlantic paradigm allows us to shed the boundaries of modern nation-states, which have served too long as the contexts for studying colonial North America, and replace them with the historically more appropriate frontiers of European colonial empires.[2] As these empires once covered much of North America, the move toward an Atlantic perspective allows us to take work done within national or regional containers—Canada, the United States, the Caribbean—and reposition it within a more comparative context. At the same time, the many studies of the demographic, economic, and social history and geography of early modern North America can be connected to imperial history and that academic castaway, maritime history, and used to re-examine the geopolitical issues of imperial formation and disintegration, as well as the rise of new North American nation-states. Of course, American historical geographer Donald Meinig, in *Atlantic America*, has seized some of these opportunities and given us a panoramic view of the early Atlantic world out of which emerged the United States, but there still seems room, particularly in such a new and expanding field of scholarship as Atlantic history, for a further geographical reconceptualization of early modern North America.[3]

Unlike Meinig's sweeping survey of the Atlantic basin, *British Atlantic, American Frontier* employs an Atlantic perspective to reconsider the geography of British America during the early modern period. The study focuses on that great arc of territory stretching from Hudson Bay to Barbados that came under British rule at various times from the late sixteenth to the late eighteenth centuries. Although a recent overview of Atlantic history has interpreted the Atlantic world as "one vast unit," uniting Western Europe, West Africa, and the Americas in an immense system of economic and demographic flows and cultural and political relationships, this book argues that the early British empire that once encompassed a good part of this Atlantic world was far from being a uniform space or an homogeneous system.[4] An Atlantic perspective that focuses on connections and flows, by its very nature, may highlight continuities and similarities rather than discontinuities and differences. The "entire Atlantic world," like our world today, may not have had a "unitary character." In fact, there were many Atlantic worlds, created by men, women, Blacks, whites, natives, Spanish, Portuguese, French, Dutch, and British. Moreover, even within the British Atlantic, there were major differences between areas under metropolitan influence and those under colonial control. At some point, British Atlantic history has to confront the brute fact that the British Atlantic world was rent in two by the revolt and independence of the thirteen colonies. This raises not only the long-standing question of American "exceptionalism" or difference, but also the question of why other parts of the British Atlantic world remained within the empire.[5] To begin to answer such questions, we need to step back from the shores of the Atlantic and consider a set of ideas about the European shaping of North America.

Any consideration of the European settlement of North America should begin with the foundational ideas of American historian Frederick Jackson Turner and Canadian political economist Harold Adams Innis. Both understood that European immigrant societies, in different ways, had interacted with the environment of the continent, and, in the process, had created two distinct North American societies. In the southern half of the continent, Turner argued that a frontier of agricultural settlement explained the development of the United States; in the northern half, Innis demonstrated that staple trades in fish and fur led to the emergence of Canada.[6] Writing just as the American frontier was officially closing in the early 1890s, Turner argued that the nature of American society—its democratic values, egalitarianism, and practical ingenuity—was explained by the availability of free land on the frontier. Exactly how the frontier shaped American society never was demonstrated adequately and has been the subject of much debate and controversy ever since, but Turner's insight that the early history of the United States somehow was enmeshed in land, territory, and an agricultural frontier was fundamental. Writing just as the influence of Turner's frontier thesis was reaching its height in the 1930s, Innis argued that European settlement in Canada had more to do with the exploitation of staples such as fish, fur, and lumber, than with freely available agricultural land. European merchants used water transport around the coasts and along the rivers to reach into the northern half of the continent to tap the immense natural resources of the interior. For Innis, the development of early Canada owed nothing to an agricultural frontier but depended on its own unique mix of staple trades, waterways, and metropolitan influence.

Since they were first put forward, these two interpretations have stimulated an immensely suggestive body of scholarship. Among work done by historians and geographers, four themes stand out. First, there is general agreement that the frontier and staples theses outline major spatial components of the early North American economy. In some parts of the continent and the Caribbean, there existed a market-driven economy of commercial staple trades comprising sugar, tobacco, wheat, fish, rice, and indigo; in other areas, particularly in poorer agricultural regions such as New England, there was a "Malthusian" or population-driven agricultural frontier.[7] Economists have shown that these two types of economy generated linkages to other sectors, which, in turn, led to economic diversification and the creation of more mature and integrated economies.[8] Although Turner and Innis did not consider the third major economic setting that existed in early North America, namely commercial towns, staple theory has been used as a general model of urban growth.[9] Geographers and historians have argued that towns in North America developed because of trade in staple products, and that linkages from particular staples facilitated the development of urban manufacturing and services.[10]

Closely related to this economic interpretation, a second theme focuses on land and staples as media of social transformation. Generalizing from the Canadian context, Canadian historical geographer Cole Harris has put forward a persuasive model of European settlement in North America.[11] According to Harris, there were three types of European colonial settlement, each with its own particular social and economic characteristics. In areas that had commercial resources, European capital quickly moved in, acquired land, imported labor and technology, and established specialized areas of production tied by long-distance transportation to overseas markets. Societies soon became stratified, providing opportunity for merchants and planters, but not for indentured servants and slaves. In settings where no ready marketable product was found, there was a weak market for land and more opportunity for labor.[12] In such places, European immigrants had a chance to acquire land and the means to achieve a standard of living higher than most peasants enjoyed in Europe. Compared to rural societies in Europe or those associated with staple trades in the Caribbean and North America, these New World agrarian societies were remarkably egalitarian. There was little support for commercial capital or aristocratic wealth. Only when markets improved or population increased did the demand for land grow and property become more valuable. In such circumstances, rural societies became increasingly differentiated between those people with land and those without. In the third type of setting—commercial towns established along the coasts—stratification was more common because almost all towns were ports tied to the trading circuits of the Atlantic world. In these urban societies, a clear demarcation existed among merchants, professionals, artisans, laborers, and slaves. Although the massive mercantile and aristocratic wealth of the major European cities was absent, the towns in North America and the Caribbean came closest of any of the three types of settlement to replicating social conditions in the Old World.

A third theme considers the relationship between the metropolis and the frontier, or what Innis called the center and the margin. In his poetic, almost mystical way, Turner saw the distinctiveness of the United States not in its European origins but in its emergence from the crucible of the American frontier. Innis, on the other hand, traced the development of early Canada directly back across the Atlantic to its European roots. For Turner, the American frontier was all important; for Innis, the connections to Europe were most significant. Nearly fifty years ago, Canadian historian Maurice Careless suggested that these two, seemingly opposite points of view could be reconciled.[13] In an influential article, Careless argued that the frontier was simply the outer edge of metropolitan influence. The central forces shaping Canada were, at first, the European cities of Paris and London, and then, later, the Canadian cities of Montreal and Toronto. Far from being "independent and self-reliant," the Canadian frontier was dependent, constantly requiring "metropolitan aid and control."[14] More recently, American historian William Cronon, drawing heavily on the work of Innis and Careless, has suggested that Turner should be read backwards.[15] In the American West, Cronon argues, the lines of economic control also flowed from city to countryside.

Nevertheless, the relationship between metropolis and frontier depended to a great extent on the nature of the link. Innis, more so than Turner, knew that economic and political power was projected across space through systems of transportation and communication. Indeed, Innis's later work on empire and communications, which seemed on the face of it so unrelated to Canadian economic history, grew out of his earlier studies of staple trades and transportation systems.[16] Recognizing that preindustrial societies had enormous difficulties projecting power over land, Innis realized that power could be projected very effectively over water. Seaborne power allowed the expansion of Europe across the world's oceans, while the waterways of the St. Lawrence–Great Lakes system provided a route into the heart of North America.[17] With the transition to industrial technology, Innis further recognized that the railroad transformed the spatial reach of power by providing, for the first time, an effective means of overland transportation and communication.

Innis's insights are essential for understanding Turner and the relationship between frontier and metropolis. In early America, the westward-moving frontier, separated from seaboard towns by hundreds of miles of difficult, overland travel, was far more detached from metropolitan influence than the comparable frontier in Canada. As Careless knew, the Canadian frontier was the outer rim of the European metropolis simply because a relatively efficient system of waterborne communications connected them. Cronon, too, recognized that the extensive railroad net across the American West had transformed the spatial reach of power. In that specific context, Turner can be read backwards. Yet in preindustrial America, Turner's insistence on the independence of the frontier carries some weight. Across the eastern half of the continent, there was a clear division in the spatial reach of power, with much more metropolitan oversight in the northern half and far more frontier autonomy in the southern. Relations between frontier and metropolis have to be related to technologies of transportation and communication.

A fourth theme derives from Innis's work and concerns the geographic context of power. Innis realized, most obviously from his own extensive canoeing expeditions in the Canadian north as well as from his monumental studies of staple trades, that different systems of economic power created their own distinctive human geographies. In his examinations of the cod fisheries and the fur trade, Innis argued that the successful reach of European power into the northern half of the continent depended on water transportation, first, in the fishing vessels of the Newfoundland fishery; second, in the canoes traversing the St. Lawrence–Great Lakes waterways. Each economic system created its own spatial pattern.[18] The fishery had virtually no landward penetration in Newfoundland, remaining confined to the coastal margin. Similarly, the fur trade depended on a network of canoe routes that connected distant fur posts to points of trans-Atlantic shipment on the seaboard. In both staples, European power was not territorial but linear. Innis further recognized that power projected linearly was inherently vulnerable, particularly when confronted by territorial force. In the face of an advancing American settlement frontier, the boundaries of the fur trade, which had once encompassed a good part of the continent, retreated. Over the course of two hundred years, a European commercial staple gave way to American agricultural settlement: what Innis so memorably called the "hostility of beaver and plough."[19]

Turner and Innis identified two fundamentally different geographical systems of power operating in the Atlantic world during the early modern era.[20] Turner delineated an expansive continental system of colonial agricultural settlement; Innis described an equally dynamic oceanic system of metropolitan-dominated staple trades. Within each of these systems, there existed what sociologist Michael Mann has called "multiple overlapping and intersecting sociospatial networks of power."[21] According to Mann, the most important networks were economic, political, military, and ideological, although it seems appropriate, given the importance of population growth in North America, to also include demographic power. These various networks interlocked together and interacted with the environment to create the two dominant spatial patterns of European power in North America.[22]

Drawing on these various ideas, I argue in *British Atlantic, American Frontier* that early modern British America was divided into two principal kinds of spaces. The first was an oceanically oriented periphery or marine empire comprised of Newfoundland, the West Indies, and Hudson Bay, from which the British derived enormous commercial wealth from fish, sugar, and furs. This space I call the British Atlantic. The second was a territorially oriented periphery or settler empire that extended along the eastern seaboard of North America from Maine to Georgia. This space I call the American Frontier. Yet lying between these two spaces was a third area that Turner and Innis did not consider. This intermediate space comprised the continental staples and port towns along the eastern seaboard that had links to both the continental interior and the world of Atlantic trade. These continental staples included the New England fishery, the Chesapeake tobacco plantations, and the Carolina rice plantations. In many ways, this intermediate space of staple production and urban development served as the point of connection, articulation, and friction between the larger oceanic and continental spaces.

All three spaces were shaped by distinctive geographical configurations of power. In the Atlantic arena, Britain created a seaborne empire dominated by staple trades, circulation of labor, naval force, and metropolitan authority. In the continental arena, American colonists developed a territorial empire based on agricultural settlement, population growth, and local autonomy. In the intermediate space lying along the eastern seaboard, American planters and merchants developed territories of staple production marked by hierarchical societies, Atlantic trade, and considerable colonial control. The expansion of these various spaces led to friction and conflict, and, eventually, the sundering of British America.

Of these three spaces, the English commercial empire of the Atlantic emerged first (chapter 1). After England won control over parts of the Atlantic from Spain in the late sixteenth century, London and West Country merchants launched a commercial outreach to the flanks of North America. In the late sixteenth and early seventeenth centuries, this outreach was to the cod fishery in Newfoundland and the sugar islands in the Caribbean; in the late seventeenth century, it was extended to the fur trade through Hudson Bay (chapter 2). All three areas were nodes of production or trade tied to the metropolis by circuits of trans-Atlantic maritime commerce that were increasingly protected by British naval power. These three areas were economically so specialized that there were few or no alternatives to their respective staples. In Newfoundland and Hudson Bay, scarcely any potential existed for an agricultural frontier; in the West Indies, land was so valuable that it was more economical to import provisions than to produce them locally. With no backcountry, elites had no local competition for power, and labor had no alternative occupation. In such circumstances, metropolitan capital reigned supreme.

Along the American eastern seaboard, two different spaces emerged. Although metropolitan commercial capital invested in the staple economies of New England, the Chesapeake, and the Carolina low country, this outreach was soon disrupted by the continent's dramatically different terms of land and labor (chapter 3). As metropolitan capital withdrew, colonial elites of merchants and planters emerged. Over time, these elites developed significant territories of staple production along the eastern seaboard. Meanwhile, a massive agricultural frontier, created more by flows of labor than capital, wrapped itself around these staple territories and pushed west (chapter 4). As this frontier expanded into the interior, the staple areas increasingly had backcountries and alternative economic opportunities. Janus-faced, American elites looked back across the Atlantic to Britain for commercial credit and forward across the Piedmont to the frontier for speculative profit. The New England fishery and the Mid-Atlantic grain trade also supported urban growth along the seaboard, creating the metropoli of Boston, New York, and Philadelphia, further settings of colonial control (chapter 5).

Eventually, the expansion of the American land-based economic and demographic system led to conflict with the other great metropolitan power on the continent, the French (chapter 6). Like the British, the French had extended a seaborne commercial empire across the Atlantic to Newfoundland, and then, in the form of the fur trade, reached up the St. Lawrence–Great Lakes corridor and into the Mississippi drainage basin. The great clash between these two geographical systems—an expanding linear French commercial system based on waterways and the territorial advance of an American agricultural frontier based on land—took place in the upper reaches of the Ohio Valley. The French and Indian War (1754–1763) was the outcome. In the end, British sea power overwhelmed the French, and the vulnerable commercial system of New France was destroyed at Quebec. Yet the great irony of the war was that the British acquired a waterborne empire from the French rather than another territorial base on the continent. The fit between the metropolitan commerce of the British Atlantic (Hudson Bay, Newfoundland, the West Indies) and the St. Lawrence system was much closer than that between the British Atlantic and the American colonies. Conflict between the two systems soon developed. Given the relative strengths of sea power and land power, Britain eventually lost control over the seaboard colonies, and a new nation-state emerged. In sum, the fracturing of empire was along structural faults embedded in the continent's evolving human geography.

Chapter 1

Creating an English Atlantic, 1480–1630

On the eve of the first English voyages into the western Atlantic in the late fifteenth century, England lay physically and economically on the outer rim of a European world centered on the Mediterranean. The country's principal export of cloth was shipped largely through London to markets in the Low Countries, while lesser trade flows passed through the outports to markets in the Baltic and the Iberian peninsula. Few English vessels ventured past the Straits of Gibraltar into the Mediterranean Sea, or south along the Moroccan coast. Around the British Isles, the English had a variety of coasting trades, as well as fisheries in the North Sea, the Western Approaches, and more distant waters around Iceland. Overall, England's maritime reach was limited, confined to the coastal periphery of western Europe.

By the early seventeenth century, England's relative geographic position in the European world had changed completely. In the wake of English and Dutch voyages to the Americas and Indies, the economic center of gravity in Europe had shifted northwards, away from the Mediterranean toward England and the Low Countries. From being on the periphery of western Europe, England was increasingly at the center of a European world economy that encompassed the Atlantic and Indian oceans. English shipping ventured west to the Caribbean, the eastern seaboard of North America, and the western Arctic; south to the coasts of Africa; and east to Russia, the Levant, and South Asia. The traditional English trade in cloth had been supplemented by New World trades in codfish and tobacco, as well as the Indies' trades in silks and spices. Along the Atlantic front of North America, the English had mapped coastlines, named prominent physical features, and gained some knowledge of the flora, fauna, and native peoples. The first permanent English settlements had been made along the eastern edge of the continent and in the Caribbean. In little more than a century, the geographic horizons of the English, like those of many Europeans along the Atlantic coast, had enlarged enormously. The old maritime world of the North Sea, English Channel, and Baltic had expanded to encompass the new world of the Atlantic.

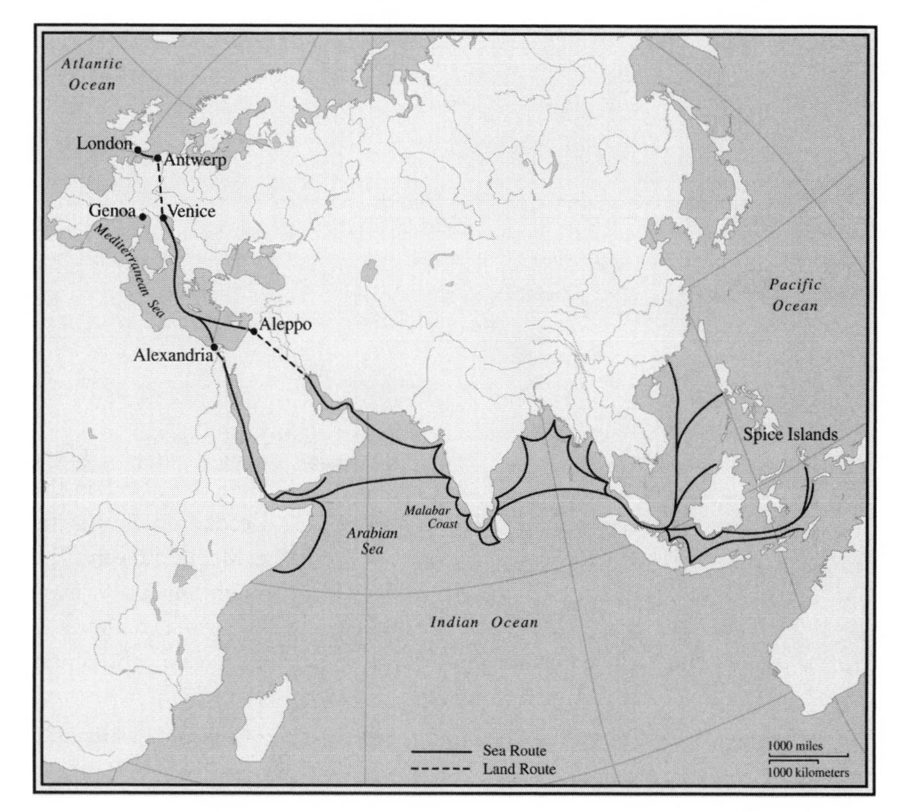

FIGURE 1.1. Spice routes from Asia to Europe. After Kernial Singh Sandhu and Paul Wheatley, eds., *Melaka*, vol. II (Kuala Lumpur: Oxford University Press, 1983), 502.

"The Indies of England"

In the late fifteenth century, Venice and its great Mediterranean empire dominated the European economy. From the time of Marco Polo in the late thirteenth century, Venice had served as the major entrepôt or clearinghouse between the trade circuits of Europe and Asia.[1] Through Alexandria and Aleppo, the city's merchants imported high value goods—silks, spices, perfumes, dyestuffs, and precious stones—from Asia, selling them profitably to merchants from all over Europe (figure 1.1). With command of this great long-distance trade, Venice had grown rich and powerful, fending off competing city-states such as Genoa, and becoming the center of the European economy. Yet Spanish and Portuguese attempts to break Venetian control of the spice trade by finding their own way to the Indies began to bear fruit in the 1490s. Although Columbus did not find Asia, he laid the basis for a glittering Spanish empire in the Americas. Meanwhile, Portuguese probing along the west coast of Africa eventually led to the break out into the Indian Ocean in 1498 and the trading for pepper along India's Malabar Coast. The Venetian monopoly on the spice trade had been broken. By the early 1500s, the Iberian powers, through their conquest of the Atlantic and Indian oceans, had

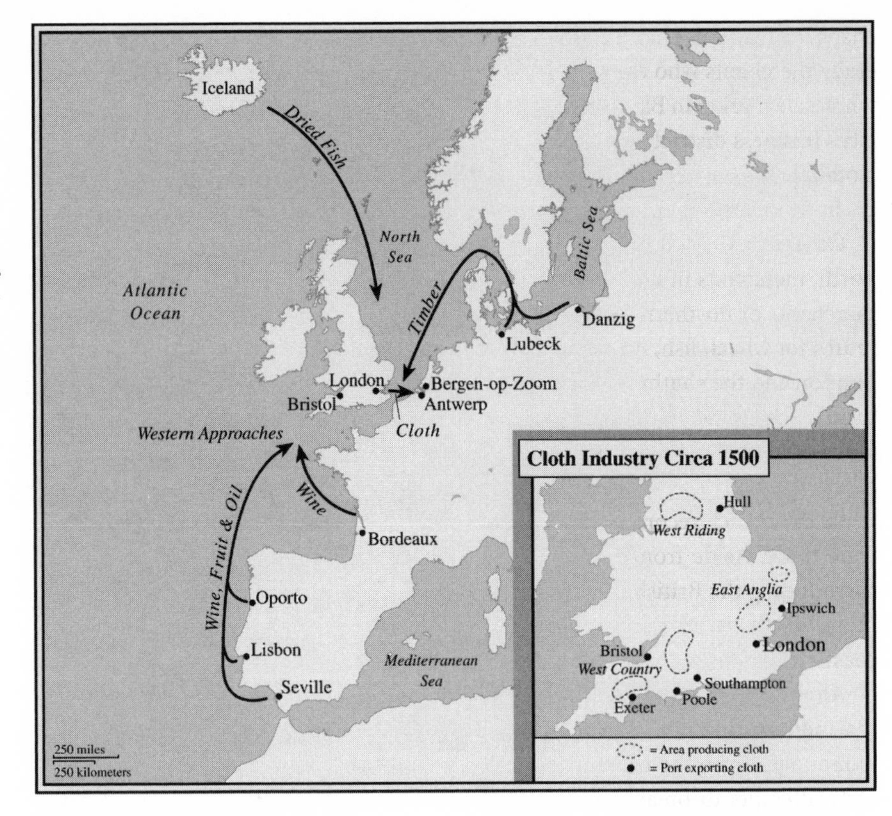

ng-
stry
and overseas trade in the early sixteenth century. After Peter J. Bowden, *The Wool Trade in Tudor and Stuart England* (London: Macmillan & Co., 1962), 46; E. M. Carus-Wilson, *Medieval Merchant Venturers* (London: Methuen, 1967), xxxii.

opened up an entirely new trade circuit to the Americas and interloped into the old Venetian trade to Asia. The wealth of these trade circuits began to pour into western Europe, much of it passing through Antwerp, the gateway city for the markets of northern Europe.[2]

The English tapped into this Iberian-controlled commerce by exchanging their cloth in Antwerp for high value goods from the Indies (figure 1.2). From about 1450, the cloth trade had surpassed wool as England's principal export, and by the early 1600s made up more than 75 percent, perhaps even 90 percent, of the country's exports. For contemporaries, the cloth trade was "the flower of the king's crown, the dowry of the kingdom, the chief revenue of the king . . . the gold of our Orphir, the milk and honey of our Canaan, the Indies of England."[3] Much of the cloth was produced in the West Country, East Anglia, and the West Riding of Yorkshire. These areas had plentiful supplies of water, needed for cleaning the wool and powering the fulling mills, as well as large part-time farm labor forces available for work. Local clothiers organized the industry on the putting-out system, supplying raw wool (and sometimes the loom) to textile workers, and then purchasing the finished cloth. The clothiers, in turn, sold the cloth to merchants in the ports. Much of the cloth trade flowed through London, with lesser amounts shipped through Southampton, Exeter, and Bristol. The principal London merchants handling the cloth trade formed the Merchants Adventurers Company (literally merchants who engaged in foreign trade); they purchased cloth at the great wholesale market in Blackwell Hall in Basinghall Street, situated in the heart of the city's business district, and then forwarded it to their agents resident at the major cloth fairs in Antwerp and Bergen-op-Zoom.[4] In return, the Merchants Adventurers imported the spices, silks, and precious stones of the Indies.

On either side of this main axis of trade were lesser shipping routes. To the north, merchants in London and the main east coast ports traded with Hanseatic merchants of northern Germany, the Baltic, and Scandinavia, exporting cloth in return for wheat, fish, furs, timber, and wood products; to the south, merchants in London and the southwest coast ports traded to southern France, Spain, and Portugal, exchanging cloth for wine, fruit, olive oil, and salt. Both these trade routes, however, were subject to periodic disruption. By the late fifteenth century, much of the Baltic trade was in the hands of Hanseatic merchants resident in London, while the French reconquest of Gascony displaced the English from the profitable wine trade. Aside from these maritime trades, the English exploited the waters surrounding the British Isles and in the North Atlantic, catching cod and herring in the North Sea, sardines and herring in the Western Approaches, and cod off the coast of Iceland.[5]

After the Portuguese and Spanish opened trade to the Indies and the Americas, merchants in London and the outports explored ways of breaking the Iberian monopoly. Given the relative economic and military weakness of England, these early attempts to break into the Portuguese and Spanish empires were modest and largely unsuccessful. In 1480 and 1481, at the time that the Portuguese were exploring the coast of West Africa, Bristol merchants financed two voyages into the Atlantic in search of the legendary island of Brazil, thought to be somewhere west of Ireland. Both expeditions were failures, and it was not until Columbus's presumed discovery of the Indies that Bristol merchants again backed exploration into the Atlantic. In 1497, the Italian navigator John Cabot, patronized by Henry VII and financed by Bristol merchants, sailed from Bristol in search of the Indies.[6] Cabot coasted the northeastern foreland of North America, somewhere between Cape Bauld (the northern tip of Newfoundland) and Cape Breton Island, finding a rocky coast, spruce forest, and plentiful codfish, but hardly the fabled riches of the east. Although Cabot had found one of the greatest fisheries in the world, the English, their domestic markets well supplied from the Icelandic and North Sea cod fisheries and unable to sell fish in the protected markets of France and Spain, did not capitalize on his discovery. Instead, other European fishermen—principally Bretons, Basques, and Portuguese—seized on the knowledge of Cabot's voyage and swung westward to exploit the new fisheries.[7]

For much of the early sixteenth century, the English showed only intermittent interest in the New World. After Verrazano's coasting of the eastern seaboard of North America in 1524 revealed that a continuous land mass lay between the Spanish possessions in the Caribbean and Cabot's discoveries in Newfoundland, the

English made an attempt to find a sea route around the continent to the Indies. In 1527, John Rut, a mariner employed in bringing the king's wine from Bordeaux to London, was commissioned by Henry VIII to sail across the Atlantic and search for a northwest passage. After making landfall in Labrador, Rut sailed southward along the entire eastern seaboard, appearing, much to the surprise of the Spanish, at Santo Domingo. Although Rut gained valuable information about the east coast of North America, he had not found a passage to the Indies, and the king lost interest in further trans-Atlantic ventures.

Meanwhile, English merchants tried to break into the Atlantic commerce of Spain and Portugal. Not permitted to trade with the Spanish empire from England, a handful of English merchants moved to Seville and traded to the Caribbean during the 1520s and early 1530s, only withdrawing in the face of a hostile Spanish Inquisition. Between 1530 and 1542, merchants from Plymouth and Southampton broke into the Portuguese Brazil trade, and also made a foray along the Guinea Coast of West Africa, but no long-term trade resulted. During a period of tension between England and Spain in the mid-1540s, English privateers preyed on Spanish and Flemish ships. Yet these various nibbles around the edges of the New World and the Atlantic trades of the Spanish and Portuguese amounted to little. The English outports, rather than London, were involved, and only a few vessels were devoted to these ventures.[8] At mid-century, English overseas trade depended overwhelmingly on cloth shipments from London to Antwerp, a city under Spanish control. England remained an economic satellite of the continent, its Atlantic world hardly larger than in 1480.

The Significance of Southern Europe for English Enterprise in North America

Yet in the late sixteenth century, English interests overseas broadened considerably. During those years, the English ranged around the Atlantic and the Mediterranean, pushing trade into Russia, the Levant, and into West Africa, as well as exploring the Caribbean, the eastern seaboard of North America, and the western Arctic (figure 1.3). The traditional explanation for this expansion has been that economic difficulties in the cloth trade in the early 1550s prompted English merchants to search for new cloth markets outside northern Europe. It now seems, however, that the depression in the cloth trade was less important than previously thought, and that much of the decline in cloth exports was due to the expulsion of Hanseatic merchants from London in 1552. With the removal of the Hanse, the Merchants Adventurers were left in complete control of the cloth trade, and actually increased their cloth exports to the continent during the late sixteenth century. Instead of looking for new export markets, English merchants made even greater efforts to bypass the middlemen in Antwerp and import goods directly from Southern Europe, the Levant, and the Indies.[9]

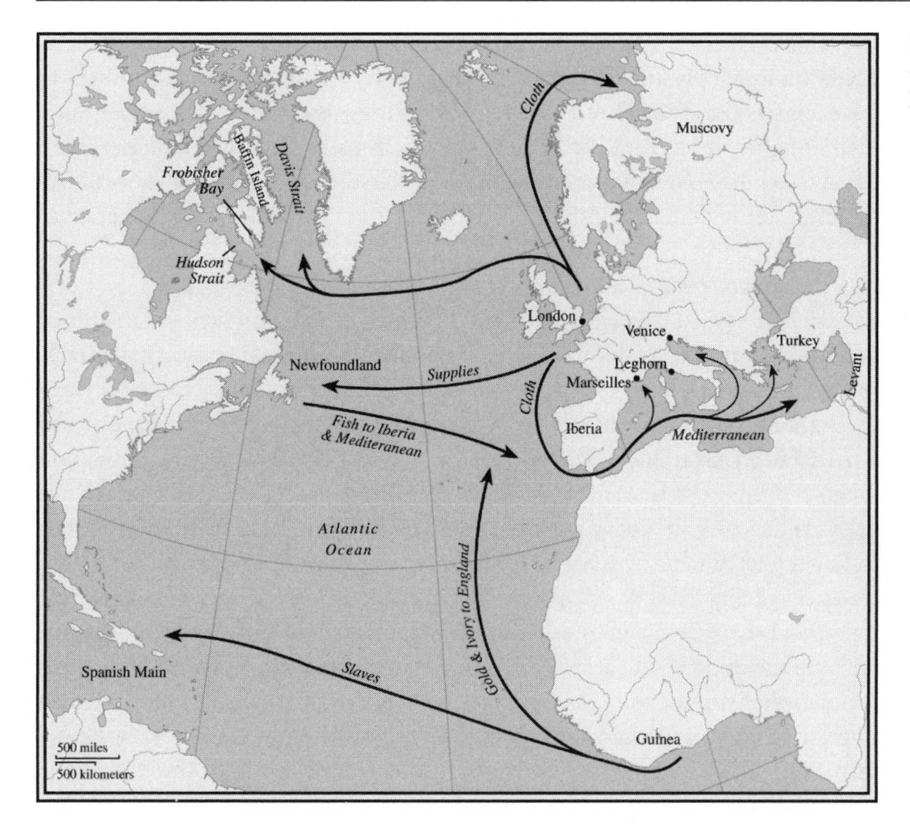

FIGURE 1.3. English overseas trade in the late sixteenth century.

The beginning of this Elizabethan expansion can be traced to attempts by London merchants to find a northeastern route to the Indies. In 1553, London merchants, as well as the Crown, backed an expedition around the North Cape of Norway to Russia with the intention of opening an overland route to Asia. With the success of this venture, the Muscovy Company was formed in 1555, financing several expeditions to Persia, as well as trading with Russia, during the late sixteenth century. Further impetus to open new trade routes came after the Spanish closed the port of Antwerp in 1569. This action not only forced the Merchants Adventurers to move their cloth market to northern Germany and then to the Netherlands, but also compelled English merchants to look elsewhere for their Indies goods. The two major markets that the English concentrated on were the Mediterranean and Spain.[10] In 1571, the Venetians scored a Pyrrhic victory over the Turks at the Battle of Lepanto, but were left too exhausted to protect their maritime empire. Taking advantage of Venetian weakness, the English moved quickly into the Mediterranean. One thrust was into the western basin, where English merchants traded cloth and dried fish in Marseilles and Leghorn (Livorno), a free port that became the main entrepôt in that part of the Mediterranean;[11] another thrust was into the eastern basin, where the English sold cloth, tin, and lead to Venice, Turkey, and the Levant. After the Spanish lifted an embargo on trade with the English in 1573, London merchants formed the Spanish Company and began trading to Spain, importing products from the Spanish empire, as well as wine, olives, and dried fruit from Spain itself.[12] In 1580–1581, Spanish Company merchants, as well as Muscovy Company merchants, who wanted a shorter route to the Near East, joined forces and formed the Turkey Company to handle all English trade with that country. Two years later, the Venice Company was created. In 1592, the Turkey and Venice companies joined to form the Levant Company with a monopoly of all English trade in the eastern Mediterranean and Near East.[13] Outside the cloth trade with northern Europe, the Levant Company managed perhaps the most profitable English overseas trade in the late sixteenth and early seventeenth centuries.

The shift in English trade from Antwerp to Spain and the Mediterranean had a crucial bearing on English involvement in the New World.[14] Although they had long supplied cloth to Spanish and Mediterranean markets, English merchants needed other items to help pay for the Indies goods formerly acquired in Antwerp. In looking for these new trade goods, English merchants turned westward to the resources of the New World, particularly the cod fishery of Newfoundland. Up until the early 1570s, France and Spain had dominated the Newfoundland fishery, but with the opening of Iberian and Mediterranean markets, the English had the stimulus they needed to exploit the fishery of the western Atlantic. The first great English colonial trade had begun.[15] According to Anthony Parkhurst, a Bristol merchant engaged in the fishery, thirty English vessels fished at Newfoundland in the early 1570s and from then on numbers increased rapidly.[16] By 1604, perhaps 150 vessels were making the voyage. Unlike the Icelandic fishery, which was mainly in the hands of fishermen from East Coast ports and supplied London and the domestic market, the Newfoundland fishery was developed by West Country merchants who had long-standing trade ties with the Iberian peninsula. By the mid-1580s, if not earlier, a triangular trade linked the West Country, the Newfoundland fishery, and the markets in Iberia and the Mediterranean.[17] So important was the fishery to the Spanish trade that Newfoundland was later "accompted [accounted] the India to the West of England."[18]

The significance of the Newfoundland fishery for English maritime power was immense.[19] Unlike the Merchants Adventurers' short-distance, cross-Channel cloth trade, the fishery was England's first long-distance, trans-Atlantic enterprise, and needed numerous vessels of large tonnage crewed by experienced seamen. During the 1570s and 1580s, the tonnage of English vessels began to increase significantly, most likely due to the demands of the fishery as well as the Levant trade.[20] As armed merchantmen were pressed into service as privateers and naval warships during the sixteenth and early seventeenth centuries, the growth of the Newfoundland fishing fleet strengthened English sea power. Moreover, the fishery was seen as "a nursery of seamen" who could be impressed into the navy in times of war.

The Newfoundland fishery also had a major impact on the English mercantile establishment. Until the rise of the fishery, the trades to northern Europe, the Mediterranean, and the Levant were largely in the hands of merchants, resident in London, who belonged to chartered companies with regulated trading monopolies. The fishery, however, was a resource open to all comers, and one that individual merchants and ship captains could exploit. As a result, West Country merchants moved quickly into production, organizing fishing enterprises, as well as trading fish to Iberian and Mediterranean markets.[21] Given the importance of dried fish to the Spanish trade, London merchants were themselves quick to enter the fishery, most likely financing West Country fishing voyages and certainly contracting to buy fish in Newfoundland, where it was loaded onto "sack" ships for markets in southern Europe.[22] For these London merchants, the dried fish trade was their first involvement in New World enterprise. Yet despite their financial resources, London merchants could not monopolize the fishery in the same way as the cloth or Levant trades. In finding their own "Indies," West Country merchants were beginning to challenge London's trading dominance.[23]

Compared to the exploitation of the Newfoundland fishery and the profitable trades with southern Europe and the Levant, English ventures to other parts of the Atlantic rim in the late sixteenth century were much less successful. In the 1550s, the English exploited a temporary weakness of the Portuguese and established a small trade to Morocco, exchanging cloth and linen for sugar, molasses, and dates. Farther south, English merchants launched several trading and raiding expeditions along the coast of Guinea, bringing back cargoes of gold, ivory, pepper, and redwood, as well as slaves for sale to the Spanish in the Caribbean. But interloping into the Spanish Main carried risks as well as rewards, and eventually became too expensive and unreliable for London merchants. After these early efforts, the Caribbean was left to freebooters—Hawkins, Drake, and others from the West Country—who soon switched from slaving to privateering and raiding.[24]

The great prize of direct trade with the Indies still eluded the English. Not yet powerful enough to challenge the Portuguese in the Indian Ocean, the English looked for an alternative way to the riches of the East and renewed their search for a sea route around the northern flank of North America. In 1576, merchants in the Muscovy Company backed Martin Frobisher, a veteran sea captain with experience in the Guinea trade, to find a northwest passage. Frobisher sailed into Arctic waters as far as southeastern Baffin Island, explored the bay that now bears his name, and returned with rock thought to contain gold (figure 1.4). The lure of gold led to Frobisher's two subsequent voyages in 1577 and 1578, which yielded much more worthless rock and an awareness of a strait (Hudson Strait) between Baffin Island and the mainland.[25] A financial disaster for the backers, the search for a northwest passage resumed only in the mid-1580s, when John Davis, a Devon mariner, won support for the venture from a mix of royal courtiers and merchants in London and the West Country. In the course of three voyages (1585, 1586, 1587), Davis explored the strait (Davis Strait) between Greenland and Baffin Island, saw "a great sea" (Baffin Bay) to the north, and confirmed Frobisher's discovery of the

FIGURE 1.4. *Englishmen in a Skirmish with Eskimos*, by John White, 1577. © Copyright The British Museum.

entrance to Hudson Strait.[26] The Indies, however, still lay tantalizingly beyond the ice, mountains, and fog.

The War for the Atlantic, 1585–1604

In his great portrait of the Mediterranean world in the sixteenth century, French historian Fernand Braudel saw the years 1578 to 1583 as "the turning point of the century."[27] During those years, Philip II of Spain disengaged from his long-running wars against the Ottoman Turks in the eastern Mediterranean and shifted his attention to the Atlantic. In 1580, he invaded Portugal and assumed the royal crown, thereby uniting the two Iberian powers and their global empires. Philip II was now master of the greatest empire that the world had ever seen, an empire that stretched east to India, Malaya, China, and the Philippines, and west to the Caribbean, Mexico, Peru, and Brazil. As the Spanish were proud to point out, it was the first empire on which the sun never set.[28] Nevertheless, the English were beginning to cast some shadows. On the continental front, Elizabeth I, anxious to preserve the balance of power in Europe, supported Dutch Protestants in their revolt against their Spanish rulers; on the Atlantic front, English freebooters, keen to break the Iberian monopoly on the New World, attacked Spanish shipping. Relations between the two countries deteriorated during the early 1580s, and eventually broke out into open warfare in 1585. Although few realized it at the time, the outbreak of the Anglo-Spanish war marked the beginning of the "great battle for control of the Atlantic and world domination."[29]

For the Spanish, an invasion of England would win the war and solve the English problem once and for all; for the English, an invasion of Spain was out of the question, but the war did offer immense opportunities to roll back the edges of the Spanish empire overseas. Over the course of the conflict, both strategies were pursued. In 1588, Philip II launched a great invasion armada against England, only for it to fail in the face of English naval attacks and westerly gales.[30] Two more armadas were sent against England, but they too were disrupted by bad weather. Meanwhile, the English, through a combination of naval expeditions and privateering, attacked the peripheries of the Spanish Atlantic empire—Newfoundland and the Indies route—as well as the Spanish Main, the treasure fleets, and Spain itself. After Philip II instigated the war by embargoing Dutch, German, and English shipping in Iberian ports, Elizabeth I, fearing that she would lose a good part of the country's naval strength when the English fishing fleet in Newfoundland sailed with their catches to Spain and Portugal, dispatched Bernard Drake to Newfoundland to warn the fishermen and attack the Spanish and Portuguese fishing fleets. Drake drove the Spanish and Portuguese off the Avalon peninsula in southeastern Newfoundland, raiding some twenty vessels and capturing about 600 men.[31] At the same time, Sir Francis Drake undertook a long-planned attack on the Spanish Main, sacking Santo Domingo, Cartagena, and St. Augustine (figure 1.5).[32] Further naval expeditions were sent against Corunna in 1589 and Cadiz in 1595.[33]

Even more effective than these organized campaigns was English privateering. At the outbreak of war, many of the merchants who had been engaged in the Spanish trade and the Newfoundland fishery refitted their vessels as privateers.[34] Between 1589 and 1591, nearly half of all English privateers were based in West Country ports, many no doubt using vessels and seamen formerly employed in the Newfoundland fishery.[35] Over the course of the war, English privateers captured at least a thousand Spanish and Portuguese vessels and their cargoes.[36] The "constant stream of booty" into England, totaling as much as £100,000 to £200,000 per year, was roughly equivalent to the pre-war trade with the Iberian peninsula.[37] The small group of merchants in the West Country and London who invested in privateering ventures garnered enormous windfall profits, leaving them well placed to finance new trading ventures after the war.[38] Meanwhile, English privateering, as well as wartime requisitioning of vessels by the Spanish Crown, devastated Spanish and Portuguese shipping. Although Philip II protected the critical treasure *flota* from New Spain, the *carreira da India* and the fishing fleets in Newfoundland suffered greatly. Portuguese carracks sailing to and from India were regularly lost to storm and capture, while the Spanish fishing and whaling fleets, based in the Basque country and the core of the country's merchant marine, were run down.[39]

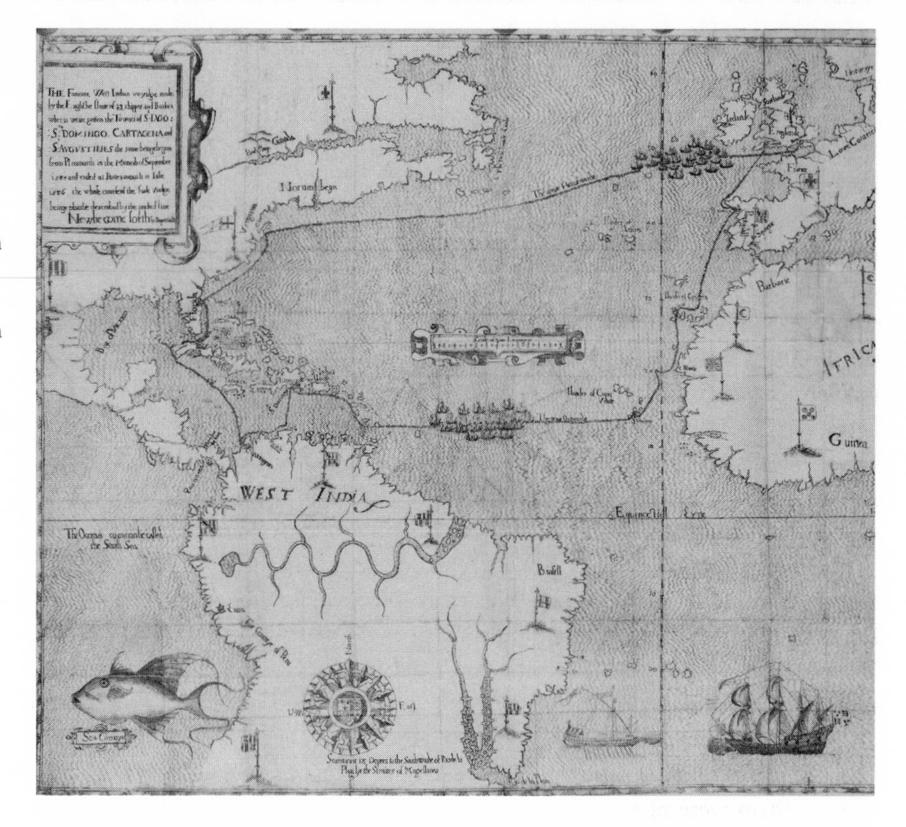

FIGURE 1.5. *The Famous West Indian Voyadge Made by the Englische Flect*, by Baptista Boazio, 1589. This map shows the course of the English fleet around the North Atlantic, the attacks on Spanish possessions in the Caribbean and Florida, and the English settlement at Roanoke in Virginia. By permission of the British Library, Shelfmark G6509.

Taking advantage of Spanish and Portuguese weakness, the English expanded into the Atlantic. During the war years, English merchants concentrated on taking over the Spanish and Portuguese fisheries in Newfoundland, while charted companies attempted to plant colonies in the New World. In the short run, the dislodgement of the Spanish and Portuguese from part of the fishery was the most significant development; according to Harold Innis, "cod from Newfoundland was the lever by which [England] wrested her share of the riches of the New World from Spain."[40] Moreover, exploitation of the fishery by individual merchants rather than by chartered companies set a pattern of commercial development that eventually would be adopted in Virginia and the West Indies in the early seventeenth century. The fishery also placed the English in an important strategic position athwart the great circle route across the Atlantic to North America; it seems inconceivable that later English ventures in Virginia and New England could have been sustained for long if the English had not already cleared the Spanish from the northern sea route to the continent.

The beginnings of this colonization came in the early 1580s. As a direct challenge to Spanish claims to the New World, English soldiers of fortune, as well as younger sons of landed gentry, lured by the potential for raiding the Spanish and the prospect of free land, attempted to plant colonies in North America. The first venture was launched by Humphrey Gilbert, a swaggering West Country man with connections to Raleigh and Drake, who became interested in overseas colonization through his brutal military campaigns against the inhabitants of southern Ireland. Awarded a massive land grant in North America by Elizabeth I, Gilbert gained support for his colonization plan from prosperous English Catholics who were becoming increasingly isolated in English Protestant society and were looking to recreate their gentry life overseas. Although Gilbert got to Newfoundland, which he claimed for the English Crown in 1583, he was lost on the voyage home and his scheme for English colonization of the New World was taken over by his half-brother, Walter Raleigh, who acquired the charter for settlement. In 1584, Raleigh sent the first of three expeditions to settle the new territory of Virginia (see figure 1.5). Rejecting more northerly latitudes, Raleigh aimed to establish a colony between 35° and 40° latitude, sufficiently south to grow Mediterranean crops and near enough to the Caribbean to serve as a base for raiding the Spanish. His expedition reconnoitered the sandy shores of the Outer Banks of the Carolinas, settling upon Roanoke Island in Pamlico Sound as a potential site for the colony; the island could be protected from Indian attack and was hidden by the Outer Banks from preying Spanish warships. Colonists were sent out the next year. But the shortcomings of Roanoke, particularly its lack of a good anchorage and decent soil, soon became apparent, and the colonists looked elsewhere to settle. During their searches, the English explored the Chesapeake Bay and realized that it was a far more suitable place for settlement. But before the colonists could remove there, most of them were picked up by Sir Francis Drake, returning from his raiding expedition in the Caribbean (figure 1.5), and taken back to England. A second group of colonists was sent out in 1587. Although they planned to settle in the Chesapeake, they were landed at Roanoke. Attempts to supply the colony from England were disrupted by the Spanish armada in 1588 and it was not until 1590 that relief ships once more entered Pamlico Sound. By then, the settlers had disappeared and with them English hopes for the Roanoke colony.[41]

English Expansion in the Early Seventeenth Century

After peace was signed with Spain in 1604, the English were well placed to expand into the farthest reaches of the Atlantic. London and West Country merchants had accumulated sufficient capital from privateering during the war that they were able to invest in overseas trading ventures, notably the Newfoundland Company, the Virginia Company, and the East India Company. The success of the Newfoundland fishery showed that the resources of the New World could be exploited and sold profitably in European markets, particularly those in the Iberian peninsula, and that further profit could be made on the Spanish and Mediterranean trades back to England.[42] The English also had the technology and knowledge to launch trans-Atlantic expeditions. London and West Country ports had plentiful ocean-going vessels and experienced seamen, as well as a growing number of explorers' accounts, promotional tracts, and maps of the ocean. Such knowledge circulated through the country's mercantile and aristocratic communities, making them much more aware of North America and its rich potential for commerce and colonization.[43]

At the intersection of this world of merchants, aristocrats, and texts in late Elizabethan England was the great chronicler and advocate of English colonization overseas, Richard Hakluyt the younger. Although resident at Oxford University during the 1570s and early 1580s, Hakluyt received financial support from the Clothworkers' Company in London, which was directly involved in trade with Spain and the Mediterranean.[44] In an attempt to counter the dominant influence of the Merchants Adventurers and their cross-Channel commerce, the Clothworkers encouraged Hakluyt's efforts to publicize longer-distance trades. In 1584, Hakluyt championed the colonization efforts of Gilbert and Raleigh in Virginia by publishing a *Discourse of Western Planting*; five years later, Hakluyt published his massive compendium of *The Principall Navigations, Voiages, and Discoveries of the English Nation*. During the same years, firsthand accounts of exploration and settlement were also appearing. Among these was Thomas Hariot's *A Briefe and True Report of the New Found Land of Virginia* (1588), an eyewitness account of the second voyage to Roanoke, which included compelling descriptions of North Carolina Algonquian native culture. An illustrated edition, with engravings of John White's watercolors of Algonquian people, was published in 1590.[45] These accounts, as well as accompanying maps, illustrations, and ships' logbooks, marked the beginning of what has been called the "imperial archive," a fund of knowledge that helped shape England's expansion overseas.[46]

With the Newfoundland fishery and Spanish/Mediterranean trade well established by the early seventeenth century, the English began to consolidate their position in other parts of the Atlantic. As the Spanish still commanded much of the Caribbean and the French claimed northeastern North America, the English concentrated on exploring the western Arctic for a route to the Indies, as well as promoting colonization of the eastern seaboard, southeastern Newfoundland, and the outer chain of Caribbean islands (figure 1.6). Among the first ventures was a renewed search for a passage to the Indies. Although the English East India Company had been founded in 1600 and had sent its first vessel around the Cape of Good Hope to the Spice Islands the following year, the Company was concerned about the great distance and expense and looked for a shorter route around North America. In 1602, the Company financed George Waymouth to search for a northwest passage. With an illuminated letter from Elizabeth I to the Emperor of China, Waymouth sailed to the western Arctic, pushing some 300 miles into Hudson Strait before turning back.[47] The Muscovy Company also maintained its interest in a polar route to the Indies, sending Henry Hudson on two voyages in 1607 and 1608 to explore eastern Greenland and the waters north of Spitzbergen. After these voyages revealed no passage through the ice, Hudson turned his attentions

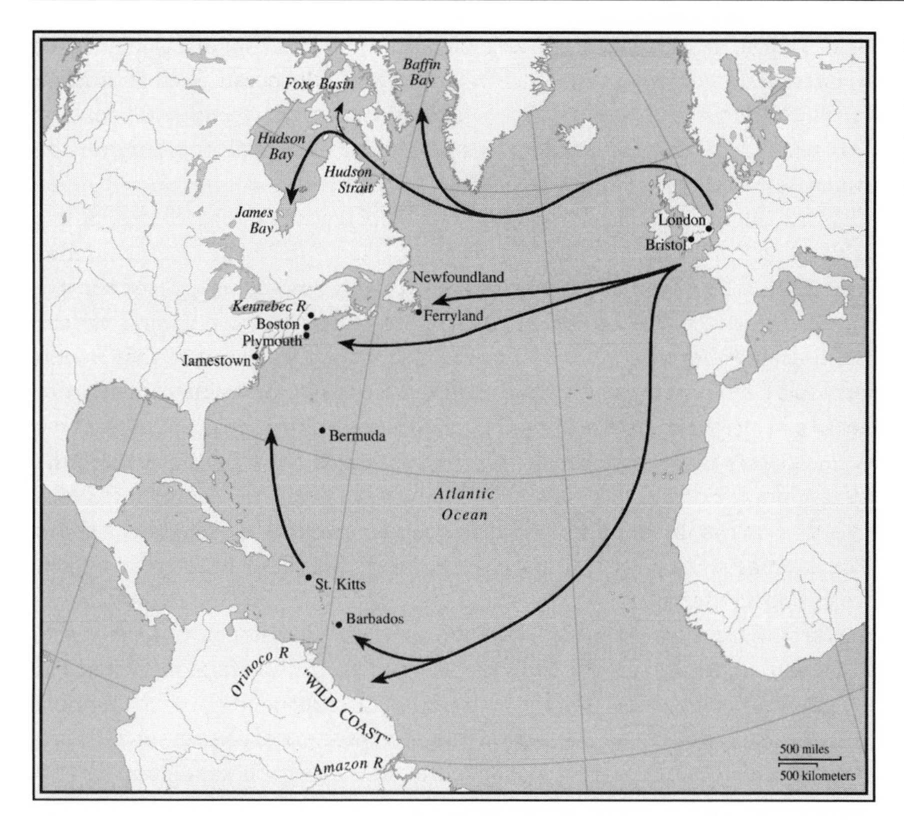

FIGURE 1.6. English exploration and settlement in North America and the Caribbean in the early seventeenth century.

westward. Sponsored by the Dutch East India Company, Hudson explored the coast north of the Chesapeake Bay in 1609, entering what is today New York Harbor and the Hudson River. With growing recognition of his seafaring expertise, Hudson won backing from merchants in the Muscovy, East India, and Virginia companies for another search for the northwest passage. In 1610, he sailed through Hudson Strait and into Hudson Bay, then along the east coast to James Bay, where he overwintered. Although a mutinous crew cast Hudson adrift, his vessel limped back to England with news that the northwest passage had been discovered. The Northwest Passage Company was soon formed in London to sponsor further voyages, which charted the west coast of Hudson Bay in 1612 and Baffin's exploration of the western end of Hudson Strait and entry into Foxe Basin in 1615. Discouraged by not finding an obvious passage to the Pacific, Baffin returned the following year to explore Davis Strait, establishing that it ended in a bay (Baffin Bay). Although these failures led to the collapse of the Northwest Passage Company, the search for the elusive passage was revived by Bristol merchants in 1631. They backed James's exploration of James Bay and Foxe's push through Foxe Channel into the southern part of Foxe Basin. Again disappointed, the search for the northwest passage to the Indies was abandoned.[48] Apart from considerable knowledge of the geography of Hudson Bay and expertise in sailing arctic waters, the English had not gained much from an investment spread over more than fifty years. Yet this scaffolding of expertise and knowledge eventually would support the launching of the English fur trade through Hudson Bay later in the century.

London and West Country merchants also focused their attention on the eastern seaboard of the continent, hoping to find gold or other worthwhile products. In 1602, Bartholomew Gosnold, probably financed by London merchants, sailed along the Maine coast and into Cape Cod Bay;[49] the following year, Martin Pring, backed by Bristol merchants, coasted the same waters; and in 1605, George Waymouth, who earlier had searched for the northwest passage, gained support from Plymouth merchants for a voyage along the mid-Maine coast. The London and West Country merchants were sufficiently encouraged by these voyages to take over Raleigh's charter to Virginia, which had lapsed in 1603, and to form the Virginia Company in 1606. As the charter covered all the seaboard between the Spanish possessions in Florida and the French claims in Acadia, the merchants divided responsibility for settlement. The London merchants gained the right to plant a colony south of latitude 41°, while the West Country merchants were given the area north of 38°.

Learning the lessons of Roanoke, the London-based Virginia Company promoted settlement around the Chesapeake Bay. Colonists were sent out in 1607, settling on a defensible bend of the James River, the most southerly and accessible river flowing into the Chesapeake Bay. With no experience of pioneering and riven by social and personal discord, the colony at Jamestown was kept going only by food supplied by local natives.[50] By 1609, the situation was so desperate that the colony was almost abandoned, saved only by the arrival of new supplies and settlers from England. A lack of merchantable products from the colony also left the Virginia Company in dire financial straits. With title to land its only asset, the Company began selling off parcels in 1612. About the same time, tobacco plants, imported from the Orinoco Delta in South America, were raised successfully, providing settlers with a crop marketable in England. Yet continuing mismanagement and an Indian massacre of settlers in 1622 led to the dissolution of the Company and the appointment of a royal governor in 1624. In spite of these setbacks, the first permanent English settlement on the eastern seaboard was in place.

Meanwhile, the Plymouth-based Virginia Company sent Pring in the summer of 1606 to reconnoiter the Gulf of Maine for a suitable site for settlement. After a good report of the mouth of the Kennebec River, the Company sent out a group of colonists to the mid-Maine coast in 1607. But a bleak winter and the death of the leaders led to the abandonment of the colony the following fall. Although no further attempts at settlement were made until 1620, West Country merchants extended the trans-Atlantic cod fishery into the Gulf of Maine, setting up fishing stations on the coastal promontories and offshore islands, much as they were doing along the coast of southeastern Newfoundland.[51] During the 1610s, the Gulf of Maine became the westernmost edge of the Atlantic cod fishery.

The first permanent English settlement of the northern part of Virginia, which became known as New England after the publication of John Smith's map in 1616, owed much to the London-based Virginia Company's desperate attempts to develop its Chesapeake Bay colony. In 1619, the Company began granting land to subsidiary joint-stock companies, which were responsible for recruiting and planting settlers. One of these grants was made to Thomas Weston, a small London merchant who had interloped into the cloth trade to the continent and was now looking to expand his trading connections with Virginia.[52] Aware that a group of English religious dissenters—the Pilgrims—resident in Leyden, Holland, had been negotiating with the Virginia Company for land in the Chesapeake, Weston took out his own charter from the Company and agreed to finance the Puritan plantation. In late fall 1620, the Pilgrims crossed the Atlantic and arrived off Cape Cod. Abandoning thoughts of sailing south to the Chesapeake Bay so late in the season, they established a settlement at Plymouth on Cape Cod Bay. Although Weston received little return on his investment, the Plymouth colony survived through the tight bonds of religious community and the establishment of family farming.

Despite the lack of financial return from the Plymouth colony, West Country and London merchants maintained their interest in the Gulf of Maine, promoting several settlement schemes associated with the fishery during the 1620s. Although none of these succeeded, mercantile and gentry interests from the West Country and London combined in the late 1620s to form the New England Company, later incorporated as the Massachusetts Bay Company in 1629. Drawing some of its support from merchants and gentry with Puritan sympathies, the Company was established not only to promote trade with New England but also to create a colony for persecuted Puritans.[53] In 1630, the Company sent the first of thousands of Puritans to the Massachusetts Bay Colony, many of them settling at Salem and Boston.

Apart from Jamestown and Plymouth, the London-based Virginia Company also had a hand in the settlement of Bermuda. The Company became involved in the island after one of its supply ships, bound for Jamestown, was wrecked there in 1609. Awarded the island as part of its Virginia grant in 1612, the Company found itself so stretched supporting Jamestown that it made over Bermuda to a subsidiary joint-stock company. Following the example of the Virginia Company and its Jamestown colony, the Bermuda Company sent out settlers in 1614 to colonize the island. With tobacco beginning to be cultivated successfully in Virginia, the settlers took up tobacco growing and by the mid-1620s were part of a tobacco economy encompassing the mainland and several Caribbean islands.[54]

Colonization efforts were also directed toward Newfoundland. In 1610, John Guy, a former sheriff of Bristol and member of the Bristol Society of Merchant Venturers, received backing from London and Bristol merchants to establish a colony at Cupids Cove, Conception Bay; six years later, he made another settlement nearby at Bristol's Hope (Harbour Grace). During the 1620s, the Stuart courtier and Roman Catholic George Calvert attempted a similar venture along the southeastern shore at Ferryland. But competition from fishermen, depredations of pirates, and the harsh climate—what Calvert called the "sadd face of wynter"—led to the collapse of all these schemes, leaving only a handful of settlers scattered along the shores of the Avalon Peninsula.[55]

Far to the south, the English were exploring and settling the "Wild Coast," that part of South America stretching from the Orinoco Delta to the Amazon. During the 1590s and early 1600s, as an outgrowth of their privateering ventures in the Caribbean, English merchants prosecuted a contraband trade in tobacco with the Spanish colonies, first from Cumaná and Caracas and then, after 1607, from Trinidad and the Orinoco River. Despite James I's counterblast against the weed, smoking had become popular in England and the domestic market for tobacco boomed. By the early 1600s, English planters had joined Dutch, French, and Irish settlers in raising their own tobacco along isolated parts of the coasts of Guiana and the Amazon delta. Yet this encroachment on Spanish and Portuguese territory, as well as Raleigh's doomed venture up the Orinoco in search of the fabled city of El Dorado in 1616, provoked the Spanish to clear the coast of northern Europeans in the early 1620s.[56]

Although English ventures in South America had ended in failure, the transfer of Orinoco tobacco plants to the Chesapeake was crucial to the long-term economic viability of the Virginia colony. Moreover, some of the English settlers evicted from Guiana turned their attention to settling the outer arc of Caribbean islands, well away from the Spanish, and suitable, so the English thought, for growing tobacco. The first English colony in the West Indies was established at St. Kitts (St. Christopher) in the Leeward Islands in 1624, and the success of that venture led to the settlement of Barbados three years later. By 1632, further islands in the Leewards—Nevis, Montserrat, and Antigua—had also been settled, footholds on the outer rim of the Spanish Main.[57]

Although the British Isles were geographically well positioned for oceanic exploration and commerce, there was nothing inevitable about the English expansion into the Atlantic. Despite Cabot's voyage, the English were latecomers to the New World. For much of the sixteenth century, the English only dabbled at the edges of the Atlantic; it was not until the 1570s, with the opening of Mediterranean markets, that the English moved more forcefully across the ocean. The Mediterranean provided the impetus for exploiting the Newfoundland fishery, which, in turn, led to the development of a deep-sea merchant marine. By the early 1580s, English sea power was sufficient to challenge Spain and thereby win access to considerable parts of the Atlantic. Over the next fifty years, English merchants and freebooters explored the western Arctic and Hudson Bay, cleared the Spanish and the Portuguese out of parts of Newfoundland, and established English claims to parts of the eastern seaboard and the West Indies. During those years, the foundations of an English Atlantic empire were laid.

Yet compared to the extensive English trade across the North Sea to the continent and the ventures beyond Europe to the Levant and the Indies, English enterprise across the Atlantic to North America was still a small affair. Apart from the cod fishery, English mercantile investment in the New World had been disastrous, with considerable sums lost on the search for a northwest passage and the Virginia Company's promotions in the Chesapeake. With the exception of the Bermuda Company, the New World had been a graveyard for chartered, monopolistic companies.[58] Nevertheless, individual English merchants, many of them outside the great London companies, had established a commercial fishery in Newfoundland, while others supported a nascent plantation economy in Virginia. Meanwhile, several utopian religious communities settled around Massachusetts Bay were surviving on family farming. By the early 1630s, the basic patterns of settlement and economic development that would define the spaces of English America were in place.

Chapter 2

Atlantic Staple Regions: Newfoundland, the West Indies, and Hudson Bay

Over the course of the seventeenth century, the English established a string of settlements along the eastern flank of North America, stretching from Hudson Bay to Newfoundland, New England, the Mid-Atlantic colonies, the Chesapeake, the Carolinas, and the West Indies. Almost from the beginning of settlement, these areas specialized in different forms of production and trade. Some regions quickly became producers of staple goods for markets around the North Atlantic, while others became frontiers of semisubsistent farming. The areas of staple production encompassed the fur trade through Hudson Bay, the cod fisheries in Newfoundland and New England, the tobacco plantations in the Chesapeake, the rice plantations in the Carolina low country, and the sugar plantations in the West Indies. The frontiers of semisubsistent farming developed in New England, the Mid-Atlantic, and, in the eighteenth century, the southern backcountry.

Although the staple-producing areas along the eastern edge of North America and in the Caribbean operated within a web of official regulations and imperial defense, the staple regions soon developed their own particular trajectories. Shortly after the beginnings of English settlement, a basic structural divide emerged between those areas of staple production dominated by metropolitan interests in England, and those controlled by local interests in the colonies. The staple regions under predominantly metropolitan control included the cod fishery in Newfoundland, the sugar islands in the West Indies, and, after 1670, the fur trade through Hudson Bay. The staples under colonial control encompassed the fishery in New England, the tobacco plantations in the Chesapeake, and, in the early 1700s, the rice plantations in the Carolina low country. The metropolitan staple areas—Newfoundland, the West Indies, Hudson Bay—comprised nodes of production or trade scattered around the western Atlantic; the colonial staple areas—New England, the Chesapeake, South Carolina—formed territories of production within a larger continental settlement frontier. Although the metropolitan staple regions were mere specks on the Atlantic rim, they were among the

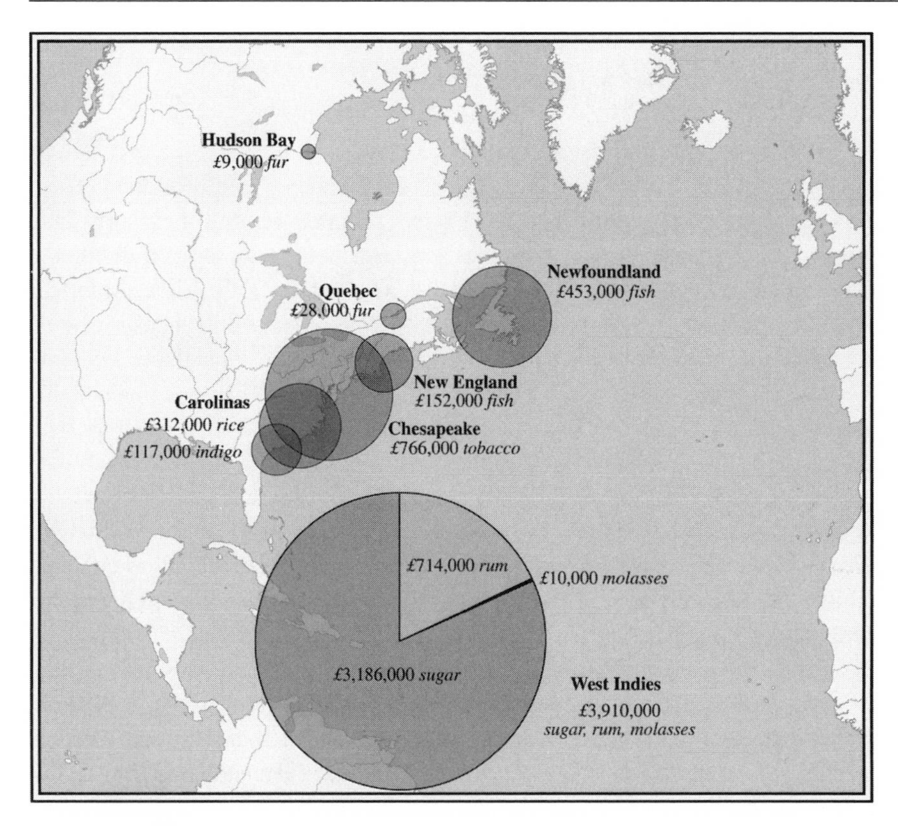

FIGURE 2.1. Staple regions in North America and the Caribbean, 1764–1775. Average annual export values are shown for the West Indies, the Carolinas, the Chesapeake, and New England for the years 1768 to 1772; for Newfoundland, 1764 to 1774; for Quebec and Hudson Bay, 1770. Based on data in John J. McCusker and Russell R. Menard, *The Economy of British America 1607–1789* (Chapel Hill: University of North Carolina Press, 1985), Table 7.3; and James F. Shepherd, "Commodity Exports from the British North American Colonies to Overseas Areas, 1768–1772: Magnitudes and Patterns of Trade," *Explorations in Economic History* 8 (1970–1971); and "Staples and Eighteenth-Century Canadian Development: The Case of Newfoundland," in *Explorations in the New Economic History*, ed. Roger L. Ransom et al., Table 5.5 (New York: Academic Press, 1982).

most valuable regions in colonial British America (figure 2.1). Sugar exports from the West Indies alone were worth more than all the exports from the mainland colonies combined. With codfish from Newfoundland and furs from Hudson Bay, the value of exports from the Atlantic areas averaged about 60 percent of total exports from British America between 1768 and 1772.[1] Moreover, these three areas developed patterns of settlement, society, and culture that set them apart from the staple regions on the continent, and helped create a distinctive British commercial empire of the Atlantic.

The Cod Fishery in Newfoundland

From the 1570s until the rise of the tobacco trade in the 1620s, the cod fishery in Newfoundland was the principal English enterprise in North America.[2] Even after the development of other areas of staple production, the fishery remained of major importance. According to customs returns for 1764 to 1774, dried cod from Newfoundland was the third-most-valuable export from British America, surpassed only by sugar and tobacco.[3] Although another cod fishery developed in New England in the 1630s and eventually expanded up the Atlantic coast of Nova Scotia and out to the Grand Banks, the Newfoundland fishery remained the most significant producer. Of the £614,310 worth of dried fish that was exported on average each year between 1764 and 1774, 73 percent came from Newfoundland, 25 percent from New England, and 2 percent from Nova Scotia.[4]

After the defeat of the Spanish in the late sixteenth century, the English competed with the French for control of the Newfoundland fishery. In the seventeenth century, the English confined their efforts to the coasts of the Avalon and Bonavista peninsulas in southeastern Newfoundland, leaving much of the rest of the island to the French (Figure 2.2).[5] During those years, the English fishery was seasonal and migratory: Merchants and fishermen from the English West Country arrived on the English Shore in the spring, fished during the summer, and returned to Europe with cargoes of dried fish in the fall. After the defeat of the French in the War of Spanish Succession (1702–1713), the British gained title to all of Newfoundland, allowing them to extend the migratory fishery westward along the South Coast and northward toward Notre Dame Bay.[6] During the late 1600s and early 1700s, a resident fishery also developed alongside the migratory fishery, particularly along the original English Shore. As the population increased during the eighteenth century, the resident fishery expanded in two directions: westward into St. Mary's Bay and Placentia Bay in the early 1700s, and then along the South Coast, reaching Cabot Strait and St. George's Bay on the west coast by the 1760s; and northward to Fogo Island in the 1720s, and from there along the old French Shore on the Petit Nord (the northern peninsula of Newfoundland) and into the Strait of Belle Isle by the 1760s.[7] At the outbreak of the American Revolution, the British had an enormously productive cod fishery extending more than a thousand miles around the coast of Newfoundland and along the shore of southern Labrador.

Despite the sway of London merchants over much English commerce during the late sixteenth century, West Country merchants quickly dominated the country's first trans-Atlantic trade.[8] They were close to the open Atlantic and had longstanding trade ties with the Iberian peninsula and the Mediterranean, the principal markets for Newfoundland cod. West Country merchants also had the resources—capital from the cloth industry, plentiful labor, numerous vessels, a wealth of provisions, and the technology—to sustain a trans-Atlantic fishery. Although Bristol merchants financed the Cabot voyages and the plantation at

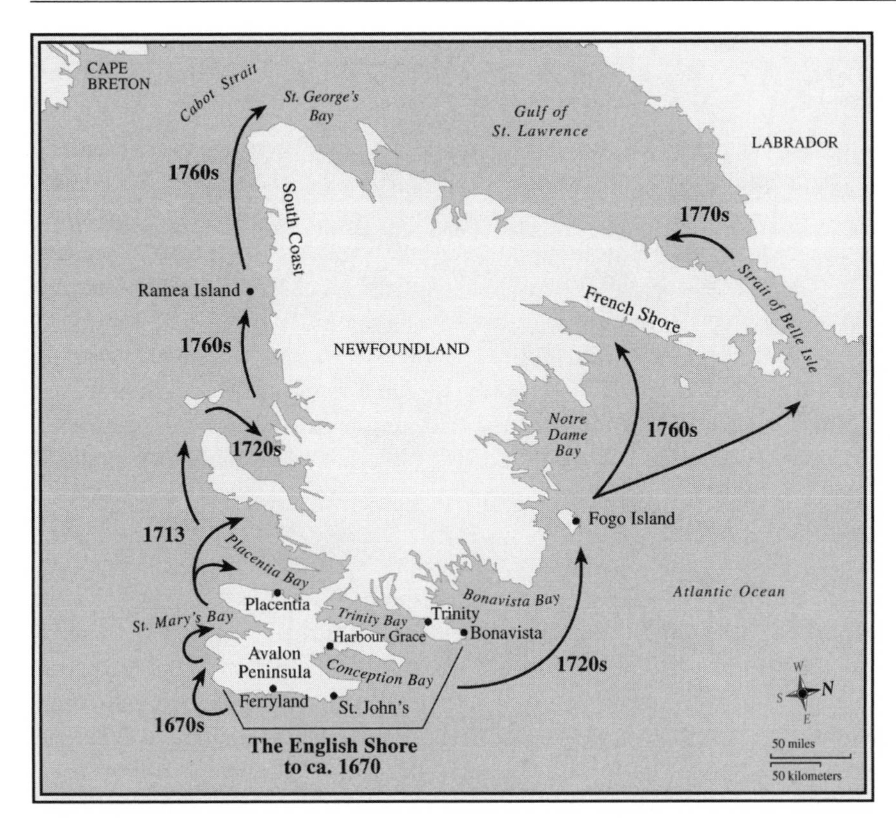

FIGURE 2.2. Expansion of the English fishery around the coast of Newfoundland, 1570–1770. After John Mannion, Gordon Handcock, and Alan Macpherson, "The Newfoundland Fishery, 18th Century," in *Historical Atlas of Canada*, vol. I: *From the Beginning to 1800*, ed. R. Cole Harris, plate 25 (Toronto: University of Toronto Press, 1987).

Bristol's Hope (Harbour Grace), the city's merchants were not a major force in the fishery, preferring to invest in the more profitable sugar, tobacco, and slave trades. Instead, merchants in Bideford, Plymouth, Dartmouth, Topsham, and Poole dominated the fishery during the sixteenth and seventeenth centuries. As the industry increasingly became concentrated in Dartmouth, Topsham, and Poole during the eighteenth century, the participation of merchants in Bideford and Plymouth waned. However, merchants in other ports also became involved, particularly those in Waterford in southeastern Ireland and in Guernsey and Jersey in the Channel Islands.[9]

Unlike the London-based chartered companies that had opened up trade to Russia and the Levant, the West Country, Irish, and Channel Island ventures to Newfoundland were individualistic enterprises, comprising many merchants and merchant partnerships in numerous ports (figure 2.3). Many of these merchants began in the Newfoundland trade by captaining vessels across the Atlantic, serving as agents (factors) at the fishing stations, and then returning to England to set up their own businesses or to take over family firms. Typical of such merchants was Benjamin Lester of Poole. Lester entered the Newfoundland trade in the late 1730s as an agent for a prominent Poole merchant, and within two decades had set up his own mercantile business in partnership with his brother Isaac. Lester's main base

FIGURE 2.3. *Arthur Holdsworth Conversing with Thomas Taylor and Captain Stancombe by the River Dart*, by Arthur Devis, 1757. Arthur Holdsworth (seated) was one of the leading Dartmouth merchants operating in the Newfoundland fishery and the Portugal wine trade. Captain Stancombe (standing on right) points to one of Holdsworth's vessels coming up the River Dart. Paul Mellon Collection, Image © 2003 Board of Trustees, National Gallery of Art, Washington.

in Newfoundland was Trinity, from where he organized a fishery that extended around Trinity Bay, up the coast to Bonavista Bay, Fogo Island, and Labrador, as well as out to the Grand Banks. Meanwhile, brother Isaac remained in Poole, arranging the recruitment of men, shipment of fishing supplies and provisions, and marketing of fish. By the 1770s, the Lester business owned at least twelve vessels, and was among the biggest in the Poole-Newfoundland trade. After Benjamin Lester returned to England in 1776, he served as the town mayor and local member of parliament, offering advice to government committees concerned with the fishery and American trade.[10]

For these English, Irish, and Channel Island merchants, the Newfoundland

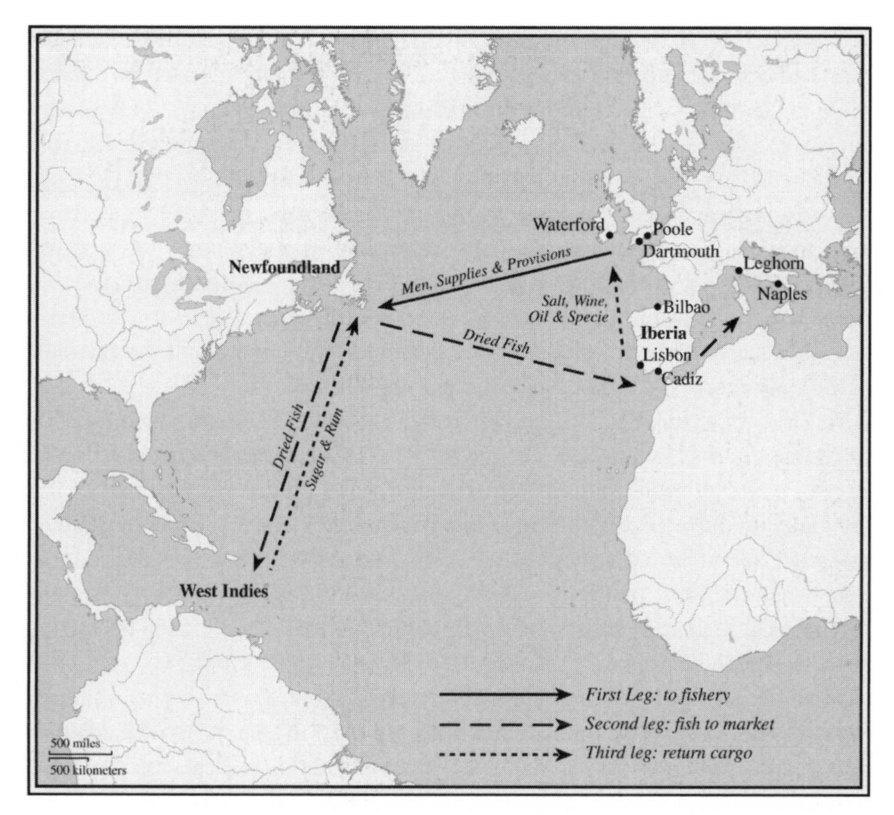

FIGURE 2.4. Newfoundland dried fish trade in the eighteenth century. After Rosemary Ommer, "The Cod Trade in the New World," in *A People of the Sea: The Maritime History of the Channel Islands*, ed. A. G. Jamieson, 259 (London: Methuen & Co., 1986).

fishery provided a resource that could be traded around the North Atlantic (figure 2.4). Most of the dried fish produced in Newfoundland was lightly salted and air-dried, producing a prime merchantable product suitable for the best markets in southern Europe. Despite periodic disruptions caused by war, the Iberian and Mediterranean markets remained the main outlets for merchantable cod throughout the early modern period. As the Spanish and the Portuguese and then the French withdrew or scaled back from the fishery, so the English expanded their markets. By the 1770s, dried cod from Newfoundland was going to all the major ports from Bilbao in northern Spain to Naples in southern Italy.[11] At that time, dried cod was the second-most-important English export to the Mediterranean, surpassed only by woollens.[12] In return for these shipments, fish merchants purchased wines and olive oil in the Mediterranean markets for sale in Britain, as well as salt for the fishery. These cargoes failed to cover the costs of dried fish imports, however, and the balance was increasingly paid by specie. Spanish gold from Central America and Portuguese gold from Brazil were used to pay for dried fish from Newfoundland.[13] In effect, two of the greatest products of the British and Iberian empires were exchanged. By this means, Britain garnered sufficient gold to purchase high value goods from the Indies. Apart from Iberian and Mediterranean markets, small quantities of refuse fish were shipped south to the West Indies as cheap food for slaves on the sugar plantations. Return cargoes to Newfoundland consisted of sugar and rum. During the years of the migratory fishery, merchants could make profits on two legs of the trade: from Newfoundland to the Mediterranean, and from the Mediterranean to Britain. As the resident fishery developed, they could also make profits on goods sold in Newfoundland.[14]

As fishing was a labor-intensive business, fish merchants needed a large, cheap work force. Although the Spanish had coerced native populations to work in the gold and silver mines of the Caribbean, Central, and South America, the available native labor in Newfoundland was small and largely avoided contact with European fishermen. Apart from some trading between the English and the Beothuk in the early 1600s, the Indians kept to themselves, content to scavenge European fishing rooms abandoned during the winter months. The gradual take-over of native fishing sites by Europeans, the growth of a permanent, frequently hostile English population, the spread of European diseases, and the withdrawal of the Beothuk into the interior of Newfoundland, where food resources were limited, all took their toll on the native population. By the late eighteenth century, few Indians were left; in the early nineteenth century, the Beothuk became extinct.[15] Although the means of destruction were vastly different in Newfoundland than in the Caribbean, the Beothuk eventually suffered the same fate as the Arawak.[16]

With no labor in Newfoundland to draw upon, West Country merchants were faced with the task of shipping fishing crews from England across the Atlantic each year to catch the cod. The French, Spanish, and Portuguese had faced the same problem much earlier, and the English merely adopted their practices. West Country merchants recruited labor that was to hand, mostly from the port towns and neighboring farms and villages. Merchants in Dartmouth hired most of their fishermen from within a radius of fifteen miles, mainly from the town itself and the surrounding parishes of Kingswear, Stoke Fleming, and Dittisham; the Torbay area; and the inland market town of Newton Abbott. Merchants in Poole also drew from their own town, as well as from a wide area of west Hampshire (Christchurch, Ringwood) and Dorset (Dorchester, Bridport, Sturminster Newton).[17] The recruits were male, mostly young, between the ages of sixteen and thirty, and from three principal occupations: the largest group were farm laborers; a second group comprised artisans and tradesmen such as carpenters, blacksmiths, masons, shoemakers, and butchers; and a third group consisted of maritime tradesmen such as mariners, fishermen, and ship artisans.[18] From the lower rungs of society, these men worked in the fishery in the hope of accumulating savings to advance their conditions back in England or to move on to the mainland colonies.

Fishermen, like agricultural laborers, were hired for one or more seasons, and usually participated in the fishery for no more than five years.[19] Much of the recruitment was done at spring hiring fairs, by contractors touring through the countryside, and through advertisements in local newspapers. The Newfoundland historian Judge Prowse recorded that in Devon the "regular place for shipping was Newton Abbott, in . . . 'The Dartmouth Inn and Newfoundland Tavern.' Here the engagement was 'wetted' with cyder, strong beer, and the still more potent Jamaica [rum]."[20] Fishermen were hired at rates set by the labor market in England, where labor was plentiful and cheap. In North America, the situation was exactly the opposite: labor was scarce and expensive, and the availability of cheap agricultural land provided an attractive alternative to wage labor in an industry such as the fishery.[21] Although Newfoundland had virtually no good agricultural land and its climate mitigated against farming, the English settlement of New England in the 1620s opened up an agricultural alternative to the Newfoundland fishery. To prevent labor in the fishery seeping away to New England, the fish merchants bound their labor through contracts. In this way, merchants could insulate the price of labor in Newfoundland from the realities of the New World. In effect, the English labor market had been extended two thousand miles across the Atlantic to Newfoundland.

A further way of insulating the price of labor was to discourage settlement in Newfoundland. With the New England experience in mind, West Country merchants lobbied against any permanent settlement, fearing that the migratory fishery would eventually succumb to a sedentary fishery. During the early seventeenth century, the Western Adventurers, a loose coalition of West Country merchants, opposed colonization schemes at Cupids and Ferryland. Both settlements were beset by problems and were eventually abandoned, leaving only a few colonists ("planters") on the island. By the latter half of the century, the Adventurers renewed their complaints, this time against the widespread practice of merchants leaving fishermen behind at the end of the season to overwinter on the island. The Adventurers put forward several arguments against settlement, pointing out that conflict would arise between migratory and permanent fishermen over the best fishing sites or rooms; migratory fishermen were being "debauched" in taverns owned by residents; and that the appointment of a governor would have to be supported by taxation, which could not be born by the marginally profitable fishery. Furthermore, the Adventurers argued, the West Country would lose its valuable provisions and supply trade to its New England competitors, and England would not have a "nursery for seamen," which would weaken the navy. Against these arguments, proponents of permanent settlement reasoned that the sedentary fishery was more efficient than the migratory because there was no need to shuttle fishermen back and forth across the Atlantic, the removal of the existing settlers would be expensive, and settlers were needed to counter the French threat to the Avalon Peninsula. After government enquiries in the 1670s, a compromise was reached: The small population of resident planters was allowed to stay, but was not accorded colonial status and thus had no official representation.[22]

From the 1570s to the 1770s, the heyday of the English migratory fishery, hundreds of vessels and thousands of men converged on the coasts of Newfoundland each year to fish for cod. The favored locations for the fishery were the peninsula capes that punctuate the coast of southeastern Newfoundland, particularly the Atlantic face of the Avalon Peninsula (from Cape St. Francis to Cape Race),

FIGURE 2.5. Lester fishing station, Trinity, Newfoundland, 1790s. This busy scene shows merchant vessels and fishing boats, a stage head, flakes, and stacks of dried fish. Dorset Natural History and Archaeological Society, Dorset County Museum, Dorchester, England.

the west side of Conception Bay, the west side of Trinity Bay, and around Cape Bonavista.[23] These coasts were well placed for intercepting cod migrating from the spawning grounds on the Banks to the feeding grounds in inshore waters. Inland from the coasts, however, Newfoundland was strikingly barren. Compared to the prosperous farming country, market towns, and villages of the West Country and southeastern Ireland, or the expanding farms and plantations of the American eastern seaboard, the interior of Newfoundland comprised an immense area of glaciated rock, coniferous forest, and innumerable lakes and bogs that Europeans barely knew. Unlike the temperate, maritime climate that marked the southwestern British Isles, a harsh, continental climate, consisting of long winters and short summers, characterized Newfoundland. Well after the spring rites had been celebrated in the villages of Devon and Kilkenny, fog, snow, and icebergs still plagued the shores of the Avalon.[24]

Arriving after the pack ice had cleared the coast in late April or May, English fishing vessels were unloaded in sheltered harbors and temporary fishing stations constructed. These consisted of one or more fishing "rooms," each comprising stages for unloading and processing fish, flakes or platforms on which to dry the cod, bunkhouses for the fishermen, and stores for salt, fish, and dry provisions (figure 2.5).[25] Stages consisted of a short wharf projecting into the sea, and a covered shed in which to process the cod.[26] Close by, an open, square box was used as a vat for making train oil from cod livers. Flakes were built about waist high, and were usually long and narrow. Depending on the amount of dried fish made by a station, flakes could cover several acres. Fishermen's cabins or "tilts," as they were called later, were simple, one-room dwellings, constructed from spruce poles (with the bark left on) set "earthfast" in a trench and buttressed by poles set at an angle. Any gaps in the walls were caulked with moss or mud. The roof was also made from spruce poles resting on a longer ridge pole and covered with bark, wood, or sod; a hole in the roof allowed smoke to escape from a fireplace built on the packed earth floor.[27] As the fishery was seasonal, these were simple, temporary structures, usually torn down at the end of the summer. So much wood was used each year for building and fuel that the spruce forests that once covered the coasts of Newfoundland were quickly depleted.[28]

Fishing was conducted in inshore waters from small, maneuverable boats, known as shallops, similar to the whale boats of the nineteenth century. The boats were open, with one or two masts, and a crew of three. As long as the weather was fine, bait available, and the fish striking, the fishermen set out every day, usually early in the morning, and returned in the evening. Each fishermen had two or three baited lines for catching cod, and in a good day could catch several hundredweight of fish. Returning to the stage, the fishermen used pitchforks to "pew" the fish onto the landing, an operation similar to forking hay onto a rick (figure 2.5). On the stage, a processing crew comprising a header, a splitter, and a salter received the fish, and quickly set about heading, gutting, splitting, salting, and washing the cod. After the filleted fish had been lightly salted and washed, they were placed on flakes to dry. Given good weather, cod took about ten days to dry completely, but rain, fog, and cloudy conditions slowed the process. Too much sun could also ruin the flesh. After the fish were properly cured, they were collected and piled into large stacks, like miniature hay ricks, and covered with old sail cloth weighted with heavy stones to squeeze any remaining moisture out of the fish. The dried fish were then moved to a store until it was time to load the sack ships that would take the fish to market in late summer or early fall. At the same time, the migratory fishermen returned to England.

The work forces at the fishing stations were specialized and hierarchical. Men and boys had different jobs—men fished and processed the catch, boys spread the fish on the flakes—and wage rates varied accordingly. At a comparable fishing station established by Plymouth merchant Robert Trelawny on Richmond Island in the Gulf of Maine in the 1630s, the work force was paid annual wages as well as a share of the season's catch. Trelawny's agent was paid £40 and one share (worth £11 9s. 6d.); four fishermen received from 20s. to £4, plus a share each; two other fishermen got £2 10s. and £3 5s. respectively, plus one half-share each; and a servant (probably a boy) was paid a flat £5 per year.[29] Payment of shares to fishermen provided an incentive above and beyond regular wages. Such payments represented labor's link to the resources of the fishery, a much different arrangement to the majority of farmers in New England and the Mid-Atlantic colonies who owned their own small share of New World land. In overall command of the fishing station was the merchant or his agent. He was responsible for the fishing station, supervised the fishing crews, made sure that dried fish was of merchantable quality, kept in written contact with the head office, and dispatched vessels to market.

Surrounded by sea, forest, rock, and potentially hostile native peoples, the English fishing stations were solitary settlements set along the Newfoundland coast. Apart from the fur trading posts on the rim of Hudson Bay, no other type of settlement in British America was quite so isolated. Beyond the fishing stations, there were no roads, farms, plantations, villages, or towns. Tied to a marine resource and dependent on a maritime trading system, the stations were the merest landward extensions of a British Atlantic economy.[30] With families left behind in Britain and few or no permanent residents in the fishing harbors and coves, the migrant fishermen had little to do except work. In essence, the migratory fishery was a productive system extended from the southwestern British Isles to the fish resources of Newfoundland each summer and then withdrawn to Europe in the fall.

Nevertheless, a resident fishery eventually developed out of the migratory fishery and the early colonization efforts. As competition for fishing rooms increased, merchants began to leave men behind to take care of their premises during the winter and to secure them for the following season. The colonization ventures also scattered a few families along the English Shore.[31] From these two sources, the resident population began to build slowly. From the 1670s to the 1770s, the resident population increased from about 1,700 in 1675 to 3,500 in 1730, reached 7,300 in the 1750s, and about 12,000 in the 1770s.[32] As late as the 1750s, the summer population, numbering perhaps 8,000 men, was larger than the year-round population. Yet during the 1760s and 1770s, the resident population gradually overhauled the migratory population, and was dominant by the time of the American Revolution.

The resident population was drawn from the same pool of labor and geographic areas as the migrant fishermen. West Country merchants recruited servants from the counties of Devon and Dorset, while Irish merchants hired from Kilkenny, Wexford, Waterford, and Tipperary.[33] Migration channels developed between particular ports in the British Isles and specific outports in Newfoundland. Merchants from Dartmouth in Devon, for example, had establishments in the St. John's area, and brought many migrants from that county to St. John's and neighboring Torbay. Although the trans-Atlantic migrations to Newfoundland created a complicated braid of channels, some general patterns of settlement had emerged on the island by the late eighteenth century. English migrants dominated Conception, Trinity, and Bonavista bays, as well as the South Coast; the Irish were concentrated mostly in St. John's, along the southern part of the English Shore, and around St. Mary's and Placentia Bays. Many outports were completely dominated by one or the other ethnic group.

In spite of their earlier fears about the growth of a resident fishery, West Country merchants became the chief organizers of the new economy. As the permanent population increased, the merchants shifted their interests from the migratory fishery to the resident fishery, becoming less involved in production and more involved in supplying goods and provisions to resident fishermen (planters) and marketing their fish.[34] At the time of the American Revolution, the transition was far from complete, with West Country merchants participating in both the migratory and resident fisheries.[35] But the outbreak of the revolutionary war with France in 1793 made trans-Atlantic shipping so perilous that the migratory fishery collapsed, and many West Country merchants either abandoned the fishery altogether or relocated their businesses to Newfoundland.

To coordinate the resident fishery, West Country merchants established permanent bases on the island. The merchants either managed these premises themselves, shuttling back and forth across the Atlantic each year, or appointed agents to look after them. Clerks and bookkeepers were also sent out from England to take care of the accounts. A stream of written communication—letters, journals, ledger books—detailing the business activities of the fishing stations were sent back by the agents to the merchant houses in England each year. Mariners and fishermen were also hired to crew the trading vessels and fishing boats. In a method of production similar to the putting-out system in the English cloth industry, West Country merchants advanced supplies and provisions to local planters in return for fish. As the merchants controlled both the import of supplies and the export of fish, they had great power over prices, particularly if they had a monopoly over trade in an outport. It only needed a bad fishing season for planters to get into debt and become virtual vassals of the merchants. In 1765, Governor Sir Hugh Palliser observed that the inhabitants were "no better than the property or Slaves of the Merchant Supplyers to whom by Exhorbitant high Prices on their Goods they are all largely in Debt, more than they can work out during life."[36]

Despite their control over the developing resident fishery, West Country merchants could not insulate the economy completely. A critical shortcoming of permanent settlement was the lack of arable land and the short growing season, which meant that residents depended to a great extent on imported food. At first, English merchants controlled the provisions trade; each spring, they shipped bread, biscuit, flour, salted meat, butter, and peas to the island, some of it coming from Irish suppliers in Waterford.[37] Yet as the resident population grew and demand for provisions expanded, American merchants increasingly shipped foodstuffs to the island. Merchants in Boston, Newport, New York, and Philadelphia needed bills of exchange on English merchant houses in order to pay for their imports of British manufactures. One way to gain such credit was to trade to Newfoundland. American merchants hawked their "Continent goods" along the coast, as well as established retail outlets in St. John's.[38] Provisions were exchanged either for bills drawn on English houses or for dried fish, which could be sold, depending on its quality, in the Mediterranean or West Indies markets. Although they lost some of the trade in provisions, West Country merchants continued to control the trade in manufactured goods and fishing supplies. Some of these supplies, such as rope, beer, and cider, came from manufacturers in southwestern England and were an important source of local employment.

While merchants organized the trade in supplies, provisions, and dried fish, planters and servants caught the cod.[39] Some planters resided in Newfoundland permanently, others for a number of years before returning to England.[40] Similar in their social and economic status to yeomen and tradesmen back in England, planters owned their own fishing room and boats, and hired servants as fishing crew.[41] In the late seventeenth century, the majority of planters owned one to three boats, while a handful had as many as six or seven.[42] In 1675, a planter with one boat had an average of four servants, while those with three boats had nearly fifteen servants.[43] By the 1720s, after years of poor seasons and depredations caused by war, very few planters had more than five or six servants.[44] Wives and children also helped process and dry the fish. Apart from fishing, planters exploited other local resources, including harvesting nearby forest for wood, fuel, and construction materials; clearing patches of land for rough pasture and small vegetable gardens; and making inshore boats for their own use and for sale to the migratory fishery.[45] The availability of these resources and activities lessened the planters' dependence on the fishery but scarcely provided an alternative to fishing. The inshore fishery and some meager land provided a modest living for planters but little more.

Much of the labor in the resident fishery was provided by servants. These were mostly young men and boys recruited from the West Country and, increasingly, by the late eighteenth century, from southeastern Ireland, and shipped out on fishing vessels to Newfoundland.[46] There, they were indentured for a season or two to the planters. Like the fishing crews in the migratory fishery, the resident servants were differentiated between the skilled and unskilled, between the fishermen and the processors, and the boys who spread fish on the flakes and did menial jobs. Compared to merchants and planters, servants were the most mobile of the residents, spending perhaps two summers and a winter in Newfoundland, and then moving on. "Soe longe as their comes noe women," observed an English naval officer, "they are not fixed."[47] For servants, the Newfoundland fishery provided a stepping-stone to the New World or some temporary employment before settling back in the British Isles.

The settlement pattern that emerged in Newfoundland during the eighteenth century reflected this new economic order. At the top of the settlement hierarchy was a handful of regional centers, such as St. John's, Ferryland, Placentia, Harbour Grace, Trinity, and Bonavista, which served as points of contact between the trading circuits of the Atlantic and the fish production of Newfoundland (figure 2.6). At these centers, West Country merchants had their establishments and organized the fishery, importing supplies and provisions from England, distributing them to fishermen in the outports, and collecting fish for export.[48] To accommodate the inflow of goods and outflow of fish, merchant premises were usually quite extensive. At the Lester property in Trinity, the station comprised several distinct work

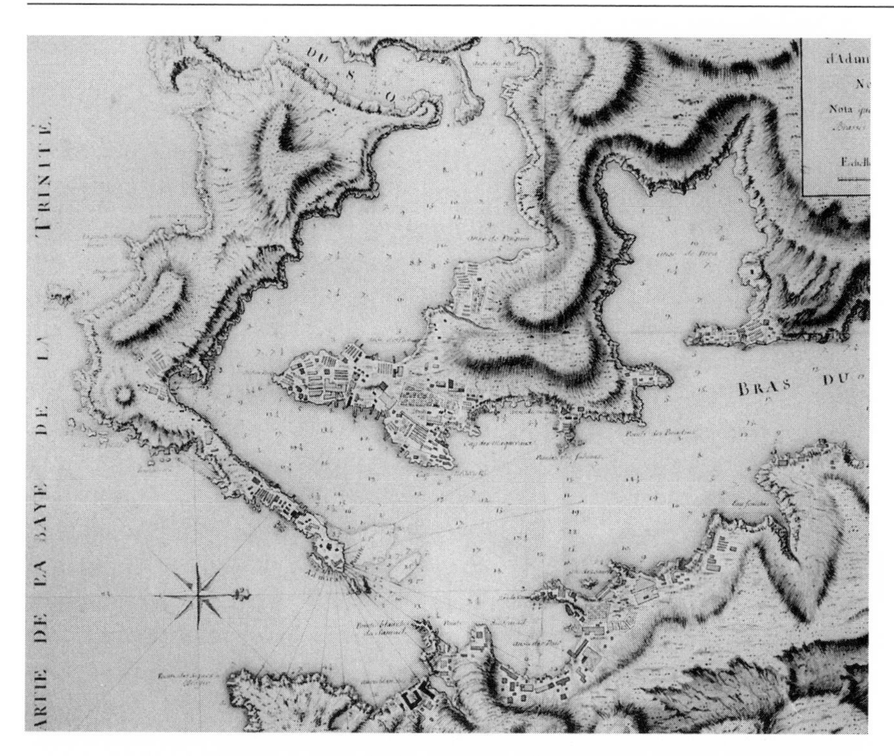

FIGURE 2.6. Detail of *Map of Trinity Harbour, Newfoundland*, by Marc Antoine Sicre de Cinq-Mars, 1762. During their brief occupation of Trinity in 1762, the French collected enough information to create this detailed topographic map. The map shows the central settlement of Trinity and the numerous fishing rooms scattered around the arms of Trinity Harbour. Individual dwellings, fish flakes, stages, and gardens are carefully delineated. The map graphically illustrates the littoral pattern of settlement associated with the fishery, and the complete absence of any agricultural hinterland. Archives du Génie, Chateau de Vincennes; copy held by Provincial Archives of Newfoundland and Labrador.

and living spaces (figure 2.7). Around a small cove were wharves and stages for unloading supply vessels and fishing boats, as well as stores for putting up salt, dried fish, seal skins, and provisions. Behind the wharves and warehouses lay a manager's house, and a shop and store for supplying planters and servants. In addition, there was a shipyard, with launch way, sail loft, forge, steamhouse, workshop, and cookroom. Many of these buildings were made to last, using sawn lumber for framing and boards and shingles for covering walls and roofs. Apart from these buildings, there was rough grazing for horses and livestock, and a small, fenced garden for growing roots and vegetables.[49]

If the larger centers were points of contact with the Atlantic world, the outports were little more than enclaves of production scattered along the Newfoundland coast. Sited around a sheltered harbor or cove, the typical outport comprised several fishing rooms, each consisting of a stage, a store for fish, salt, and gear, a dwelling, and a patch of cleared land. The various structures were made out of

FIGURE 2.7. The properties of Benjamin Lester and Company, Trinity, Newfoundland, circa 1800. Dorset Natural History and Archaeological Society, Dorset County Museum, Dorchester, England.

wood, and had a makeshift, utilitarian appearance. Houses were constructed from sawn lumber or spruce poles, much like the "tilts" of the migratory fishermen. One visitor to Newfoundland in 1680 described the planters' houses as "low and simply built, the best sort with sawd plancks from the foundation up, roof and all, others with the whol timber joyned together, standing stable wise and the roof covered with the rinde or Bark of trees, wth green turfs cast over them."[50] A few acres of cleared land were devoted to rough pasture and vegetable gardens; beyond lay spruce forest, bog, and rock.

With the fishery completely dominating the economy and settlement of the island, Newfoundland supported only a thin sliver of English and Irish society. The staple economy maintained merchants, planters, and servants, as well as a handful of professionals—doctors and clergy—in the larger outports, but there was no support for other social classes, such as gentry, farmers, artisans, or even the wandering poor. Without an agricultural base, there could be no aristocracy or gentry, as there was in England, or a large settler population, as there was in the mainland colonies, to provide a counterweight to the merchants. West Country merchants had little competition in Newfoundland for economic power. In control of the fishery and holding the principal government positions on the island, West Country merchants reigned supreme.

Yet like other metropolitan elites in the British Atlantic, fish merchants were temporary rather than permanent settlers. Although content to make money from the fishery, fish merchants did not view Newfoundland as home. Much of the wealth that they accumulated from the Newfoundland fishery was transferred back to the British Isles and invested in government bonds, trading ventures, mortgages, and real estate. The Georgian town houses in Poole, Dartmouth, and Waterford were the most conspicuous signs of mercantile success.[51] Perhaps the clearest symbol of this colonial relationship was the two dried fillets of cod carved on the marble fireplace in the dining room of Benjamin Lester's Mansion House in Poole (figure 2.8). The wealth of the Newfoundland fishery made fine living in England possible.

FIGURE 2.8. Fillets of cod carved on a mantlepiece in Benjamin Lester's Mansion House, Poole, Dorset, circa 1778. English Heritage, National Monuments Record, © Crown copyright. NMR.

While the mercantile elite participated in a trans-Atlantic world centered on Britain, the resident planters created a mosaic of local, vernacular cultures in Newfoundland. Unlike other settlement frontiers in North America, there was little mixing of different ethnic groups: Catholic Irish and Anglican English rarely intermarried. As a result, individual outport populations usually comprised one or the other ethnic group. As generations passed, these outport populations became increasingly thick with kith and kin, helping to maintain Old World folk cultures.[52] Irish and West Country English dialects were spoken in many communities, and much Old World folklore was preserved.[53] English and Irish names helped domesticate the landscape.[54] Despite these cultural differences, the material circumstances of life were fairly uniform. Almost all the resident population depended on the fishery and the rhythms of the sea. Agriculture was minimal, comprising the growing of potatoes and other roots, and the raising of cattle. Game and berries provided further foodstuffs. Nearby forest was cut for fuel, fences, and buildings; wood-framed structures replaced the stone and cob dwellings common in the southwestern British Isles. Divided by religion and ethnicity and hemmed in by poverty, the residents were among the poorest white settlers in British America.

As economic and political power over Newfoundland and the fishery lay in Britain, there was only the barest institutional framework on the island. During the seventeenth century, colonial proprietors scarcely provided any administration for their settlers, while the English government was content to let the fishery take care of itself. The system of fishing admirals, devised in the early days of the migratory fishery, continued throughout the century and into the early 1700s. According to this arrangement, the commander of the first English vessel to reach a harbor in the spring had responsibility for maintaining law and order. Such a system was open to many abuses, and tended to favor the strong over the weak. Disputes between fishing admirals were supposed to be adjudicated by the naval commander responsible for convoying fishing vessels across the Atlantic, but his authority was difficult to enforce, even in St. John's, his residence during the summer. In the outports, lawlessness was common.

As competition between migratory and resident fishermen for good fishing rooms became intense in the early eighteenth century, lawlessness became endemic. At last, in 1729, the British government appointed the naval commander as governor and commander-in-chief. By placing a naval officer in charge of civil affairs, the government achieved three aims: First, it ensured that the metropolitan government had direct control over the administration of the island; second, it saved the British government or the West Country merchants the expense of financing a civil governorship; and finally, naval ships and their crews provided a flexible government infrastructure.[55] For his part, the governor appointed magistrates, usually drawn from the principal merchants, who were responsible for law in the outports.[56] Such merchants worked closely with the governor and junior naval officers, and served as representatives of the British state at the local level. In 1750, courts were established on the island to hear capital crimes.[57] These courts were held in the principal settlements during the fishing season when naval vessels were cruising the coast. Junior naval officers served as surrogate judges, brigs on the vessels were used as lock-ups, and marines acted as police. Although a practical and efficient legal apparatus and infrastructure had been put in place, there was no legislative assembly on the island and all laws continued to be made in Britain. As several merchants served as MPs in the British parliament, West Country merchants effectively governed Newfoundland through Westminster rather than through St. John's.[58]

Other forms of metropolitan power developed on the island. As the Newfoundland fishery was one of the most important staples in British America, a major contributor to the Spanish trade, and a significant nursery of seamen, Britain had considerable interest in protecting the island and the West Country fishing fleet. By the late seventeenth century, the navy was convoying fishing vessels back and forth across the Atlantic, patrolling the coasts of Newfoundland, and protecting English commercial interests in the Mediterranean.[59] In the eighteenth century, fortifications were built at St. John's and Trinity.[60] Metropolitan religious institutions also had a presence. The Anglican Society for the Propagation of the Gospel and the Roman Catholic church sent missionaries to Newfoundland in the eighteenth century, although there were few of them and much of their pastoral work was in St. John's.[61] Moreover, the Anglican Church was associated closely with the establishment of government, military, and merchants, rather than with fishermen and servants.

Over the course of two hundred years, the British extended a metropolitan economic, political, and military system around the coast of Newfoundland. With no local labor or agriculture to draw upon, West Country, Irish, and Channel Island merchants imported thousands of men and tons of supplies and provisions each year to harvest one of the richest fisheries in the world. Seasonal and year-round settlements were established along rocky, forbidding shores, their populations completely dependent on the fishery and trade connections around the North Atlantic. While fishermen and their families gained a minimal subsistence from the staple, fish merchants transferred much of the wealth of the island back to the British Isles. With no house of assembly on the island, the important decisions affecting Newfoundland were taken by a military governor and a British parliament that contained strong representation from the West Country. During times of war, the British navy safeguarded the fishing fleets and trade routes, and depended on the migratory fishery for experienced seaman. To incorporate Newfoundland into a "Greater New England," as some American scholars have argued, is to ignore these ties back across the Atlantic.[62] In the seventeenth and eighteenth centuries, Newfoundland was much more integrated into the British Atlantic than into the continental colonies.

The Sugar Colonies in the West Indies

If the Newfoundland fishery was the great buttress of Britain's Atlantic economy in the north, the sugar colonies in the West Indies formed an even greater support in the south. After the sugar revolution swept through the islands in the mid-seventeenth century, the West Indies became the most important staple-producing region in the Atlantic; between 1768 and 1772, the islands exported sugar, molasses, and rum worth £3,383,915 on average each year to Britain, equivalent to 54 percent of all exports from British America.[63] Moreover, the sugar colonies contributed enormously to the growth of the slave trade, provided a market for British manufactured goods and American provisions, supported a huge merchant marine, and generated massive capital accumulation in the British Isles. And like Newfoundland, the West Indies remained tightly within Britain's metropolitan grasp throughout the early modern period.

English involvement in the West Indies began with the colonization of Barbados and the Leeward Islands (St. Kitts, Antigua, Nevis, and Montserrat) in the 1620s and early 1630s. After a period of agricultural experimentation with the growing of tobacco and cotton, the cultivation of sugar cane was introduced from Brazil to Barbados in the mid-1640s and soon took hold (figure 2.9).[64] During the second half of the seventeenth century, Barbados was the principal sugar-producing island in the West Indies, and the richest English colony in the New World.[65]

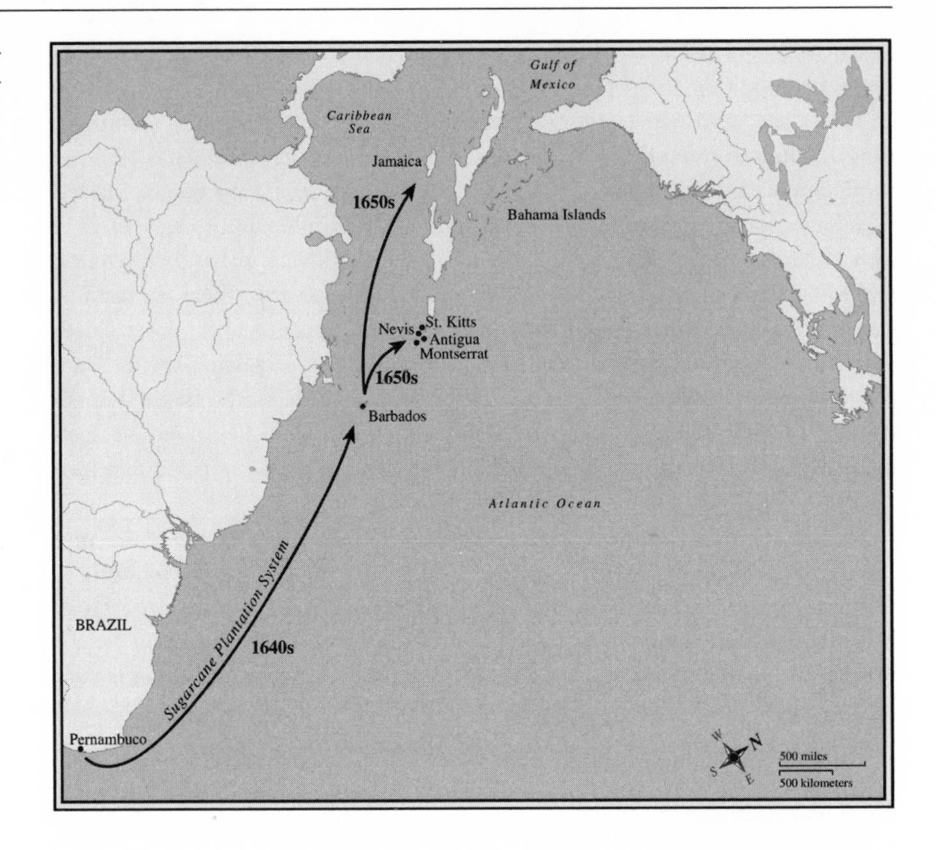

FIGURE 2.9. Expansion of sugar cultivation in the English West Indies, mid-seventeenth century.

Meanwhile, sugar cultivation was gradually extended to the Leewards, and these islands took over from Barbados as the main sugar producers after about 1710. Looking to expand their plantation frontier, the English seized Jamaica from the Spanish in 1655, and turned it into another, much larger, sugar-producing island. By the early 1740s, Jamaica had replaced the Leewards as the principal sugar producer, and became the richest colony in British America.[66] After the defeat of the French in the French and Indian War (1754–1763), Britain acquired the Windward Islands of Grenada, Dominica, St. Vincent, and Tobago. Although the British maintained the existing plantation economies on the islands, the Windwards were minor sugar producers until the 1770s.[67] By the outbreak of the American Revolution, the British controlled a sugar-producing region that encompassed one of the Greater Antilles (Jamaica) and most of the Lesser Antilles. Outside the Caribbean, the British also governed the Bahamas, a colonial offshoot of Bermuda, first settled in the 1640s and a struggling economic backwater.

English demand for sugar lay behind the expansion of the sugar economy in the West Indies. In the early 1600s, sugar was a luxury commodity produced in the Mediterranean, the Canaries, and Madeira; only small amounts were imported into England for consumption by the wealthy classes. As production spread to

Brazil and then to the West Indies, imports into England increased, prices fell, and consumption rose. In the early 1650s, only a few thousand tons of sugar were exported from the West Indies to England; by 1700, the figure had reached 24,000 tons; and by the 1770s, about 88,000 tons.[68] By then, sugar had become a dietary staple in virtually all homes in the British Isles, particularly taken in conjunction with that other great colonial staple: tea.[69] Sugar had also become the leading import into the country. In 1700, sugar imports were second in value only to linens and calicoes from the East Indies; by the 1770s, sugar was the most valuable of all England's imported goods.[70] In return, manufactured goods and provisions were shipped to the sugar islands. By the late eighteenth century, the trade between England and the West Indies comprised between one-sixth and one-fifth of the country's foreign commerce.[71] The West Indies also had important trade links with the continental colonies and Newfoundland, exchanging rum, sugar, and molasses for dried fish, lumber, and provisions.

Merchants in London and Bristol directed much of the early settlement and economic development in the English West Indies. After the failure of the Virginia Company in 1624, the great London merchants involved in the chartered companies shied away from investing in trans-Atlantic ventures, leaving the field open to domestic tradesmen, shopkeepers, and ship captains.[72] Following the precedent set by West Country merchants in the Newfoundland fishery, London and Bristol merchants moved quickly to exploit the rich opportunities in the West Indies. The merchants were instrumental in developing the early tobacco and sugar plantations on the islands, supplying capital and labor (indentured servants and slaves), as well as a transportation connection to the English market. In some cases, merchants went into partnerships or "mateships" with planters in order to develop plantations; in other cases, merchants invested directly in plantations, creating a powerful group of merchant-planters.[73] By investing in production as well as circulation, merchant-planters followed a pattern similar to that in the Newfoundland fishery, where West Country merchants were involved in both fishing and transporting and marketing of fish. As the sugar economy became established, the early partnerships gradually were replaced by a new arrangement: A class of large planters looked after production, consigning their sugars to merchants, who took care of transportation and marketing. Acting on behalf of the planters, the merchants sold the sugar on a commission basis, ordered provisions and goods, and provided credit. In effect, the merchant acted as the planter's agent or factor, as well as banker. One result of this partnership was to create a powerful coalition of interests—the West India lobby—in London;[74] another result was to bind at least some of the planters completely to the merchants. Although the great sugar planters were much richer and more powerful than the fishermen planters in Newfoundland, like the fishermen, the sugar planters eventually came to rely too heavily on merchant credit and became entrapped in debt. By the late eighteenth century, some of the greatest profits from the sugar industry were not accruing to planters in the West Indies but to merchants in Britain.[75] The wealth of the West

Indies, like that of Newfoundland, was to be found not on the islands of the Caribbean but in the shires and cities of the British Isles.

Soon after the English settlement of the islands, a class of large planters began to develop. In Barbados, the Earl of Carlisle, the first proprietor, leased 10,000 acres to a syndicate of London merchants, as well as making smaller leases, ranging from 50 to 1,000 acres, to incoming settlers. Even before the sugar revolution of the mid-1640s, a group of large planters had emerged on the island, with some controlling plantations of several hundred acres.[76] As sugar replaced tobacco as the staple crop, the costs of production increased enormously. Unlike the relatively straightforward drying of tobacco leaf, sugarcane had to be crushed, boiled, and cured, a complex process that required skilled hands and substantial infrastructure. To recover the cost of a sugar works, planters had to increase production, which, in turn, required more land and labor. As a result, the industry needed considerable amounts of capital, which favored wealthy planters. As the sugar boom drove up the price of land in the 1640s, small holders sold out to larger planters already on the island.[77] Meanwhile, new investors from England, many from gentry and commercial backgrounds, purchased plantations.[78] By the 1660s, a class of large planters had emerged.[79] According to a census of 1680, 175 of the biggest planters in Barbados, drawn from 159 families and constituting 7 percent of all property holders, controlled 54 percent of the land and 60 percent of the slaves. Of these 159 families, 62 had been landowners in 1638, while 34 had arrived between 1640 and 1660.[80]

As the sugar revolution spread from Barbados to the Leewards and Jamaica, the rise of the planter class was repeated. A distinct group of large planters began to appear in the Leewards in the 1680s, and was fully developed by the early eighteenth century.[81] In 1707, the governor of the Leewards described Nevis as "a rich little Island, but there are few people, the Island was devided amongst a few rich men that had a vast number of slaves, and hardly any common people, but a few that lived in the town."[82] Much the same could have been said of Antigua, St. Kitts, and Montserrat. During the same years, the planter class emerged in Jamaica. As early as the 1670s, there was a small group of very large planters on the island, and this group increased as the island was settled and plantations developed in the last quarter of the seventeenth century.[83] By the early 1700s, the big sugar planters completed dominated the economy of the island.[84]

As plantations began producing substantial profits, an increasing number of planters entrusted the day-to-day running of their holdings to estate managers and returned to England to live off their wealth. As generations passed and estates were inherited by people living in England, absenteeism became an entrenched way of life. The bankruptcy of many planters in the late eighteenth century and the take-over of their plantations by their creditors — the sugar merchants — resident in Britain further consolidated the pattern of absentee control. Such absenteeism was most common among large planters in the Leewards and Jamaica, less so in Barbados. The prevalence of absenteeism had several effects: First, absenteeism detached planters from the source of their wealth. This weakened their control over daily operations and, in many cases, led to problems of management and breakdowns in the running of plantations. Second, absenteeism removed a significant element of the planter class from the West Indies, thereby stunting the development of an indigenous elite in the colonies. Finally, absenteeism helped ensure that the West Indies remained economically, politically, and ideologically bound to Britain.[85]

Among the planters who exemplified many of the trends of the West Indian planter class were the Pinneys of Nevis.[86] Originally from west Dorset, the West Indian line of the Pinney family began with young Azariah, who became involved in Monmouth's ill-fated rebellion in 1685 and, for his troubles, was shipped out to Nevis. After serving as a planter's agent for a number of years, Azariah gained control of a small plantation and a handful of slaves. Through careful management, he accumulated further holdings on the island. Meanwhile, his son married a sugar heiress with properties in Antigua and adjoining the original Pinney estate in Nevis. Both father and son died within a few months of each other in 1720, and the Pinney estates—by then, among the largest in Nevis—passed to grandson John Frederick Pinney. With the sugar fortune established, John Frederick was content to live the life of a country squire in Dorset, visiting Nevis only twice to oversee his plantations. Yet these visits were not sufficient to shore up the family fortune, and, like other planters, John Frederick began to slide into debt. At his death, the property passed to a young cousin, John Pinney, who not only restored the family fortune but increased it many times over. Pinney achieved this by managing the plantations himself from 1764 to 1783, and transferring income back to England. After leaving Nevis, Pinney used his accumulated capital to set himself up as a sugar merchant in Bristol (figure 2.10). Over time, he extended loans to planters in the West Indies, making far more money from interest payments than he ever did from his own plantations. By the end of the eighteenth century, the Pinneys, like other merchant houses in England, were benefitting much more from the sugar economy than were the planters resident in the Caribbean.

The development of first tobacco and then sugar plantations required a massive labor force that was unavailable in the West Indies. Although the Spanish had made use of native labor in the early years of settlement in Hispaniola, the English found some islands uninhabited and others settled by hostile Caribs and maroons (escaped Spanish slaves). In Barbados, a combination of Spanish slaving raids and the spread of European diseases most likely depopulated the island in the early 1500s, leaving it open for later English settlement.[87] In the Leewards, however, there were remnant populations of Caribs, and these were quickly destroyed by the first settlers. In St. Kitts, the French and English, who occupied different parts of the tiny island during the early years of colonization, combined forces and massacred the local Indians. Although Jamaica had no native population left by the mid-seventeenth century, the English had to contend with maroons, who fought running guerrilla wars against the settlers well into the eighteenth century.

FIGURE 2.10. John Pinney House, Bristol, 1788–1791. Photograph by the author, 1999.

Clearly, neither the Caribs or the maroons were a suitable source of plantation laborers.

As in the Newfoundland fishery, the merchants and planters involved in developing the plantations turned to local supplies of labor in the British Isles.[88] From the 1620s to the 1750s, a mix of English, Irish, and Scots migrated to the West Indies, some of their own free will, others deported as criminals or prisoners of war. Of the migrants who went voluntarily, a great many, almost half, were indentured servants.[89] The major migration streams of English servants to the West Indies were directed to Barbados in the 1650s and 1660s, the Leewards in the 1660s and 1670s, and Jamaica in the 1680s.[90] The timing of these inflows reflected the changing demand for labor as the sugar revolution spread through the islands and as plantation work forces switched from servants to slaves. Jamaica also received substantial numbers of servants in the 1720s and 1730s, reflecting a continuing need for skilled and semi-skilled labor. As in the fishery, the overwhelming majority of servants were male, aged between fifteen and thirty, and comprised farmers, laborers, tradesmen, and youths.[91] Unlike the fishery, however, these servants were drawn from across England. As the two principal ports involved in the sugar trade were London and Bristol, the migration fields covered considerable areas; London attracted labor from much of England, while Bristol, the "metropolis of the west," drew from the southwest, the midlands, and south Wales.[92] The mixing of servants from different parts of England would make the maintenance of English regional cultures difficult, if not impossible, in the West Indies. A second major migration stream came from Ireland. During the 1650s and 1660s, the largest migrations to the English West Indies were most likely Irish servants recruited for plantations in the Leeward Islands.[93] Indeed, Montserrat quickly gained a reputation as an "Irish island."[94] After the Act of Union between England and Scotland in 1707, large numbers of Scots emigrated to the colonies. Of those who went to the West Indies, several hundred fetched up in Jamaica, serving in the professions, plantation management, and commerce.[95] Apart from free migrants, thousands of prisoners of war, criminals, and kidnap victims—"sturdy beggars, thieves, and vagabonds" according to Jamaican planter Edward Long—were shipped off to the West Indies.[96] Most of the prisoners came from Cromwell's brutal campaigns against the Irish and the Scots between 1649 and 1655. During those years, Barbados alone received some 12,000 prisoner-servants.[97] Criminals were given the choice of transportation or incarceration, and many chose the plantations. Such was the demand for labor, however, that kidnap rings or "spirits" operated in London and Bristol, kidnaping servants and selling them to merchants in the West Indies trade.[98] Given such methods of labor recruitment, the phrases "to be Barbadosed" and "to be spirited away" soon entered seventeenth-century popular parlance.[99]

Unlike the seasonal rise and fall of labor in the Newfoundland fishery, the influx of servants had a dramatic, albeit short-lived impact on the white populations of the English West Indies. In Barbados, the first island to feel the effect of immigration, the white population increased from approximately 1,200 in 1635 to 23,000 in 1655; in Jamaica, the population increased from 2,900 in 1661 to 7,400 in 1698.[100] Some of this growth was due to natural increase but much of it was the result of immigration. With men comprising most of the immigrants, the white populations in the English West Indies were unbalanced and natural reproduction was extremely slow. Yet as plantations changed over to slave labor in the late seventeenth century, the white servant class was gradually phased out. At first, servants were replaced by slaves in the unskilled laboring jobs in the sugar-cane fields; eventually, they were displaced from the skilled sugar processing work, leaving only a handful of whites in clerical and managerial jobs.[101] With little opportunity to purchase land and few white women available as potential marriage partners (a situation similar to that in Newfoundland), the redundant servants were "not fixed" on the islands. Although some stayed, forming a poor white laboring class, many looked elsewhere to settle. Between 1650 and 1666, some 17,000 whites left Barbados, mainly for Jamaica, Surinam, Virginia, and New England; in the 1670s, several hundred more left for the Carolinas.[102] Such outflows led to a fall in the white population in Barbados, from a peak of about 23,000 in 1655 to only 12,500 in 1715. In the Leewards, the fall in the white population was less dramatic, from 10,400 in 1678 to 7,700 in 1708, reflecting a lesser dependence on white servants. Settled after the other islands and benefitting from immigration from Barbados and the Leewards, Jamaica saw an increase in its white population from 7,400 in 1698 to 17,900 in 1768.[103] As the largest of the sugar islands, Jamaica offered greater opportunities than in the other islands for white immigrants to settle or to take managerial positions on the plantations.

The change from white servants to black slaves reflected the changing cost of labor in the English West Indies.[104] As the tobacco and sugar plantations developed in the early seventeenth century, indentured servants from the British Isles were plentiful and cheaper than slaves from West Africa. In Barbados, the average price of a male servant was about £10 between 1635 and 1660, while that of a male slave was about £30. Although servants were indentured for five to seven years depending on age and slaves were bound for life, the high mortality rates of both servants and slaves meant that six or seven years was a realistic working life for both groups. Given such short life expectancies, planters found servants, indentured for at least four years, much cheaper than slaves. Yet in the late seventeenth century, the situation was reversed. The population of England appears to have stopped growing during those years, forcing up wages of domestic labor and lessening the need for servants to emigrate.[105] Moreover, the brutal work conditions on the sugar plantations made the West Indies a much less attractive destination than the mainland colonies. As the supply of servants fell, particularly in the 1660s, so their price rose. During those years, merchants resorted to underhand methods to recruit servants, while planters engaged labor on ever more favorable terms. Eventually, the price of indentured servants had risen so high that slaves, always an expensive form of labor, became an economic alternative. The transition to slaves was first made during the 1650s and 1660s in Barbados, and thereafter spread to the Leewards and Jamaica, as well as to the Chesapeake and the Carolinas; in effect, Barbados became the model for the other slave-holding societies that developed in British America.

Until the sugar revolution, there had been little demand for slaves in the English West Indies and only about 20,000 were imported during the first three decades of settlement.[106] But with the establishment of large-scale sugar production in the

1650s and the growing shortfall of English servants in the 1660s and 1670s, the slave trade increased rapidly. As Edward Long observed: "the Negroe trade . . . [was] the ground-work of all."[107] Like other New World trades, the slave trade became a battleground between regulated chartered companies and independent merchants. In 1663, the Company of Royal Adventurers into Africa was chartered in London with a monopoly of all trade along the coast of West Africa. After financial setbacks, the Company was reconstituted in 1672 as the Royal African Company (RAC), again with monopoly trading privileges. Such was the demand for slaves, however, that many English merchants pressed for a repeal of the monopoly and interloped into the trade. In 1698, the Company lost its monopoly in return for a 10 percent tax on all goods exported from England to West Africa. This arrangement lasted until 1712, when the trade was thrown open to all comers. Thereafter, the Company limped along until its dissolution in 1750, while private merchants from London, Bristol, Liverpool, and such lesser outports as Lancaster, Whitehaven, and Lyme Regis took over the trade.[108]

As the English trade for slaves developed along the coasts of West Africa, two different patterns of trading emerged. The Royal African Company and some independent merchants established forts and factories through which a fort trade was conducted; other merchants cruised along the coast and conducted a ship trade. The fort trade was particularly concentrated at Gambia and along the Gold Coast, where there were as many as seventeen forts. Of these, Cape Coast Castle served as the RAC's African headquarters, staffed by an Agent-General, company servants, soldiers, and slaves. At the height of its power in the 1680s, the Company employed more than 300 white men.[109] As slaves were traded, they were held at the forts and factories in congested *barracoons* until vessels arrived to take them across the Atlantic. The ship trade was more flexible, and extended up and down the slave coasts. Slave vessels acted as mobile warehouses, exchanging British manufactured goods for human cargoes (figure 2.11).

Between 1651 and 1775, chartered companies and private merchants exported some 1,656,000 slaves from Africa to the British American and Caribbean colonies (figure 2.12). Of these, approximately 1,377,000 survived the horrendous middle passage between Africa and the Americas, while perhaps another 279,000 died during the crossing.[110] The slaves who arrived in the British West Indies comprised about 86 percent of all slaves shipped to British America. Among the West Indies–bound slaves, 538,000 or 39 percent of the total went to Jamaica, 314,000 or 23 percent to Barbados, and 326,000 or 24 percent to the Leewards, the Windwards, and the Bahamas. Tens of thousands of slaves, however, were re-exported to other parts of the Americas. With its windward location outside the Lesser Antillean Arc, Barbados served as the entrepôt for both white and black populations in the British West Indies, distributing immigrants and slaves to other Caribbean islands, as well as to the Chesapeake and the Carolinas.[111] After the signing of the *asiento* agreement between Britain and Spain in 1713, slaves were also exported to the Spanish colonies, mainly through Jamaica. Compared to the migrations of

FIGURE 2.11. *The Southwell Frigate,* by Nicholas Pocock, circa 1760. A portrait of a Bristol-owned, former privateer fitted out for a slaving voyage along the coast of West Africa. The vignettes show the ship trade: English goods exchanged for African slaves. Bristol Museums and Art Gallery.

Europeans to the mainland colonies, these slave shipments to the West Indies were far larger, representing enormous movements of people.[112] For British America as a whole, the dominant migration stream was not from Europe to North America but from West Africa to the Caribbean.

Although the migration hinterlands of white servants covered many parts of the British Isles, the source areas for black slaves were far larger and culturally more diverse.[113] The first slaves imported into Barbados came from Dutch slaving expeditions as far afield as Madagascar, the Gold and Slave coasts of West Africa, the Angolan region of Central Africa, and Brazil.[114] After the English entered the slave trade in the 1650s, slavers concentrated on the West and Central African coasts, an enormous area that stretched from the Senegambia in the north to Angola in the south. Within this area, much of the activity in the late seventeenth and early eighteenth centuries centered on the Gulf of Guinea—the Windward, Gold, and Slave coasts—and became known as the Guinea trade.[115] Slaves from Coromantine, a generic name for the Gold Coast (present-day Ghana), were considered the best; Antiguan planter Christopher Codrington considered the male slaves "really all born Heroes."[116] From the 1730s, the slavers shifted eastward and took an increasing proportion from the populous Bight of Biafra. Given such an extensive source area, slaves were drawn from numerous peoples, including the Ashanti, Fanti, Dahomey, Yoruba, Benin, and Ibo. Depending on the period of trade and preference of the buyer, the proportions of different tribal groups among

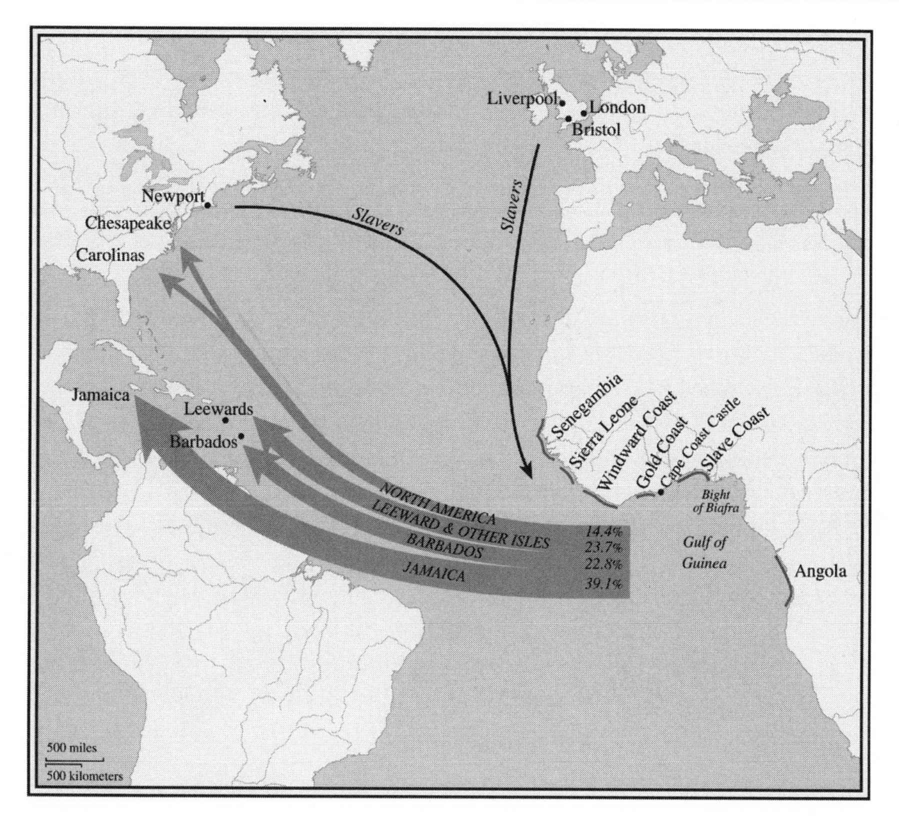

FIGURE 2.12. Slave shipments from West Africa to the Caribbean and North America, 1651–1775. Based on data in David Eltis, Stephen D. Behrendt, David Richardson, and Herbert S. Klein, eds., *The Trans-Atlantic Slave Trade: A Database on CD-ROM* (Cambridge: Cambridge University Press, 1999).

West Indian slaves varied considerably. In Barbados, a large number of slaves came from the Akan peoples (Ashanti and Fanti) on the Gold Coast.[117] Mixing different ethnic groups on plantations was deliberate policy; as one observer commented in 1694: "the safety of the Plantations depends upon having Negroes from all parts of *Guiny*, who not understanding each others languages and Customs, do not, and cannot agree to Rebel, as they would do . . . when there are too many Negroes from one Country."[118]

The greatest demand from planters was for healthy young adult males, but high prices for these slaves ensured that slavers also took adult females as well as children. Of the 62,000 slaves shipped by the Royal African Company to Barbados and Jamaica in the years 1673 to 1725, just over half were men, a third women, and the rest children.[119] Although the composition of the slave population was more balanced than that for the white servant population, the work force did not reproduce. Harsh conditions on many plantations ensured a low rate of reproduction and a constant need for new workers. Despite the costs, planters preferred to "buy rather than breed" their work forces.[120] In effect, the sugar plantations, particularly the larger ones in Jamaica and the Leewards, became monstrous consumers of people, and the West Indies became reliant on an ever-increasing trans-Atlantic trade in human beings.

The massive importations of slaves had a major impact on the demography of the British West Indies, dramatically boosting their total populations and radically changing their racial compositions. In Barbados, the population increased from about 23,000 in 1645 to 87,000 in 1773; in the Leewards, from 21,000 in the late 1670s to almost 89,000 in the early 1770s; and in Jamaica, from 18,000 in 1675 to 216,000 in 1775.[121] Blacks formed a majority of the population in Barbados around 1660, in Jamaica in the early 1670s, and in the Leewards after 1680.[122] By the early eighteenth century, the ratios of blacks to whites was four to one in Barbados, six to one (later increasing to eleven to one) in the Leewards, and eleven to one in Jamaica.[123] The higher ratios of blacks in Jamaica and the Leewards, as compared to Barbados, reflected their much larger plantation work forces. Overall, the sugar revolution had steered the British West Indies onto a demographic course almost unique among the British American colonies; only South Carolina on the mainland had a population with a black majority.

Apart from these inflows of capital and labor, the sugar economy needed supplies and provisions. Like other British American colonies, the West Indies depended on Britain for manufactured goods, such as textiles, ironware, and pottery, but the islands also needed vast quantities of foodstuffs and lumber. The spread of sugar cane cultivation across much of Barbados and the Leewards, as well as around the coastal lowlands and more accessible valleys of Jamaica, left little room for growing foodstuffs or harvesting forest for timber. Although waste land on the islands was set aside as provisions grounds for the slaves and some cattle were raised in the more mountainous areas of Jamaica, there was still a great demand for foodstuffs. The British Isles, particularly southeastern Ireland, supplied salted meat and butter, but the American colonies shipped most of the necessary foodstuffs and lumber.[124] During the seventeenth century, New England dominated the trade, exporting salted meat, dried fish, livestock, pine boards, and staves (figure 2.13). In the early eighteenth century, the Mid-Atlantic and the Chesapeake colonies entered the market, shipping bread, flour, and corn; these colonies were soon joined by South Carolina, which exported livestock and rice.[125] In return, the American colonies imported sugar, molasses, and rum. As in Newfoundland, the provisions trade allowed American merchants to enter a market that was otherwise dominated by metropolitan merchants.

As the sugar revolution swept through the West Indies, the natural landscape was transformed. On Barbados and the Leewards, much of the tropical rain forest was cleared, apart from a few inaccessible gullies and steep slopes, and replaced with a settled landscape of plantations and sugar cane fields (figure 2.14). On the much larger island of Jamaica, plantations spread around a coastal strip some 10 to 15 miles wide, leaving the rugged interior, comprising the broken and difficult terrain of the Cockpit Country and the precipitous Blue Mountains, largely unsettled.[126] Over time, it became the refuge of the maroons and other escaped slaves. Apart from Jamaica, the small size of the islands reminded visitors of parts of England, particularly the Isle of Wight. In the mid-twentieth century, the English travel

FIGURE 2.13. Detail from *View of St. Christopher's*, by Thomas Hearne, 1775–1776. This scene of the foreshore at Basseterre, the capital of St. Kitts, shows sawn lumber, open boats, and hogsheads, all imported from the American colonies, probably from New England. © Christie's Images Limited, 1994.

FIGURE 2.14. Detail from *The Island of Montserrat from the Road before the Town*, by Thomas Hearne, 1775–1776. This view shows vessels waiting for cargo, the main port town of Plymouth, the fort defending the town, sugar cane fields, dispersed plantations with windmills, forested ravines with severe gully erosion, and the volcanic mass of the Souffrière Hills. © Christie's Images Limited, 1994.

writer Patrick Leigh-Fermor likened Barbados to "a shire that had drifted loose from the coast of England and floated all the way to . . . tropic waters, its familiar fields having acquired outlandish flowers and trees on the journey, but never in great enough quantities to impair the deception."[127] Yet such a comparison missed obvious differences. The tropical heat and humidity, the torrential downpours, the ever-blowing trade winds, the devastating hurricanes, the remnants of jungle with their exotic flora and fauna, the outcrops of coral, the volcanic rumblings, the warm, translucent seas all marked the islands from the British Isles. Moreover, the pattern of settlement was much different. Even on islands as small as Barbados, St. Kitts, Nevis, and Montserrat, plantations were dispersed across the landscape, with service functions, such as churches, established at central, accessible locations. The availability of land, at least initially, ensured that the clustered villages of England with their farmhouses, parish churches, and open fields were not reproduced in the tropics.

The plantation landscape that developed in the British West Indies was one of the most specialized in early modern British America. The plantations were large units of production, usually covering several hundred acres, with the biggest estates — in Jamaica and the Leewards — encompassing more than a thousand acres (figure 2.15). Frequently pieced together from smaller holdings, plantations were usually irregular in shape, although they often contained some regular-sized fields.[128] Most plantations had a similar layout, with the sugar processing works, slave huts, slave hospital, overseer's house, and great house clustered together at a central location.[129] Around this knot of buildings lay the cane fields and provisions grounds where the slaves raised their own foodstuffs. The close proximity of the great house to the slave huts, sugar works, and cane fields allowed the planter to keep an eye on the slaves and the workings of the plantation. On his Mountravers plantation in Nevis, John Pinney instructed his manager "to have trees planted in any part of the estate that will not intercept the view of the estate from the house."[130]

Although plantation great houses were the largest and most conspicuous residences on the islands, they were surprisingly modest in size and style compared to the plantation homes erected in the Virginia tidewater or the Carolina low country during the mid-eighteenth century or the magnificent "power houses" that so dominated rural England. Apart from a few exceptional houses, such as Rose Hall in Jamaica and the Principal's Lodge, Codrington College (formerly the great house of Consetts plantation) in Barbados, plantation houses were comparable in size to small manor houses in England or planters' houses in Ireland.[131] On the outside, West Indian great houses were marked by plain Jacobean or Georgian architecture, some with piazzas to provide shade; on the inside, rooms were usually "fitted up and furnished in the English taste."[132] Houses also had gardens and shaded walks. Visiting Antigua in 1775, Scottish writer Janet Schaw observed a "superb" plantation house outside St. John's, "laid out with groves, gardens and delightful walks of Tamarind trees, which give the finest shade."[133] Although gardens could

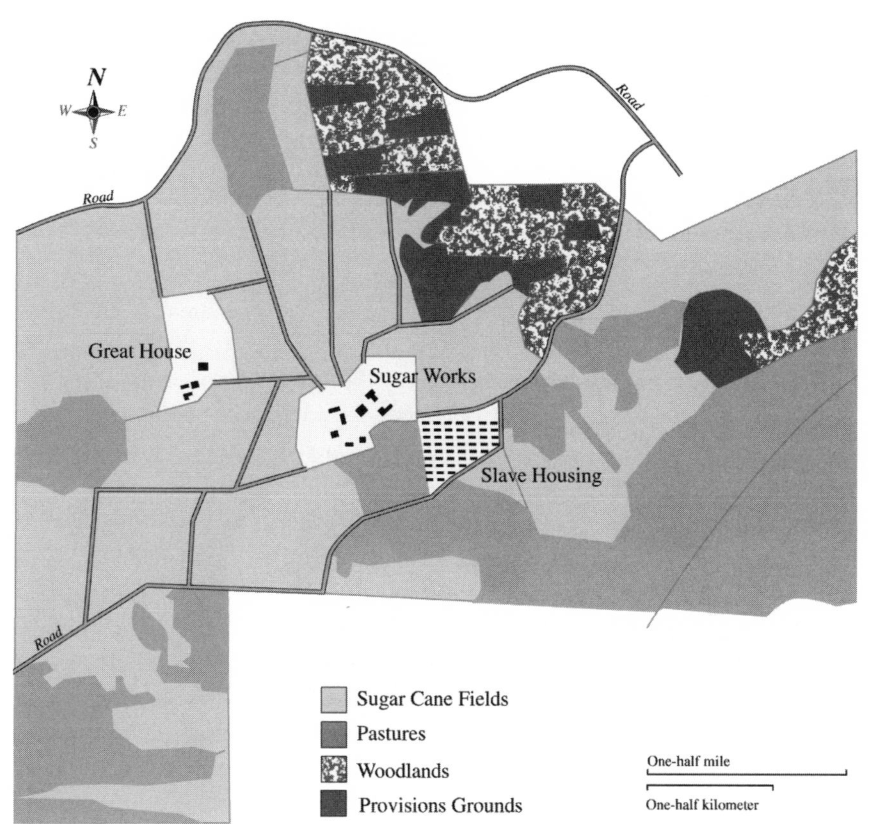

FIGURE 2.15. Plan of Carrickfoyle Estate, Trelawny Parish, Jamaica, 1793. The plan shows an irregularly shaped plantation; sugar cane fields, pasture, and provisions ground; and the distinct spaces of great house, sugar works, and slave houses. After James Robertson's plan of Carrickfoyle Estate, Trelawny, 1793 in B. W. Higman, *Jamaica Surveyed: Plantation Maps and Plans of the Eighteenth and Nineteenth Centuries* (Kingston: Institute of Jamaica Publications Limited, 1988), 36.

be maintained by elderly slaves, there was little point erecting a mansion in the West Indies when the planter class looked to Britain to settle. The great wealth of the planters, especially those from Jamaica and the Leewards, was displayed not in the West Indies but in the town houses and country estates of England.[134]

Cane fields and sugar works formed the economic nucleus of the plantations. In the tropical climate of the Caribbean, sugar cane could be grown year round, taking between fifteen and eighteen months to mature. To ensure that a crop was harvested each year, planters staggered the planting of their fields. Such a regime also meant that labor was continually employed in planting, tending, weeding, and harvesting the crops. At first, canes were planted in trenches, but severe soil erosion led to the development of cane hole cultivation. These square holes, approximately five feet in size, were laid out regularly across a field and then planted with canes. The holes prevented down wash of soil in any direction and also provided room for manure. Although extremely laborious to dig with hoes, cane hole cultivation became standard practice on plantations by the early eighteenth century.[135] After the cane was cut, it was transported by ox-drawn wagons or on the backs of mules to the sugar works for processing (figure 2.16). Like the stage in a Newfoundland

FIGURE 2.16. *Parham Hill House and Sugar Plantation, Antigua,* by Thomas Hearne, 1779. Although Hearne depicts plantation slavery as a West Indian pastoral, his view reveals much about the plantation system. In the foreground, he shows slaves cutting sugar cane under the watchful eye of an overseer on horseback; in the middle-ground, a sugar works (including windmill) and great house; and in the background, forested slopes too steep to cultivate.

fishing room, the sugar works was the heart of a West Indian plantation. The works comprised several buildings: a mill, where three rollers powered either by animals, water, or wind crushed the cane to produce juice; a boiling house, where the juice was heated in coppers until it began to coagulate and form sugar; and a curing house, where the sugars were allowed to dry. In addition, there was an animal house for accommodating the animals that powered the mill, a trash house for storing waste canes that were used as fuel in the boiling process, and a still house for making rum from the molasses removed during the curing. Essentially, the sugar works was a "factory set in a field."[136]

The demands of cultivating and processing the sugar cane led to a division of labor among the work force. Most basically, there was a division between the fit and unfit. Those able to work comprised about two-thirds of a plantation workforce, and were divided into skilled and unskilled workers. The skilled included gang leaders, boiler men, millwrights, coppersmiths, blacksmiths, carpenters, coopers, joiners, and masons, as well as a small group employed in the planter's household as cooks, domestics, butlers, washerwomen, grooms, and gardeners. The unskilled comprised the field hands, the great bulk of the slave work force. These workers were divided into two or three gangs depending on age. The first or "great gang" comprised adult men and women, and was set to do the laborious tasks of digging cane holes, carrying manure, and planting and harvesting the cane; the second or "young people gang" was composed of adolescents, who did the less-demanding tasks of weeding, gathering trash canes, and tending livestock; the third or "children's gang" comprised young children from the ages of five or six to eleven or twelve, who collected fodder for livestock and looked after domestic fowl. The unfit slaves included the elderly and infirm, pregnant women, and very young children.[137]

Although slavery marked all these workers, the division of labor led to different work conditions and life expectancies. Gang leaders, skilled craftsmen, and household domestics led easier lives than the field hands digging cane holes under the broiling sun. These elite slaves were highly valued by the planters, commanded good prices at slave auctions, and enjoyed generally better living conditions than the majority of the work force.[138] As a result, they lived longer than field hands. Skilled workers also appear to have formed the leadership among the slaves, in some cases leading slave revolts.[139] Field workers were "generally treated more like beasts of burden than like human creatures," bearing the brunt of the physical work, frequently working twelve-hour days six days a week, and much of the punishment.[140] Janet Schaw noted that "every ten Negroes have a driver, who walks behind them, holding in his hand a short whip and a long one . . . [The slaves] are naked, male and female, down to the girdle, and you constantly observe where the application has been made."[141]

The successful implementation of large-scale slavery on the sugar islands owed much to a repressive legal system, permanent and well-maintained military facilities, and a particular geographic context. All the island governments enacted slave codes, with Barbados introducing the first in 1661. Infractions were met with severe punishment. During the eighteenth century, British regiments garrisoned the larger port towns, as much to maintain internal order as to defend against external attack, while militias composed of able white men operated at the local parish level. The small geographic areas of Barbados and the Leeward Islands also helped the planters maintain control. On these well-settled islands, there was virtually no refuge for escaped slaves. Moreover, any slave revolt had to take over an entire island, a difficult proposition.[142] In Jamaica, the mountainous interior inhabited by the maroons served as a place of escape until the island's government made peace with the maroons in 1739 in return for handing back escaped slaves. From being the hunted, the maroons became the hunters. Although slave revolts occurred in Antigua in 1736 and in Jamaica in 1760, they were unsuccessful, testament to the power of the planters to control the interior spaces of the sugar islands.

The society that grew out of the sugar economy was the most stratified in early modern British America. At the top of the social hierarchy was a small group of planters; in the middle, a larger group of professionals, merchants, and petty retailers; and at the base, an enormous mass of slaves. Although many of the larger planters retired to England to live off their sugar profits, there were sufficient planters left, particularly in Barbados where absenteeism was less prevalent, to constitute a local elite or "plantocracy." The planters controlled the main source of wealth, commandeered the principal government positions, and dominated the politics of the islands. Like the merchants in Newfoundland, the planters had no backcountry farming population to deal with, and thus wielded immense power. Apart from the planters, there was a small professional group comprising Anglican clergymen, lawyers, doctors, estate managers, and clerks; a mercantile class of merchants, commission agents, tradesmen, and shopkeepers; and an artisan class of carpenters, blacksmiths, masons, tailors, butchers, and shipwrights. Many of these whites lived in the port towns. Remnants of the old indentured servant class lingered on in all the islands, forming a group of poor whites—"redlegs" in Bajan parlance—who scratched a living from small-holdings and odd jobs.[143] Although slaves were beyond free society, deprived of all rights, there were small groups of free blacks and coloreds on the islands. Some of these had never been enslaved, while others had been released from slavery (manumission) by their masters. They comprised a growing group of petty tradesmen, overseers, artisans, tavern keepers, and clerks.[144]

Like the West Country merchants in Newfoundland, the sugar planters in the West Indies were a transient elite that shuttled back and forth across the Atlantic.[145] In 1764, Lord Adam Gordon observed in Jamaica that the English "Inhabitants look[ed] upon themselves there as passengers only."[146] Whether in the Caribbean or England, the planters pursued a gentry lifestyle, passing the time drinking, horse racing, and gambling. In the West Indies, the planters attempted to recreate Georgian England in their great houses and formal gardens; in England, they purchased town houses in Bath and London and country estates in the shires. In Wiltshire and Dorset, the Beckford, Dawkins, and Drax families, drawing their income from plantations in Jamaica and Barbados, purchased extensive country estates and built substantial houses. With the largest sugar estates in Jamaica and reputed to be the richest man in Europe, William Beckford indulged in gracious living by building the Gothic folly of Fonthill Abbey and decorating it with Italian Renaissance paintings and furnishings.[147]

While richer members of the planter class reveled in the sophisticated high culture of Europe, slave populations created their own creole culture in the West Indies. Although the institution of slavery and the plantation economy dramatically changed the ways of life of Africans shipped to the New World, the huge numbers of slaves in the British West Indies provided sufficient critical mass to support the development of a creole language.[148] Along the West African coast, pidgin English, a simplified form of the language, was the *lingua franca* between English slave traders and African suppliers; it was also used aboard the slaving vessels and on the plantations in the West Indies. Pidgin also served as a *lingua franca* among the slaves, who spoke many different West African languages and dialects. Several African languages were spoken on the plantations, especially among first generation slaves, but there were never enough slaves from particular linguistic groups to preserve complete languages. As a result, elements of African languages mixed with pidgin English to create a West Indian creole. Although it appeared broken to English ears, the language developed its own grammatical construction and dialect, and formed the most distinctive regional language in early modern British America.

Apart from creating their own creole language, slaves maintained other elements of their African folk culture. West African music (particularly drumming), singing, dancing, folk tales, birth and death rituals were all transferred across the Atlantic.[149] Given the different tribal and cultural backgrounds of the slaves, the details and nuances of individual folk traditions were probably not maintained, and a simplified common culture most likely emerged in the Caribbean. The slaves also took instruments and ideas from the dominant white culture, including playing the fiddle at dances and wearing European-style masks for Christmas John Canoe parades. Elements of West African peasant agriculture and diet also survived. The slaves used hoes similar to those in Africa for cultivating sugar cane, as well as for digging their provisions grounds. In these small gardens, slaves grew a range of crops, such as yams and millet (guinea corn), that were common in tropical Africa.[150] Although African slaves had been shackled to a brutal plantation system in the West Indies, they had considerable control of those areas of their lives outside the production process. Over time, they developed a heavily Africanized, creole vernacular culture.

Unlike Newfoundland, the British West Indies were sufficiently settled and prosperous that they could support basic institutions. The principal organs of colonial government—governors, legislatures, and civil service—were established in Barbados, Antigua (capital of the Leewards), and Jamaica. As the franchise was restricted to larger property-holders, planters soon controlled the legislative assemblies and molded them to their own interests. Law courts were also established, and again the planters dominated the magistrate's bench as well as formulated, through the assemblies, most colonial laws. The Anglican Church was established throughout the islands, and upheld the social hierarchy. The islands were divided into parishes, and had parish councils or vestries to provide local government and raise taxes.[151] The vestries included the rector, local magistrates, church wardens, and elected vestrymen, and represented the planter interest at the local level. The clergy were often little more than "paid servants of the local plantocracy."[152] Indeed, several rectors themselves owned plantations and held slaves. For much of the seventeenth and eighteenth centuries, the church and the vestries resisted any missionary work by other denominations, leaving many poor whites of Irish extraction and the vast bulk of the slave population without any religious ministration. The only dissenting groups that were tolerated were Quakers and Sephardic Jews, both entrenched among the commercial classes in the port towns. With slaves and poor whites largely excluded, Anglican parish churches were a microcosm of the ruling class; their pews rented out by the vestry according to social precedence.[153] Reminiscent of English and Irish churches in their plain Georgian architecture, the parish churches of the sugar islands were as much symbols of the elite as the great houses on the plantations.[154]

Yet the West Indies, like Newfoundland, were part of a larger militarized British

FIGURE 2.17. Detail from *View of Antigua: English Harbour, Freeman's Bay, and Falmouth Harbour, Monk's Hill etc from the Hill near the Park*, by Thomas Hearne, 1775–1776. The militarized world of the British Atlantic: in the foreground is a group of British naval and army officers, including the Governor of Antigua, Sir Ralph Payne, seated second from right. This group overlooks Halfmoon Battery and Fort Berkeley, commanding the harbor entrance, on the left, and Freeman's Bay, the main anchorage, on the right. In the background is King's Yard, the principal naval dockyard in the eastern Caribbean. © Christie's Images Limited, 1994.

Atlantic. The threat of slave revolt, the menace of French or Spanish attack, and the extraordinary wealth of the islands ensured a sizeable British military presence. During the eighteenth century, the British constructed numerous fortifications on the islands, garrisoned the principal port towns (see chapter 5), built naval bases in Barbados, Antigua, and Jamaica, and stationed a squadron of warships in the Caribbean (figure 2.17).[155] Although the smaller Leeward islands were always vulnerable to invasion and exchanged hands several times during conflicts between Britain and France, the long-term security of the West Indies rested on British naval power.[156] As long as Britain commanded the sea, the West Indies and the sugar and slave trades were secure.

Over the course of the seventeenth and eighteenth centuries, Britain established a firm hold over the British West Indies. Metropolitan merchants and colonial planters imported tens of thousands of servants and slaves, as well as enormous amounts of provisions and supplies, to power a tremendously productive sugar economy. Large, highly specialized plantations spread over tropical island landscapes, producing sugar and rum for markets around the Atlantic rim. Over time, sugar became the most important colonial product imported into the British Isles, and the West Indies became a significant source of capital accumulation for planters and merchants resident in Britain. While absentee planters enjoyed some of the most lavish lifestyles in Europe, enslaved men and women created a distinctive creole culture in the Caribbean. Given the immense economic significance of the sugar islands to Britain and the influence of the West India lobby in London, the British government went to great lengths to maintain political and military control over the islands. Despite trade to the continental colonies, the West Indies remained economically, politically, militarily, and ideologically dependent on Britain.

The Fur Trade through Hudson Bay

Compared to the two great staples of the cod fishery in Newfoundland and the sugar plantations in the West Indies, the fur trade through Hudson Bay might appear of little importance to the development of British America. Indeed, American scholars have scarcely considered the significance of the trade.[157] Allowing for the relatively small value of furs—worth on average £8,236 per year between 1760 and 1775[158]—the trade was important for at least three reasons: First, it established an English presence on the northern flank of North America; second, the trade brought Europeans and natives together in a long-lasting relationship that helped preserve an indigenous presence across the northern half of the continent; finally, the trade was a particularly attenuated version of a more general seaborne commercial system that included chartered trading ventures to West Africa and South Asia. In many ways, the fur trade through Hudson Bay was the most extreme manifestation of Britain's commercial empire of the Atlantic.

The English fur trade in northern North America began in the late 1660s, when two disaffected French fur traders tipped off a group of London merchants that the route to the richest furs in Canada lay through Hudson Bay. Building on earlier English voyages to the Bay, London merchants sent out a trading ship in 1668, which returned the following year with marketable furs. In 1670, the Hudson's Bay Company was chartered, with a monopoly of trade over the entire drainage basin of Hudson Bay. The Company established trading posts around the coasts of Hudson Bay and James Bay, and for much of the next hundred years traded bayside (figure 2.18); only in the early 1770s did the Company, in response to competition from fur traders from Montreal, move inland and establish interior trading posts.

The Hudson's Bay Company was unique among English chartered companies operating in North America in that it managed to protect its trade and so remain in business. This had much to do with the severe subarctic environment. Hudson Bay was open for navigation for only a few months each summer, and there was no potential for farming at bayside posts. The high overhead costs of maintaining trading posts around the Bay dissuaded other merchants from interloping into the trade, while the lack of any agricultural alternative ensured that no rival economic activities developed. As a result, the Company could police its trade and maintain its monopoly.[159] Such control extended to prices of trade goods and furs, wages of employees, and regulations in the workplace. In its labor contracts, for example, the Company stipulated that employees were not to engage in private trade.[160] Inland from the Bay, however, the Company had little control, and the fur trade was open to all comers. Competition between the Company and traders operating

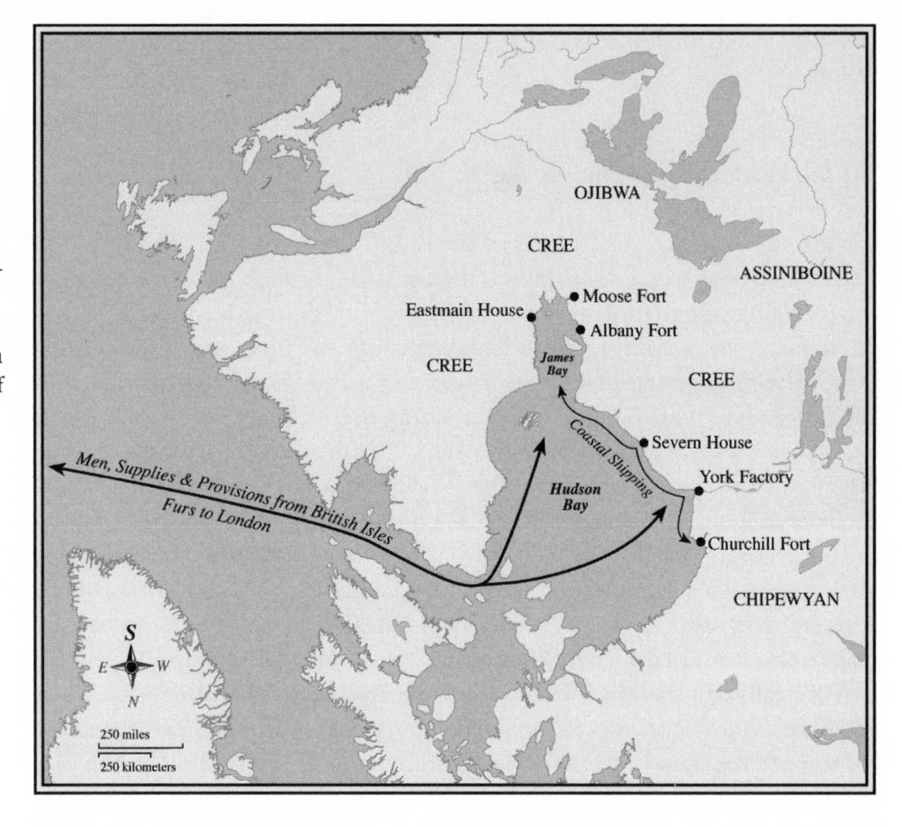

FIGURE 2.18. Fur trade through Hudson Bay, 1760s. After D. Wayne Moodie, Victor P. Lytwyn, and Arthur J. Ray, "Competition and Consolidation, 1760–1825," in *Historical Atlas of Canada*, vol. I: *From the Beginning to 1800*, ed. R. Cole Harris, plate 61 (Toronto: University of Toronto Press, 1987).

out of Montreal was intense during the eighteenth century, and continued after the British conquest of New France.

Bayside trade followed a simple and relatively unchanging pattern. After the ice melted on the rivers in early summer, "upland" Indians canoed down the rivers draining into Hudson Bay to trade furs at bayside posts. At the same time, two or three supply vessels loaded with provisions and trade goods were dispatched from London to the Bay. At the posts, the traders met the Indians, exchanged presents in elaborate ceremonies, and then began trading. Manufactured goods such as knives, hatchets, guns, shot, powder, kettles, beads, blankets, and cloth were exchanged for beaver, marten, and otter furs, as well as elk hides. Tobacco and alcohol were also traded to the Indians. Beaver fur was the main item of trade; "made beaver" became the standard against which trade goods and other furs were valued. Some trade goods were paid for directly with furs, others were purchased on credit. After trading was completed, the natives returned inland while the furs were forwarded across the Atlantic to London, where they were sold at auction to hatters for making felt hats.[161]

The Hudson Bay Company divided its trading area into the Bay and the Bottom of the Bay (James Bay). Trading posts were established at Fort Prince of Wales (Churchill), York Factory, and Severn House on the Bay, and at Albany Fort, Moose

FIGURE 2.19. York Factory, Hudson Bay, 1782. Toronto Public Library (TRL), J. Ross Robertson Collection: JRR 2329

Fort, and Eastmain House on James Bay.[162] All these posts were at the mouths of major rivers draining the immense Hudson Bay watershed, creating a radial pattern of posts and hinterlands around the southern rim of Hudson and James bays. Although Albany Fort on James Bay was the first company post, York Factory on Hudson Bay eventually superseded it in importance, and became "the capital place in Hudson's Bay" (figure 2.19).[163] Until the Company pushed inland in the 1770s, the only interior fort was at Henley House, a subsidiary of Albany Fort, built on the Albany River in 1720 in response to French competition.[164] The first major interior post was constructed in 1774 at Cumberland House on the Saskatchewan River, a strategic point across the old French route from Lake Winnipeg to the Saskatchewan and Athabasca river systems and well placed to siphon off trade from the Montrealers.[165] The posts themselves were utilitarian structures. Apart from the massive, stone-built fort at Churchill, Bay posts comprised stockaded enclosures, complete with corner bastions mounted with cannon or swivel guns against possible native or French attack, as well as numerous wooden buildings. These included separate quarters for officers and men; cook rooms; warehouses for furs, trade goods, and provisions; blacksmith, carpenter, and cooper shops; a trading house; a magazine; and ancillary store sheds.[166]

By the early 1770s, the Hudson Bay Company had almost 200 men serving in the fur trade.[167] These men were dispersed among the bayside posts, with individual staffs ranging from a handful to as many as 60 men.[168] The work forces were divided between officers and servants: the officers were responsible for overseeing the fort and the trade, while the servants were employed as clerks, skilled tradesmen, and laborers. Among the tradesmen were carpenters, blacksmiths, bricklayers, masons, gunsmiths, coopers, shipwrights, and tailors.[169] The officers were either English or Scots, and many spent their entire careers in the service of the Company. The servants, in contrast, were recruited for three or five years, and when their term expired either returned to the British Isles or signed on for further service.[170] Almost all the servants were from the Orkney Islands, off the northern tip of Scotland.[171] The vessels that provisioned the Bay posts each year took on fresh water and supplies at Stromness in the Orkney Islands before making a northerly passage across the Atlantic, and ships captains took the opportunity to recruit labor for the Bay. Labor was cheap in Orkney, and could be hired at Old World rates. As the Company preferred to hire bachelors for the spartan life at the posts, most of the men were young and single, many from Stromness and the surrounding rural parishes.[172] The social background of officers and servants varied considerably; some officers were drawn from the gentry class, while many servants had been farm laborers. Nevertheless, there was promotion through the ranks; by the eighteenth century, some employees from humble backgrounds had become officers and chief factors. An employee's position at the posts depended on his place in the Company hierarchy, rather than on his social class in Britain.[173]

Surrounded by an alien, potentially hostile environment and people, the posts were much like ships at sea, with similar command structures and below-deck culture. Fur trader Joseph Robson thought that the post officer or chief factor had "generally sea-officers principles, and exert[ed] the same arbitrary command, and expect[ed] the same slavish obedience here, as [was] done on board a ship."[174] The factor kept in written communication with the Company in London, handled relations with the Indians, and supervised and disciplined the work force. Contact with Indians was limited to a few weeks trading each summer. After the elaborate opening ceremonies, trade was conducted through a hole in the fort wall; the only Indian allowed into the post was the trading captain who observed transactions in the trading room.[175] Any fraternization or socialization with Indians was strongly discouraged; Robson noted that "to converse with an Indian [was] a great crime."[176] Nevertheless, factors were allowed to take native women as wives, although officers usually left their wives and mixed-blood children behind when they returned to England. At some forts, servants were also allowed to marry native women.[177] Relations between officers and servants were formal: the two groups had their own living and sleeping areas, much like messes on board ships. The factor maintained disciple over the work force through a variety of means, including fines and flogging, a punishment used in the navy.[178] Although Hudson Bay was not a colony, the post officers, like their counterparts in the Royal African Company and East India Company, were representatives of imperial Britain, and observed official protocols. Royal birthdays and accessions were celebrated with parades and gun salutes, the drainage basin of Hudson Bay was called Rupert's Land after Prince Rupert, and important posts were named after royalty (Fort Prince of Wales, York Factory, Albany Fort). In contrast, the servants brought with them the rural vernacular culture of Orkney, most evident in their dialect and recreation: dancing and drinking appear to have been common pastimes.[179]

In his discussion of the economic development of the northeast, Harold Innis observed that the "bitter wars of New England and the extermination of the Beothic in Newfoundland contrasted strikingly with the fur trade and its dependence on friendly relations with the Indians."[180] Indeed, the fur trade was unique in British America for its dependence on exchange with native peoples rather than on the acquisition of their land. In this respect, the trade was similar to that of the Royal African Company along the coast of West Africa and the East India Company in India. In all three cases, the companies, at least initially, were far more interested in exchange rather than production. Like the East India and the Royal African companies, the Hudson's Bay Company acquired small parcels of land from native peoples on which to erect trading posts or "factories," and left the vast hinterlands to the natives.

The fur trade brought Indians and Europeans together into a complex trading relationship that had great impact on native population, settlement, economy, society, and culture. Most likely the settlement of Europeans around Hudson Bay introduced diseases unknown to the Indians, which must have taken some toll. Certainly, the smallpox epidemic that swept up from the southern plains into the woodlands northwest of Lake Superior in 1780–1781 devastated the Ojibwa, Cree, and Assiniboine.[181] Native settlement patterns were also disrupted. Opportunities to trade with the Hudson's Bay Company encouraged native groups to move from their traditional hunting grounds toward the posts. The Cree moved closer to all the Bay posts, while the Chipewyan shifted toward Churchill and the Assiniboine toward York Factory.[182] By positioning themselves between bayside traders and interior tribes, these Indians acted as middlemen, controlling the flow of furs and trade goods. The trade in firearms also gave the Cree, Assiniboine, and Chipewyan a military advantage over other Indian tribes without ready access to European weapons; an alliance of Cree and Assiniboine fought the Dakota Sioux almost incessantly during the late seventeenth and eighteenth centuries.[183] Ties between traders and Indians also became strong. Some Cree settled in the vicinity of the posts; as "homeguard Indians," they supplied furs, meat, feathers, and other "country produce" to the traders, as well as acting as couriers and guides.[184] Apart from creating a division between groups who trapped and others who traded, the fur trade encouraged some to specialize in making birchbark canoes and supplying foodstuffs (wild rice, pemmican) to European traders. Moreover, the trade changed the technology and diet of Indians. The introduction of manufactured goods to stone age cultures led to the replacement of skin vessels with copper kettles, bows and arrows with muskets and shot, and furs with woolen blankets. Tobacco, brandy, and rum became luxury items of consumption, and were commonly used by traders as gifts. Alcohol became "so bewitching a liquor amongst all the Indians" that it soon became addictive.[185] The trade also elevated some Indians within native society; particularly effective and trustworthy suppliers were rewarded with gifts and military uniforms, and were known as trading captains. When a captain and his "gang" left a bayside post after trading, "the cannon from the Fort [was] fired and every kind of respect paid them that [was] in our power."[186]

The fur trade through Hudson Bay was the most extreme permutation of Britain's commercial empire of the Atlantic. In the harsh subarctic environment of Hudson and James bays, the Hudson's Bay Company maintained a handful of trading posts that served as points of connection between native peoples in the continental interior and British manufacturers and consumers across the Atlantic. Small numbers of officers and servants staffed the posts. For these men, there was no alternative to working at the posts, except a 3,000-mile voyage back to Britain. In this respect, life at the posts was similar to that at the fishing stations in the migratory fishery. Yet even in Newfoundland, families eventually settled and a minimal agriculture developed; on the Bay, there were only a few mixed-blood families and no possibility of farming. In many ways, the fur trade was a seaborne commercial system projected from London to the Bay, and life at the posts was a landward version of life aboard ship.

"A Grand Marine Empire"

During the early modern period, British economic, political, military, and ideological power reached across the Atlantic and integrated Newfoundland, the West Indies, and Hudson Bay into "a grand marine empire" (figure 2.20).[187] Although the fur trade was of minimal economic importance, the cod fisheries, the sugar plantations, and the slave trade were central components of the British Atlantic economy. Given their enormous commercial value, the British extended military and political power across the Atlantic, while myriad cultural links bound these colonial peripheries to the metropolitan core.

From the late sixteenth century onwards, British merchants and planters assembled enormous amounts of capital, labor, and supplies to exploit these staple regions. The Hudson Bay Company sent hundreds of men to trading posts on Hudson Bay; West Country merchants shipped tens of thousands of men to fishing stations in Newfoundland; and London, Bristol, and Liverpool slave traders transported hundreds of thousands of men, women, and children to sugar plantations in the West Indies. These were the largest seasonal and permanent migration flows across the Atlantic during the colonial period. Management, too, was sent out to oversee operations, keeping in touch with head offices in Britain through a stream of written communications. Hundreds of thousands of tons of supplies and provisions, drawn from the British Isles and the American colonies, helped keep these settlements and their populations going. The fish, sugar, and furs exported from these staple regions generated immense wealth, much of it transferred back across the Atlantic to Britain.

These regions formed islands of settlement—rather than a continuous settlement frontier—on the western edge of the Atlantic. In the north, the fishing stations along the coast of Newfoundland and the fur posts around the rim of Hudson Bay occupied littoral spaces fronting the sea, a boreal wilderness behind.

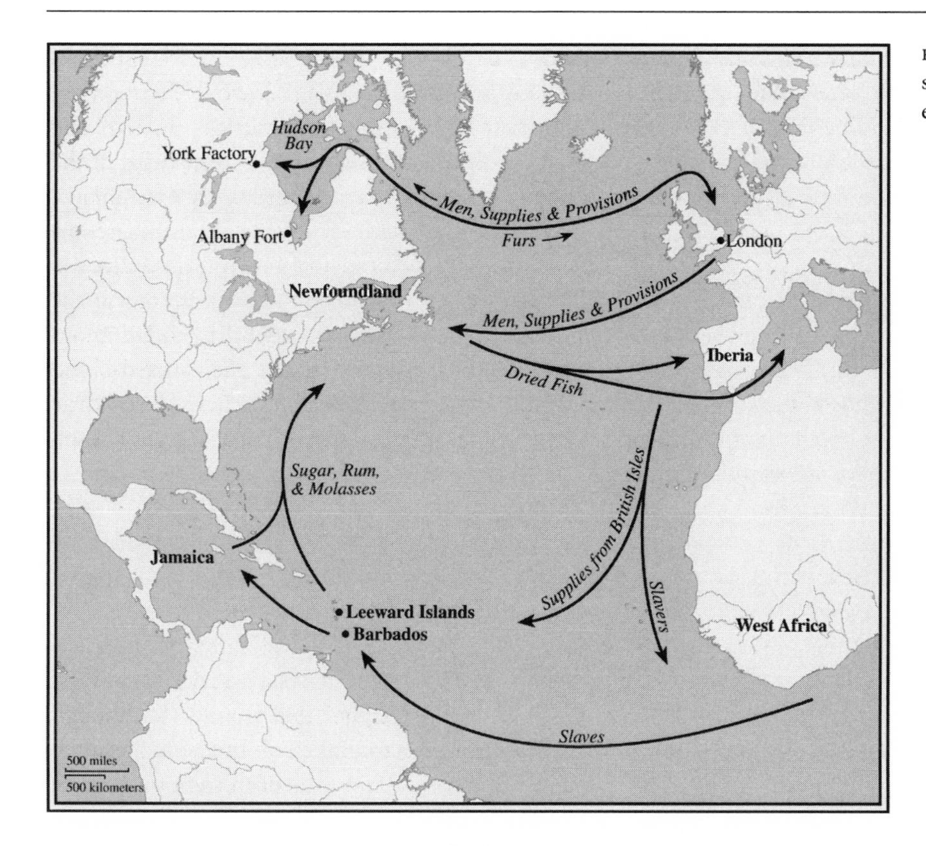

FIGURE 2.20. Atlantic staple trades, mid-eighteenth century.

In the south, the sugar plantations in the West Indies spread over small island territories and around the coastal perimeter of Jamaica. Unlike the staple regions on the continent, these staple islands had no backcountries of agricultural settlement. With no farming close at hand, metropolitan merchants and planters had no competition from colonial elites for economic power and free labor had no alternative occupation. In such starkly defined physical spaces, metropolitan capital acquired a considerable leeway.

Within these littoral spaces, a particularly attenuated pattern of settlement emerged. Compared to the diverse landscapes of farms, hamlets, villages, and market towns in the British Isles, or the farms, plantations, and port towns of the continental colonies, the settled landscapes of the British Atlantic consisted of little more than fur posts on Hudson Bay, fishing stations in Newfoundland, and plantations as well as a handful of small port towns (see chapter 5) in the West Indies. These various settlements were functional, utilitarian work places. As the elites that invested in these work places lived more or less permanently in the British Isles, there was little incentive to embellish them. The metropolitan landscapes of capital accumulation and display, represented by the Georgian town houses and country estates of the fish merchants and sugar planters, were separated from the colonial landscapes of production and toil, represented by the fur posts, plantations, and fishing stations, by the North Atlantic ocean.

The three regions were also marked by relatively small, racially distinct, and unbalanced populations. The total population of Hudson Bay, Newfoundland, and the West Indies was about 412,000 in the early 1770s, approximately four-fifths of the settler populations in New England or the Mid-Atlantic region. Although the European populations of Hudson Bay and Newfoundland were of British descent, the switch from white servants to black slaves in the West Indies created a regional population drawn largely from West Africa. As a result, the British Atlantic was racially bifurcated. In its northern circuits, it was overwhelmingly British and white; in its southern, tropical circuit, it was predominantly African and black. Yet the populations of all three regions were demographically unbalanced. In Hudson Bay and Newfoundland, the British populations were predominantly male and grew only by importing people, rather than by natural increase; in the West Indies, the African population, although more balanced, was so brutalized by the slave regime that it did not reproduce. Consequently, the trans-Atlantic circulation of servants and slaves became critically important for the functioning of these labor forces.

Compared to the rich complexities of societies in Britain or Africa, the societies that developed in Newfoundland, the West Indies, and around Hudson Bay were drastically simplified.[188] As the staple economies supported some occupations but not others, large parts of British and African societies were missing. In Newfoundland and Hudson Bay, there was virtually no farm population; in the West Indies, it was extremely limited. The societies that did develop were ordered and hierarchical. In the top echelon were tiny elites of merchants and planters: they moved back and forth across the Atlantic, and attempted to reproduce on the periphery some of the manners and mores of the metropole. Old World relationships between masters and servants were maintained, even enhanced, in the New World. In Hudson Bay and Newfoundland, the lack of alternative opportunities for labor strengthened the hand of merchants and factors over their servants; in the West Indies, planters constructed a harsh slave regime. In the middle ranks were small pockets of professionals — doctors, lawyers, clergymen — concentrated in the larger port towns. In the lowest levels of society were the staple work forces. Although fishermen, plantation servants, and fur traders were mobile, migrating back to Britain or forward to the continental colonies, an increasing number of fishermen chose to stay in Newfoundland, while slave populations remained shackled to the West Indies. Both of these groups mixed elements from the Old World and the New to create local vernacular cultures. During the early modern era, these local cultures were too underdeveloped and hemmed in by economic circumstance to mount any serious political challenge to the trans-Atlantic elites.

Although New World societies took root in Hudson Bay, Newfoundland, and the West Indies, there was little sense of new beginnings. The great majority of traders, planters, merchants, and servants were temporary, rather than permanent, migrants. With their roots in England, Ireland, or Scotland, many hoped to return to the British Isles after they had accumulated some wealth from the fishery, plantations, or fur trade. These migrants were not de Crèvecoeur's "American, this new man," products of the transforming experience of the American frontier, but English, Irish, and Scots, representatives of a British Atlantic economy. Only the African slaves were permanent migrants and they had no choice in the matter. For them, the West Indies was a new beginning, however terrible, from which they created a New World society.

Overlaying these Atlantic staple regions were significant political and military spaces. The British government had considerable political influence over Hudson Bay, Newfoundland, and the West Indies, while trade and navigation between the colonies and Britain were tightly controlled. The navigation acts of 1651 and 1660 brought colonial commerce under metropolitan regulation.[189] To encourage the growth of English shipping and to provide the government with taxation revenue from import duties, colonial products such as sugar, tobacco, rice, and furs had to be shipped to England before they could be re-exported overseas. In 1696, further measures were introduced. The Board of Trade was established in London to oversee the colonies, while the customs service in North America was strengthened and vice-admiralty courts established to adjudicate disputes over shipping and navigation. Such centralized political control was enforced through the navy. As command of the sea was essential for British trade and navigation, as well as to protect the British Isles from continental invasion, the state supported the strongest navy in Europe.[190] Although the navy could not control all the spaces of the British Atlantic, it patrolled the principal sea-lanes, convoyed merchantmen, and defended key ports and territories. As virtually all settlement in the West Indies, Newfoundland, and Hudson Bay lay along the coastal fringe, British naval power easily could be brought to bear to maintain order. Moreover, these littoral settlements were vulnerable to external attack, particularly from the French, and depended almost completely on the navy for protection.

Within this matrix of political and military power, a maritime commercial empire took shape. Unlike the agricultural empire of North America, this mercantile empire of the Atlantic was capital intensive, hierarchical, and familiar.[191] The massive amount of shipping sailing to and from Britain represented an immense amount of fixed and human capital, much of it invested by metropolitan merchants and chartered companies. In contrast to the continental farming frontier, which was open to immigrant labor, the oceanic frontier was the preserve of mercantile capital. In such circumstances, relations between capital and labor were well defined. For the tens of thousands of men crewing the merchant vessels of the British Atlantic, life was ordered, hierarchical, and disciplined, much like the worlds of the fur posts, fishing stations, and sugar plantations.[192] Indeed, the boundary between the merchant marine and the Newfoundland fishery was particularly porous. Such maritime worlds were known and familiar to people in Britain. An immense amount of information about the Atlantic as well as direct experience of the "empire of the sea" circulated through London and the outports, giving mercantile and political elites a clear sense of the economic and social contours of their "grand marine empire."[193] In contrast to the American colonies, the British Atlantic and the islands of staple production set within it were spaces dominated by metropolitan authority.[194]

Chapter 3

Continental Staple Regions: New England, the Chesapeake, and South Carolina

At the same time that they were establishing fishing stations in Newfoundland and plantations in the West Indies, the English extended their economic outreach to the eastern seaboard of North America. In the early 1600s, metropolitan merchants and proprietors began developing regions of staple production in New England and the Chesapeake, and then, in the 1670s, in the Carolinas. In time, the cod fishery in New England, the tobacco plantations in the Chesapeake, and the rice plantations in the Carolina low country would all became major parts of the colonial American economy.[1] Nevertheless, English mercantile and landed capital failed to control the development of these staple industries. Compared to the Atlantic islands of Newfoundland and the West Indies, the English encountered very different conditions along the edge of the continent. Almost limitless land, immense agricultural opportunities, and scarcity of labor all mitigated against tight metropolitan control over staple production. English merchants and proprietors soon withdrew, leaving colonial merchants and planters to develop the staples. A continental staple region gradually formed, occupying an intermediate position between the trade circuits of the Atlantic and the resources of the continental interior. By the early eighteenth century, elites in the port towns of Massachusetts, the tidewater of Virginia, and the low country of South Carolina were growing wealthy from their staple economies and enjoying a measure of economic independence; later in the century, these elites would use their economic position to challenge British political supremacy along the eastern seaboard. The early loss of metropolitan control over these key sectors of the colonial economy had fateful long-term consequences.

The Cod Fishery in New England

The English established a cod fishery in coastal New England while they were developing a colony in the Chesapeake and well before the Pilgrims sailed into Massachusetts Bay and planted an agricultural settlement at Plymouth. New

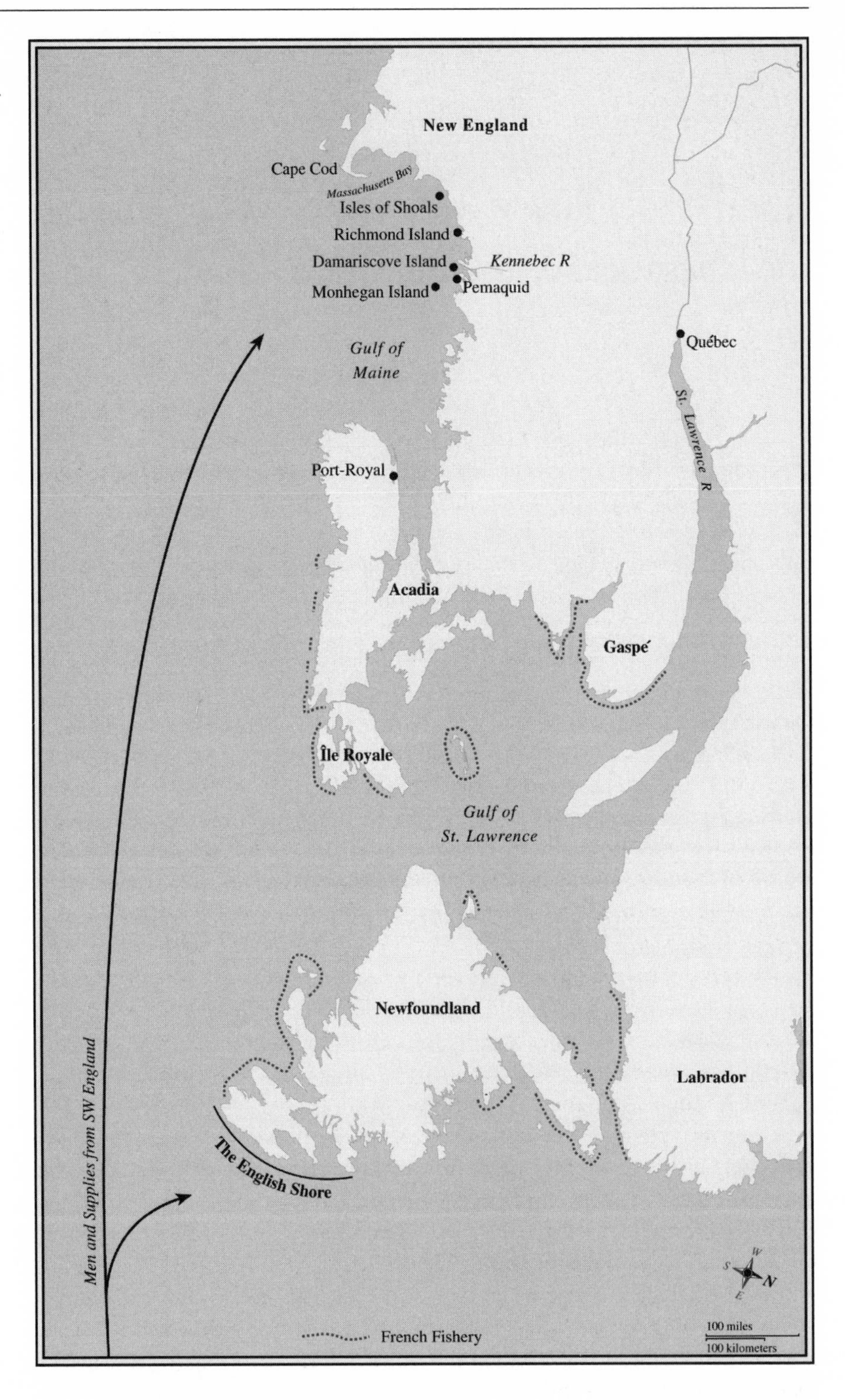

FIGURE 3.1. Expansion of the English fishery into the Gulf of Maine, early seventeenth century.

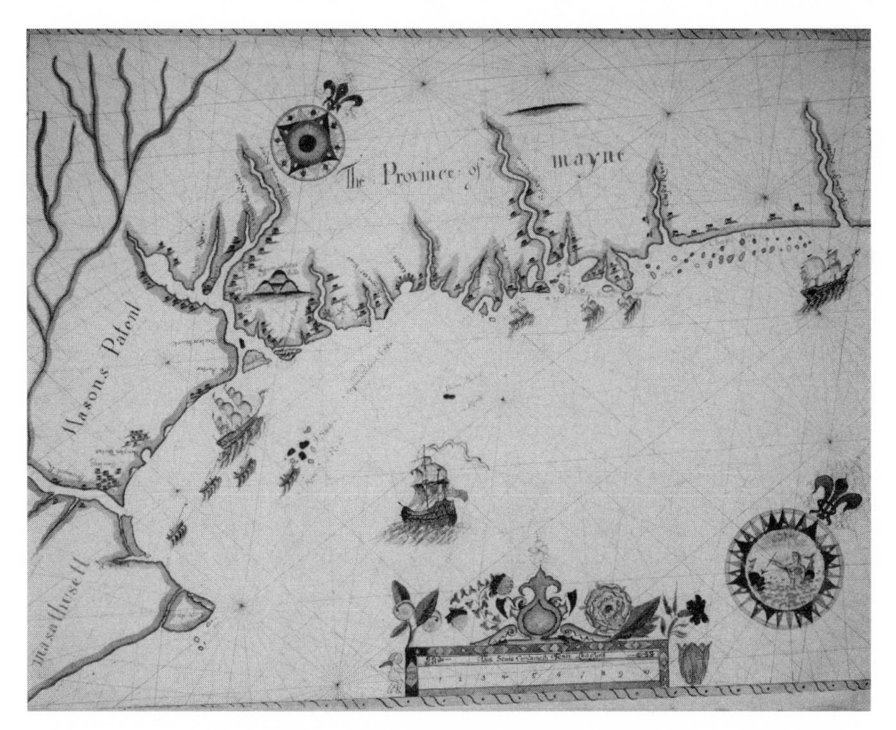

FIGURE 3.2. *Map of the Province of Mayne,* circa 1653. The map shows the coast of New England from Cape Ann (bottom left) to the Kennebec River (top right), as well as fishing boats and merchantmen near the Isles of Shoals. Baxter Rare Maps Collection, Maine State Archives.

England was thus a region of staple production before it became an English agricultural frontier (see chapter 4). Even after the expansion of agricultural settlement, the cod fishery remained the most important part of the region's commercial economy; between 1768 and 1772, dried fish made up nearly a third of all exports by value from New England, and more than half of exports from Massachusetts.[2] Moreover, the fishery stimulated the shipping, shipbuilding, and lumbering industries, as well as provided a market for local agricultural produce. More than any other product, dried cod laid the basis for New England's commercial wealth and economic independence, a fact that contemporaries recognized: In 1784, Boston merchant John Rowe presented a wooden model of the "Sacred Cod" to the Massachusetts legislature as a symbol of the importance of the fishing industry.[3] The cod still hangs over the public gallery in the House of Representatives.

West Country involvement in the New World led to the early growth of the New England fishery.[4] After the collapse of the Plymouth-based Virginia Company's settlement at the mouth of the Kennebec River in 1608, West Country merchants turned their attention to exploiting the surrounding seas. By 1610, the trans-Atlantic migratory fishery had been extended beyond Newfoundland, past the French fisheries along the coast of Acadia, and into the Gulf of Maine (figure 3.1). Unlike the barren shores, coniferous forests, and severe winter climate of Newfoundland, southern Maine had a coastline with some agricultural potential, a mixed forest, and a milder climate. As a result, the fishery could be pursued year round, with the best fishing usually in December and January; furthermore, provisions could be raised locally. The seasonal fishery in Newfoundland was thus transformed into a year-round fishery in the Gulf of Maine. Permanent fishing stations were established on the outer islands and promontories of the Maine coast, particularly the Isles of Shoals, Monhegan Island, Damariscove Island, Richmond Island, and Pemaquid Point (figure 3.2). Vessels from the West Country loaded with men and supplies arrived at these bases in the spring, an inshore fishery was prosecuted during summer and winter, and cargoes of dried fish were dispatched to markets in southern Europe in spring and fall. Almost from the beginning of the fishery, men were left to overwinter; they usually served for three years at the fishing stations, before returning to wives and families in the West Country or moving onto settlements around Massachusetts Bay. Where arable land and forest adjoined the fishing stations, a variety of cereals and vegetables were grown, cattle were raised, and pigs were left to forage in the woods. Some of the fishing stations were self-sufficient in basic foodstuffs, and needed only manufactured goods—clothing, fishing gear, weapons, earthenware—from England.

Yet the establishment of Puritan settlements around Massachusetts Bay in the 1620s and 1630s, as well as the booming tobacco colonies in the Chesapeake, undermined the West Country fishery in the Gulf. A combination of cheap land and high wages on these agricultural frontiers proved irresistible to landless labor working in a fishery at wage rates set in England. For the Plymouth merchant Robert Trelawny, who operated a fishing station on Richmond Island near Cape Elizabeth during the 1630s and early 1640s, the presence of these alternative economic opportunities made hiring and retaining fishermen immensely difficult. Trelawny's agent John Winter found labor in New England expensive: "A husbandman will not serue vnder £10 or £12 p' yeare, & yett do but litle work. I Cannot Conceaue which way their masters Can pay yt, but yf yt Continue this rates the servants wilbe masters & the masters servants." Faced with the high cost of local labor and the loss of imported fishermen, Winter advised Trelawny to hire fishermen "better Cheepe at home then the[y] will be heare, to Com[e] out in the shipe for the voyage, and to agre[e] with them for the tyme the[y] shall stay heare after the voayage is ended," and to "binde them in a sumsion [an agreement] of money to performe their promyse that they make with you." Otherwise, "yf any mans servant take a distast against his master, away the[y] go to their pleasure." Some of the "runawaies" from the station, Winter noted, "gott into a ship of London bound for Malaga [Spain] . . . the other I heare ar gonn for Virginnia."[5] The outbreak of the English Civil War in 1642 further hastened the demise of the West Country fishery in the Gulf.[6] Clashes between Royalist and Parliamentary forces in the West Country disrupted trade and shipping, and some merchants, including Trelawny, were imprisoned.

In place of the West Country migratory fishery, a New England–based resident fishery developed.[7] As the West Country fishery declined, fishermen drifted away from the fishing stations and settled along the coasts of southern Maine and Massachusetts, establishing a string of fishing settlements bearing such Devon names as Falmouth, Biddeford, Kittery, Appledore, and Barnstable.[8] Out of these harbors and coves, a resident, inshore fishery was prosecuted, much like the one that developed in Newfoundland later in the century. Some fishermen operated within a few miles of the shore; others ventured into the Gulf of Maine and organized fishing stations on the offshore islands. The common fishing craft was the shallop, crewed by two or three men. After a day's fishing, the boats returned to shore, where the catch was unloaded at a stage, processed, lightly salted, and then laid out to dry on flakes or a cobble beach. Given good drying weather, the New England inshore fishery produced prime merchantable cod suitable for the best markets in southern Europe. The fishermen exchanged their dried fish for supplies from Boston merchants, who forwarded it to market. Most transactions were done on credit and, as in Newfoundland, many settlers became mired in debt.[9]

At first, London merchants financed the development of the New England resident fishery. They were involved in the sack trade in Newfoundland, and saw the New England fishery as another source of dried fish for the Spanish market. As in Newfoundland, London merchants were not involved directly in production, but provided credit, supplies, provisions, and market connections to Boston merchants, who, in turn supplied fishermen and bought their fish.[10] A triangular trade linking London, New England, and the Iberian peninsula developed, much like that connecting the West Country, Newfoundland, and Spain. Although London houses controlled all three legs of the trade, Boston merchants were keen to trade on their own account, and were soon forming partnerships among themselves to finance independent voyages to the fish markets. By the 1660s, colonial merchants had established control over bilateral trade between New England and the Mediterranean. In little more than two decades, New England merchants had accumulated sufficient capital that they had achieved a measure of economic independence from the metropole.

Much of the labor for the New England inshore fishery came from the old West Country fishery into the Gulf of Maine and the English fishery in Newfoundland.[11] At the end of each fishing season in Newfoundland, servants were faced with returning to England, struggling through a Newfoundland winter, or moving on to New England. Lured by the promise of agricultural land and higher wages, many servants found room onboard American vessels that had delivered provisions to the island and were returning to New England. Over time, the drain of men from the Newfoundland fishery to New England became so great that British naval vessels were instructed to curb the outflow.[12] Most of the migrant fishermen were from the West Country, but there were also some Irish, who had been recruited at Waterford by West Country merchants, and Channel Islanders, particularly from Jersey.[13] As the servants in the migratory fishery were mostly young, single males, the fishing settlements that developed in southern Maine and Massachusetts were demographically unbalanced. For much of the seventeenth century, Marblehead, the major fishing port in Massachusetts, had a predominance of men and a social life much like the fishing stations in Newfoundland.

In the 1670s, the New England fishery underwent considerable change. The sugar revolution in the West Indies and the adoption of large slave work forces on the plantations produced a great demand for cheap food. Realizing the opportunity to supply refuse cod to this new market, New England merchants developed an offshore or banks fishery focused on Georges Bank, the Nova Scotian banks, and Newfoundland's Grand Banks (figure 3.3).[14] As fish caught offshore had to be preserved by heavy salting, the banks fishery produced a much poorer cure, one suitable for the West Indies market. As the banks fishery developed, New England increasingly supplied the West Indies, while Newfoundland exported to southern Europe. By the early 1770s, 62 percent of New England fish exports were going to the Caribbean, the rest to the Wine Islands, the Iberian peninsula, and the Mediterranean.[15] After starting as minor participants in a triangular trade organized by London, New England merchants used their control over the inshore and offshore fisheries to establish two independent, bilateral trades: one with southern Europe, the other with the Caribbean. As capital accumulated in New England, it was invested in expanding the banks fishery, instituting new trades, developing shipbuilding, speculating in real estate, and embellishing the port towns.

Among the early settlers who made the transition from the West Country fishery to the New England fishery was merchant William Pepperrell (c.1647–1734).[16] Originally from the Plymouth area in Devon, Pepperrell most likely worked in the Newfoundland fishery for a number of seasons before moving to the Gulf of Maine and settling on the Isles of Shoals sometime during the 1670s (figure 3.3). As this cluster of rocky islands off the coast of Maine had served as a fishing station since the 1610s, they were an ideal base for Pepperrell and he probably made a small investment in the inshore shallop fishery. After some success on the islands, Pepperrell moved to the mainland in 1680 and took up land overlooking the sheltered harbor at Kittery Point. Over the next fifty years, Pepperrell established himself as one of the most prominent merchants in the Piscataqua region. He maintained his interest in the inshore cod fishery, supplying fishermen on the Isles of Shoals, as well as invested in the developing offshore fishery, sending fishing vessels to the Nova Scotia and Newfoundland banks. He also exploited the interior pine forests along the Piscataqua River. From these various areas, Pepperrell traded lumber to Newfoundland, lumber and refuse fish to Barbados and Antigua, and prime merchantable fish to Portugal. He also had a brisk coastal trade down to Boston, the emporium of New England. In return, he purchased English goods in Newfoundland and Boston, as well as direct from suppliers in the West Country and London; in addition, he imported "Barbados goods" (rum, sugar, molasses) from the West Indies. Pepperrell also became a major shipowner and shipbuilder, owning at least seven vessels in the 1690s. His wealth was reflected in the fine Georgian house that he built overlooking Pepperrell Cove at Kittery Point in the 1720s (figure 3.4).

Unlike the inshore fisheries in Newfoundland or the Gulf of Maine, the offshore fishery in New England needed a substantial infrastructure. The large, ocean-going vessels employed in the banks fishery required deepwater wharves,

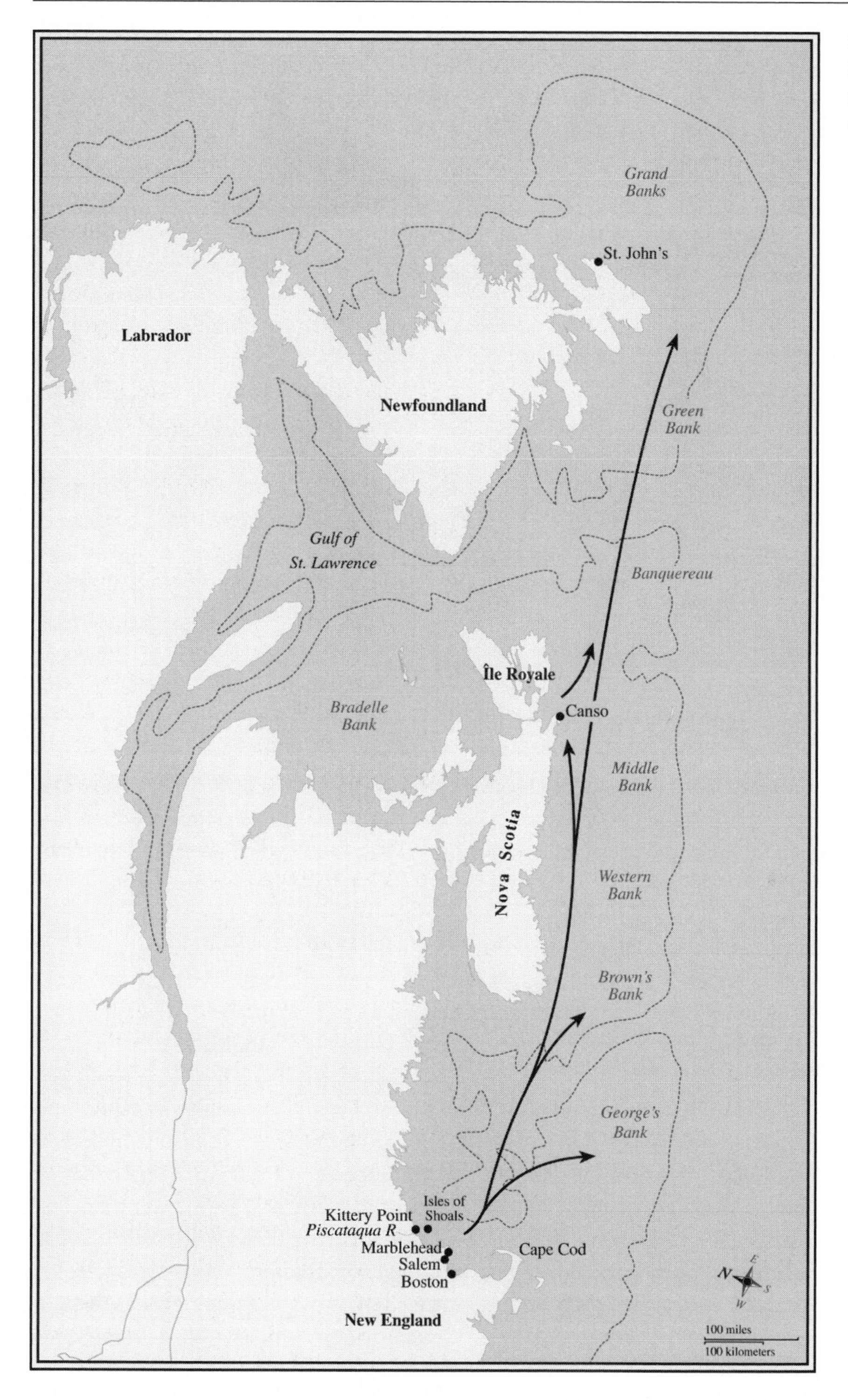

FIGURE 3.3. New England banks fishery, early eighteenth century.

FIGURE 3.4. William Pepperrell House, Kittery Point, York County, Maine, circa 1720. Situated on the northeastern frontier of New England, Pepperrell's comfortable Georgian home was built from the proceeds of the fishing and lumbering trades. Library of Congress, Prints and Photographs Division, HABS, ME, 16-KITPO, 4-1.

shipyards, ropewalks, ships chandlers, provisions, warehouses, and extensive areas of flakes for drying cargoes of salted fish. Such requirements led to considerable urban growth and development. Although Boston remained an important fishing port throughout the colonial period, as well as the regional entrepôt (see chapter 5), the expansion of the banks fishery had a major impact on the outports.[17] During the early eighteenth century, the industry developed in Marblehead and Salem on Boston's North Shore, and, in the latter part of the century, expanded to Gloucester, Manchester, and Ipswich, also on the North Shore, as well as to Plymouth on the South Shore, and Chatham and Yarmouth on Cape Cod (figure 3.5).[18] By the early 1770s, port towns in Massachusetts controlled 86 percent of the New England cod fishery.[19]

Apart from these fishing ports, New England merchants established shore stations as close as they could to the fishing grounds. After the British took over peninsular Nova Scotia from the French in 1713, New Englanders set up fishing stations along the colony's Atlantic coast, particularly at Canso, its easternmost point (figure 3.3).[20] From there, fishing vessels made several voyages to the banks each season. During the early 1760s, New England fishermen also settled along the South Shore of Nova Scotia, establishing inshore and banks fisheries at Yarmouth, Barrington, and Liverpool.[21] At the same time, merchants from the Channel Islands moved onto Cape Breton Island and established a trans-Atlantic migratory fishery. During the 1760s and early 1770s, Nova Scotia served as the meeting ground for

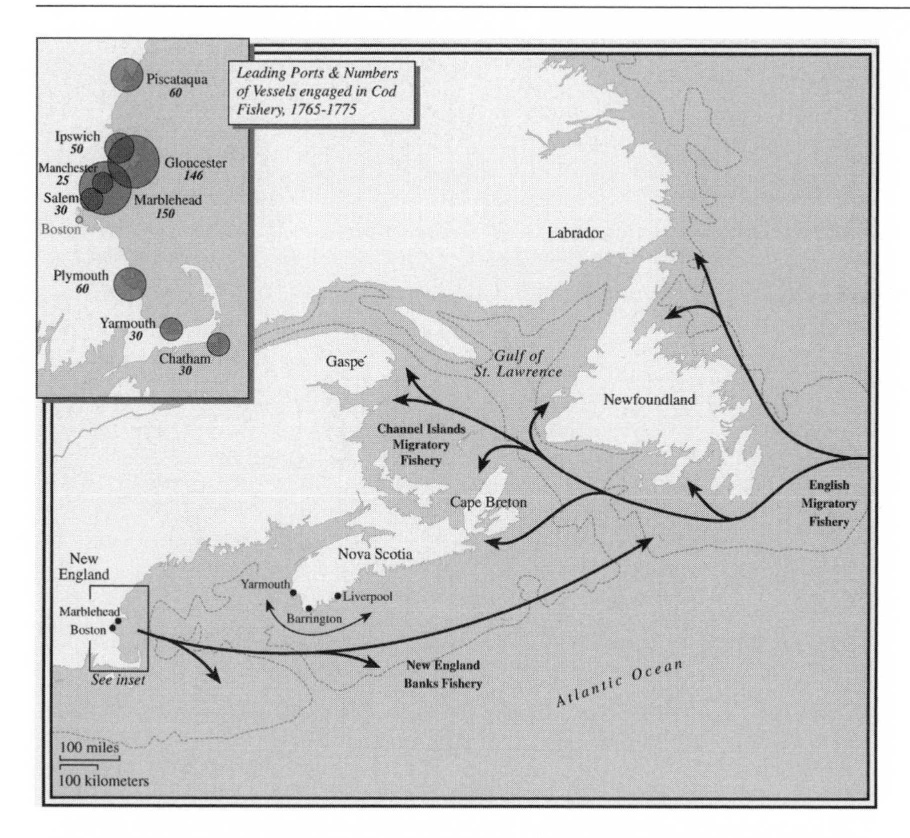

FIGURE 3.5. Western Atlantic cod fishery, circa 1770.

the two great fishing systems of the North Atlantic—the migratory fishery from the British Isles and the banks fishery from New England (figure 3.5).[22] At the outbreak of the American Revolution, the New England fishery stretched all the way from Cape Cod, along the Atlantic shore of Nova Scotia, to Newfoundland's Grand Banks.

As the banks fishery had to be conducted in open waters, several hundred miles from shore, and for long periods, large, seaworthy vessels had to be used. In the 1670s, the most common type of fishing vessel was the ketch, a fully decked, two-masted, round-sterned vessel, ranging in size from about 12 to 70 tons, although most were no more than 40 tons.[23] In the early 1700s, the ketch began to be replaced by the schooner, a New England invention, which was similar in appearance to the ketch but had a higher second mast and an increased sail area. Ketches and schooners were usually manned by six to eight men and boys.[24] The crew normally consisted of four "sharesmen," who worked for a share of the catch and did the skilled work of heading, splitting, and salting the fish, and four boys employed as apprentices.[25] Vessels fished the banks for as long as it took to fill their holds (figure 3.6). Men and boys stood beside the deck rails and fished with baited hand lines. The fish were processed and salted on deck, and then stowed in the hold. When the banker returned to port, the salted fish were washed and laid out to dry

FIGURE 3.6. Fishermen hand-lining for cod from the rail of a Grand Banks schooner, early nineteenth century. The type of vessel and method of fishing were not much different in the late colonial period. George Brown Goode, *The Fisheries and Fishing Industries of the United States*, Section 3 (Washington, D.C.: U.S. Government Printing Office, 1887).

on flakes. Because of the heavy salting needed to preserve the fish for weeks at sea, the cured fish were never as good as the lightly salted fish of the inshore fishery and were more suitable for the Caribbean market.

The growth of the offshore banks fishery generated important linkages to other industries, particularly shipbuilding, manufacturing, lumbering, and agriculture. As substantial vessels were needed for deep-sea fishing and for transporting dried cod to markets overseas, Massachusetts merchants set up shipyards in Marblehead, Salem, and Gloucester.[26] The yards turned out vessels not only for fishing and trade, but also for the export market in Britain. Shipbuilding provided employment to local shipwrights, rope makers, sail makers, and skilled artisans, and to a variety of suppliers and manufacturers. Timber for the vessels came from local lumber contractors, while pitch and tar were imported from the Carolinas. Ropewalks were set up to manufacture ropes and rigging, canvas works to make sail cloth, and smithies to turn locally produced iron into anchors and other metal work.[27] New England shipbuilders were virtually self-sufficient in lumber and materials. Crews on the fishing vessels as well as the populations of the port towns also needed provisions; salted meat and butter were supplied by local farmers.[28] In Newfoundland, virtually all of these links were transferred back across the Atlantic to the West Country; in New England, they were retained locally, giving rise to a more developed and integrated economy.

The fishery provided merchants in Massachusetts, like their counterparts in the West Country, with a resource to trade around the North Atlantic. New England fish merchants not only exported dried cod to southern Europe and the West Indies, but also imported goods from those regions and traded them up and down the eastern seaboard.[29] Imports from the Caribbean were particularly important. New England merchants imported sugar and molasses for domestic consumption, as well as for re-export to other colonies. The sale of West India products in the Mid-Atlantic colonies allowed New England to import much-needed bread and flour from Pennsylvania and New York. Caribbean molasses was also processed in New England to make rum for domestic consumption and re-export. In the early 1770s, coastal exports, consisting mainly of molasses and New England rum, comprised more than 60 percent of the value of New England's total commodity sales to overseas markets.[30] The trans-Atlantic and coastal trades established Massachusetts fish merchants among the prime traders in New England, and among the most powerful in the British American colonies. The fishery, as well as the lumber and provisions trades, provided Massachusetts merchants with an economic springboard to challenge British mercantile dominance in the North Atlantic.

The fishery and maritime trades also provided the fish merchants with wealth that could be reinvested on the continent. Leading merchants in Boston, Salem, and Marblehead invested in local real estate, buying up town lots and houses, improving wharves, and building stores and warehouses. One such merchant was Philip English. Originally from the island of Jersey, English settled in Salem in the early 1670s and soon became a major figure in the cod fishery and trans-Atlantic trade. By the 1680s, he had acquired substantial real estate holdings in the town, including a "Great House," a tavern, a warehouse, and several lesser houses.[31] Other merchants invested beyond their towns, purchasing outlying farms and undeveloped land on the frontier. Thomas Hancock, one of the leading merchants in Boston in the late eighteenth century and a major trader in provisions to the Newfoundland fishery, speculated in frontier land in Maine, investing with the Kennebec Proprietors in the 1750s.[32] Such speculative dealings were no doubt widespread among the fish merchants and strengthened their economic ties to the interior.

As the banks fishery developed, the source of labor shifted from Newfoundland to New England. Although some fishermen from Newfoundland continued to filter into Massachusetts in the late seventeenth and early eighteenth centuries, much of the labor for the fishery came from fishermen resident in the port towns. By the late seventeenth century, the populations of Marblehead, Salem, and Gloucester were balanced sufficiently that they were able to reproduce through natural increase.[33] Whereas the fishery in Newfoundland depended on a continual influx of male servants from the West Country and southern Ireland, the fishery in New England provided its own labor from the port towns. As a result, fishing communities along the North Shore were more stable, with generations of men entering the fishery and other seafaring trades. The majority of these men were young, aged between fifteen and thirty, and most worked at sea for only a few years.[34] For those who worked in maritime occupations in Salem between 1745 and 1775, about 30 percent died, usually from disease, before they reached the age of thirty, while another 15 percent quit the sea before they reached the same age. Although these figures apply to all maritime trades, they do suggest that fishing, as one particular trade, was a young man's occupation. Most likely the men who shipped out on fishing schooners from Salem and Marblehead were much like those who hired on as fishermen in Dartmouth, Poole, and Waterford: young men seeking waged employment before they married and settled down to a life ashore.

Like other staple industries, the New England fishery created a strikingly stratified society. At the apex of the social and economic hierarchy were the merchants: They financed and organized the fishery and trans-Atlantic trades, accumulated considerable wealth, and dominated town life. At first, merchants in Boston made the greatest profits from the fishery, importing manufactured goods from England, exchanging these goods for fish in the outports, and exporting fish to markets in southern Europe and the Caribbean. Merchants in North Shore ports were merely middlemen between the great Boston merchants and the fishermen.[35] As the banks fishery expanded in the early eighteenth century, however, Marblehead and Salem merchants started to trade on their own account, importing directly from England and exporting overseas. Merchants in Gloucester soon followed suit. By the 1750s, merchants in Marblehead, Salem, Gloucester, and other North Shore ports were largely independent of Boston and were accumulating their own profits from overseas trade.[36] During these years, the so-called "Marblehead Gentry" emerged, a group of wealthy merchants engaged in the fishery, overseas trade, and manufacturing.[37] Probated estates reveal this group's economic rise. Between 1716 and 1735, the richest 10 percent of property holders in Marblehead increased their share of the town's total wealth from 46 percent to nearly 60 percent, with some merchants holding estates worth more than a thousand pounds. By 1770, this group still held more than 60 percent of the town's wealth, and some had more than doubled their estates to over two thousand pounds.[38] Of these merchants, Robert Hooper, nicknamed "King" Hooper, sent "up the Straights [i.e. beyond Gibralter] 5 or 6 vessels of his own with fish every year" in the early 1750s.[39] Another merchant, Jeremiah Lee, the wealthiest person in Marblehead, left an estate valued at £45,148 in 1775.[40] Fish merchants such as Hooper and Lee were among the wealthiest men in the American colonies.

Further down the social and economic ladder, a middling group included small fish dealers, retailers, ship captains, shipwrights, and artisans. These people either had some property to their names or a skill needed in the fishery, shipping, and shipbuilding trades. Between 1716 and 1735, this middling group in Marblehead—approximately 60 percent of total decedents—held 44 percent of the town's wealth and had estates valued between £15 and £395. At the bottom of the ladder were fishermen, who sold their labor to captains of vessels owned by the merchant elite, worked hard in a frequently dangerous occupation, and earned scant reward. Although labor had been in short supply in the mid-seventeenth century and wages were relatively high, the growth of the region's population in the early eighteenth century produced more than enough labor for the fishery and helped depress wages. Annual earnings climbed only slowly from about £20 in the 1660s and 1670s to £25 to £30 in the early 1770s, and much of the increase merely kept pace with rising prices for goods, provisions, and housing.[41] Between 1716 and 1770, the poorest 30 percent of decedents in Marblehead held less than 2 percent of the town's wealth, and owned estates valued less than £22.[42] At the outbreak of the American Revolution, fishermen were scarcely better off in terms of income and material possessions than their forefathers a century earlier. The fishery produced fortunes for a few, a living for a middling group, and poverty for the bottom third.

Compared to the shifting groups of merchants, planters, and servants in the Newfoundland fishery, the populations of the New England port towns were increasingly stable by the early eighteenth century. Most merchants had no close familial ties to Britain and were developing strong commitments in New England, cementing their social and economic position in the port towns through business partnerships, joint investments, family ties, and intermarriage. Like their counterparts in the West Country, New England fish merchants were leaders of their communities, controlling town government, contributing to local churches, and providing employment directly or indirectly to most of the population; they held to a "corporate, paternalistic conception of the social order."[43] The decline of immigration from England also weakened trans-Atlantic ties among fishermen, while natural growth of the population strengthened familial bonds within the fishing communities.[44] Many of the young men who went to sea from Salem during the eighteenth century came from the same neighborhood and had grown up together: New England equivalents of the West Country fishermen in Dartmouth and Poole.[45]

Compared to the enclaves of West Country and Irish folk culture scattered along the coast of Newfoundland, the populations of the fishing towns in Massachusetts and Maine were part of a fairly uniform regional vernacular culture. After several generations of settlement and the intermingling of people from different regional backgrounds in England, many Old World cultural traits had slipped away by the early eighteenth century. To be sure, the varied origins of the English settlers continued to be reflected in the religious denominations of the port towns (Anglicans in Marblehead, Congregationalists in Salem and Gloucester) and some "idiomatic peculiarities" of West Country speech still lingered among residents of Marblehead.[46] Nevertheless, the populations of the port towns were part of the larger New England culture region, and this was reflected in foodways and dialect. A New England diet, comprising salt fish, corn, rye bread, molasses, and rum, had replaced English foodways, while a common accent, derived from Boston, was most likely developing among many inhabitants along the North Shore.

A further contrast to Newfoundland was the greater prosperity of the New England fishing ports. During the eighteenth century, Salem and Marblehead both

FIGURE 3.7. Jeremiah Lee House, Marblehead, Essex County, Massachusetts, 1768. One of the most magnificent town houses in colonial America, the Jeremiah Lee House was the New England equivalent of the great merchant houses built in the port towns of England during the eighteenth century. Library of Congress, Prints and Photographs Division, HABS, MASS, 5-MARB, 55-1.

built new institutional buildings, including town houses and churches, while local merchants built splendid residences, symbols of their economic standing, social status, and cultural sophistication.[47] Although New England merchants looked across the Atlantic to Britain for architectural inspiration, the residences that were built along the North Shore were not replicas of English Georgian town houses; instead, they were hybrids of English design and New England materials (figure 3.7).[48] The grandest town houses in Salem and Marblehead drew on England for Georgian and Palladian designs, which circulated in the form of pattern books, and were decorated with English furniture, fabrics, and wall hangings. As a result, the houses looked like their trans-Atlantic counterparts in their symmetrical facades, formal rooms, and fine furnishings, but in terms of construction they were completely different. Although a few houses were built out of brick, like Georgian houses in England, most North Shore merchant homes were framed structures covered with clapboards. Even Jeremiah Lee, who wanted his Marblehead mansion to look like a stone house, had to settle for boards cut to look like ashlar. Moreover, the greater amount of space available for building allowed many of these houses to stand detached from their neighbors, giving the North Shore port towns a more spacious feel than the cramped towns of Dartmouth, Topsham, or Poole. In one respect, though, New England merchants were similar to their West Country counterparts. Just as the Lesters had fillets of cod carved on the marble mantlepiece of the Mansion House in Poole, so several North Shore merchants displayed the sources of their wealth in their houses: Jeremiah Lee had the magnificent staircase of his mansion in Marblehead built out of mahogany from Santo Domingo, while Benjamin Pickman had each riser on the staircase of his house in Salem decorated with a gilded cod.[49] Such buildings gave the New England port towns a much greater air of permanence than the Newfoundland outports and symbolized the growing prosperity and cultural confidence of the New England fish merchants.

The institutional context was also much stronger in New England than in Newfoundland. After the creation of the Massachusetts Bay Colony in 1630, the fishery developed within a local government and legal framework. Unlike Newfoundland, Massachusetts had a colonial legislature and law courts, and the leading fish merchants used these institutions, rather than Parliament in London, to control the fishery. Leading fish merchants sat in the Massachusetts legislative assembly, and several provided leadership to the growing independence movement that developed in the 1760s and early 1770s. Moreover, religious denominations had a much greater presence in New England than in Newfoundland, contributing to greater social control. The Congregational Church was established in Salem and Gloucester, while the Anglican Church and Society of Friends (Quakers) were important institutions in Marblehead.

Although part of an immense fishing region in the northwest Atlantic, the fishery in New England developed differently from that in Newfoundland. New relationships among land, capital, and labor in early New England defeated the West Country migratory system and allowed the establishment of a colonial, resident fishery. Over time, New England merchants accumulated sufficient capital to allow them to trade independently of British merchants, as well as to develop other trans-Atlantic trades and shipbuilding. Although New England port towns relied heavily on Britain for manufactured goods as well as on the Mid-Atlantic colonies for bread and flour, such dependence was considerably less than in Newfoundland. An extensive backcountry supplied lumber for shipbuilding, while local farms provided provisions for the fishery. Furthermore, the New England port towns were economically far more diverse and independent than the outports in Newfoundland. With their fishing and overseas trade, social and economic stratification, urban form, and agricultural hinterlands, the port towns on the North Shore were most similar to their rivals in Devon, Dorset, and the Channel Islands.

The New England cod fishery grew out of the West Country fishery to Newfoundland in the early seventeenth century, developed its own base in the New World during the middle of that century, and became a serious rival to the West Country by the early eighteenth century. Just as the inshore fishery of Newfoundland provided a resource for West Country merchants to trade to southern Europe, so the banks fishery provided Massachusetts merchants with a product to sell in the West Indies. And just as the Newfoundland fishery allowed West Country merchants to challenge the dominance of London, so, in turn, the New England fishery permitted Massachusetts merchants to defy the West Country. Open access to one of the great natural resources of the New World allowed merchants on the periphery of England, then North America, to challenge the economic control of the core. As early as the 1660s, English mercantilist Sir Josiah Child noted that New England was the "most prejudicial plantation to this Kingdom."[50] A little over a century later, New England's economic challenge would be transformed into political, military, and ideological defiance of Britain.

Tobacco Plantations in the Chesapeake

After the cod fishery, the second great staple developed by the English in North America was tobacco. In the early seventeenth century, tobacco was the principal agricultural product of Virginia and Maryland, as well as Bermuda and Barbados. Although the sugar revolution of the mid-1640s displaced tobacco from Barbados, the cultivation of tobacco continued to dominate the Chesapeake, and remained the most important agricultural staple on the mainland throughout the colonial period; between 1768 and 1772, tobacco comprised 72 percent of exports by value from the Chesapeake, and 29 percent of all exports from the continental colonies.[51] Moreover, the growth of the tobacco economy created one of the most dynamic agricultural regions on the mainland. Although early settlement schemes funded by the London-based Virginia Company and English aristocrat Lord Baltimore failed, the region emerged during the mid- and late seventeenth century as the principal farming frontier on the eastern seaboard. As the tobacco economy expanded in the early eighteenth century, the Chesapeake became the wealthiest agricultural region in the thirteen colonies, marked by prosperous planters owning large plantations and numerous slaves. Yet the growth of the tobacco trade—second only to sugar in importance—led to tighter metropolitan economic control over the region. By the late eighteenth century, relations between British merchants and Chesapeake planters had become so sour that some of the leading tidewater gentry threw their support behind the growing American independence movement. As in New England, the Chesapeake had become one of the principal battlegrounds along the eastern seaboard between continental and Atlantic economic systems.

Almost from the beginning of colonization, the Virginia Company's attempts to settle at Jamestown ran into difficulties. After friendly encounters with the Powhatan, English relations with the Indians deteriorated and turned into bitter conflict between 1609 and 1614.[52] Although a truce brought a measure of stability between the two groups, English expansion beyond Jamestown provoked an Indian uprising and massacre of settlers in 1622. Apart from losing colonists to warfare, English settlers also suffered an extremely high death rate from disease. The location of Jamestown coincided with the zone of transition on the lower James between saline water of the Chesapeake Bay and fresh water of the James River. This zone provided not only salt-contaminated drinking water but also a perfect incubator for typhoid and dysentery. These diseases soon took their deadly toll.[53] Finally, the Company failed to find readily accessible staples such as minerals or furs and soon found itself in economic difficulties.

As early as 1610, the Virginia Company started to explore ways of raising additional funds. As its principal asset was land, the Company began selling off large tracts, an especially attractive strategy as the tobacco economy developed. Land was sold to Company shareholders, some of whom combined together to form sub-companies, which were entrusted with developing the tracts through recruitment of settlers and establishment of plantations or "hundreds" (an English term for a collection of parishes). Among these groups was the Society of Martin's Hundred, which was composed of several wealthy City of London merchants, including Sir John Wolstenholme, a member of the East India Company and a backer of Henry Hudson's ill-fated voyage to Hudson Bay in 1610 and of William Baffin's exploration of Baffin Bay in 1615.[54] The Society purchased land downriver from Jamestown, sent out 220 settlers in 1618, and created the plantation of Wolstenholme Towne. Yet Martin's Hundred and other company settlements along the lower James were largely destroyed in the Indian uprising of 1622, and the system of company development was abandoned. The Virginia Company also granted land to indentured servants who had completed their terms of work and decided to stay in the colony. Despite these initiatives, the Virginia Company struggled to survive; in 1624, the charter of the Company was revoked and control over the colony passed to the Crown.

A decade later, Cecil Calvert, the second Lord Baltimore, began the proprietary colonization of Maryland. Cecil was the son and heir of George Calvert, who had been a shareholder in the Virginia Company and the East India Company, and a backer of the short-lived plantation at Ferryland on the Avalon Peninsula in Newfoundland. After the failure of the Newfoundland settlement, George Calvert was granted a proprietary colony in the Chesapeake in 1632, but did not live to take up the grant. Instead, Cecil Calvert took over the enormous land grant north of the Potomac, and set about colonization.[55] A Roman Catholic in an England increasingly hostile to Catholics, Calvert aimed to create a refuge overseas for persecuted Catholics, as well as to re-create a feudal landed society. As a solution to the problem of financing colonization, which had so bedeviled the Virginia Company, Calvert made large land grants or manors to shareholders, provided they brought out settlers.[56] These settlers were to be tenants on the manors, paying rent to their feudal lords. Calvert was attempting to transfer a landowning system from the Old World to the completely different context of the New. In England, land was in short supply and expensive, ensuring high rents for landlords, but in the Chesapeake, land was plentiful and cheap, almost worthless in its uncleared state. As a result, landlords in Maryland had to keep rents low in order to attract tenants. The lack of rental income, however, undermined the landholding system, and it soon fell into disuse. Meanwhile, the proprietary ownership of the Calverts was coming under legal challenge, especially during the English Civil War and Puritan Commonwealth (1642–1660), with the result that the government of Maryland was frequently suspended.[57] Although the Calverts were confirmed in their title at the Restoration of Charles II in 1660, the hierarchical society of feudal landlords and tenants had irretrievably broken down.

The failures of the Virginia Company and of Lord Baltimore represented the first significant retreats of English commercial and landed capital from direct control of colonization along the American eastern seaboard. The withdrawal of the big London merchants from Virginia in the early 1620s and the collapse of Baltimore's feudal experiment in Maryland in the 1630s were soon followed by the departure of the West Country fish merchants from the Gulf of Maine in the early 1640s. Commercial capital had been defeated by the high cost of labor, while landed capital had been frustrated by the availability of land. The different relationship between land and labor in the New World had confounded the best-laid plans of Old World entrepreneurs. By the outbreak of the English Civil War in 1642, metropolitan merchants largely had shifted their investment away from settlement and staple production toward trade: shipping goods and servants across the Atlantic and marketing colonial produce in European markets. As metropolitan merchants withdrew from the eastern seaboard, the way was open for the emergence of a class of local entrepreneurs in the colonies. In New England, the removal of the West Country merchants allowed colonial merchants to take control of the fishery; in the Chesapeake, the collapse of the Virginia Company and Baltimore's feudal society permitted numerous small English merchants and thousands of settlers to pour in and reshape the settlement, economy, and society of the region.

After the great metropolitan merchants pulled out of the Chesapeake, many lesser merchants, ship captains, and petty retailers in London and the English outports took over the tobacco trade.[58] Learning the lesson of the Virginia Company, these small merchants and traders contented themselves with exchange: shipping servants across the Atlantic, provisioning and providing credit to planters, and marketing tobacco. Such traders were more like the merchants operating in the West Indies than the West Country men in Newfoundland, although the money to be made from the tobacco trade was considerably less than in the sugar business. In the early years of the tobacco trade, English merchants formed syndicates or "adventures," entrusting goods to a ship captain or supercargo, who would exchange them for tobacco from planters scattered along the creeks and inlets of the Chesapeake. As this system tied up a vessel in the tidewater for considerable periods, a new system of trading developed. By the middle of the seventeenth century, English merchants increasingly were sending out salaried factors to handle the tobacco trade. The factors were responsible for buying tobacco and selling English goods, and usually stayed in the Chesapeake for only a few months. Some factors, however, settled in the Chesapeake and handled the English accounts on a commission basis. By the end of the century, a handful of them had accumulated sufficient wealth through their dealings and investments in land that they formed a nascent merchant-planter class.[59]

During the seventeenth century, growth in the tobacco economy depended on the English market. "The trade of this province," the Maryland assembly declared in 1697, "ebbs and flows according to the rise or fall of tobacco in the market of England."[60] The first tobacco had been imported into England from the Spanish empire during the 1590s, and smoking quickly became a craze.[61] With the successful cultivation of tobacco in the Chesapeake, England had its own source of supply and the domestic market boomed. Between the early 1620s and the late 1670s, tobacco exports expanded enormously, rising from some 65,000 pounds to 20 million pounds. As production and exports increased during the 1620s, prices in the English market fell but planters still managed to make a profit by expanding productivity.[62]

The dramatic growth of the tobacco economy during the seventeenth century attracted an enormous number of English immigrants to the Chesapeake. Between 1625 and 1680, some 90,000 white migrants arrived in Virginia and Maryland, by far the largest European migration to any part of the eastern seaboard in the seventeenth century and second only to the influx into the English West Indies.[63] In terms of their employment, age, marital status, and gender, the migrants were similar to the servants who went to the West Indies, or hired on with West Country merchants for the Newfoundland fishery. Between two-thirds and three-quarters of the migrants were indentured servants, their passages across the Atlantic paid by merchants, who recouped their costs by selling them "for the price of their passage" to planters in the Chesapeake.[64] The remainder were free migrants, who paid their own way across the ocean and arrived with some capital. Whether indentured or free, the migrants were overwhelmingly young, single males. About a quarter of the migrants were women, and they, too, were mostly young and single. The migrants were from a variety of backgrounds. Although a few appear to have been younger sons of gentry seeking their fortunes in the New World, most were from the middling and lower levels of society, drawn from the ranks of farmers, laborers, artisans, and youths. With little stake in English society and scant chance of acquiring an independence, they looked for work overseas, hoping to establish themselves on the land in the colonies.

As with the migration to the West Indies, much of the English migration to the Chesapeake was funneled through London and Bristol, the major tobacco ports in England during the seventeenth century. The migrants who left from London were from all regions of the country, although the majority appear to have been from London itself and surrounding areas, particularly the southeast and the Thames Valley. The migrants who passed through Bristol were from the city, southwestern England, the Severn Valley, and South Wales.[65] In general, the migration field stretched across southern England, from London to Bristol. Although this ensured that the migrants were mostly from the south of the country, rather than from the north, east, or west, there were substantial cultural differences between migrants from the London area and the West Country.[66] This was a far more diverse migration than the one to Newfoundland, and the varied regional backgrounds of the

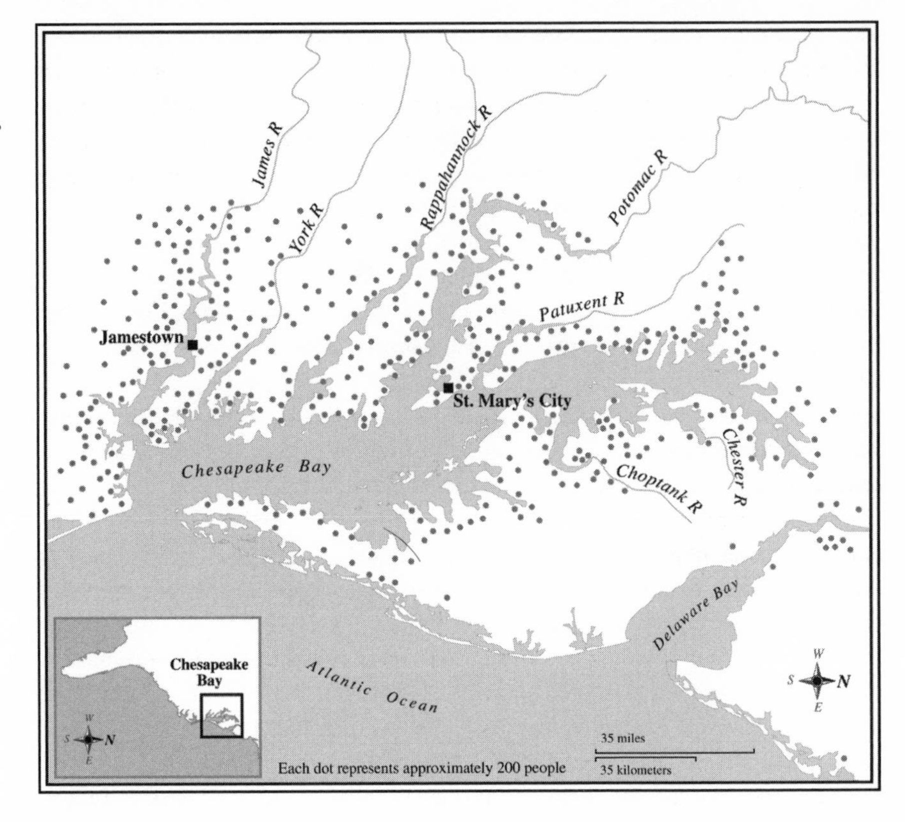

FIGURE 3.8. English settlement in the Chesapeake, 1675. After Herman R. Friis, *A Series of Population Maps of the Colonies and the United States 1625–1790* (New York: American Geographical Society, 1968).

migrants ensured that considerable cultural mixing would occur on the Chesapeake settlement frontier.

The English migration to the Chesapeake had a more lasting impact on the population of the region than had the corresponding migration to the West Indies. Whereas the migration to the sugar islands had a comparatively short-lived effect on the white population, the flow to the Chesapeake laid one of the two foundations for the region's demographic future. Between 1620 and 1680, the white population of Virginia and Maryland increased from about 1,000 to approximately 56,000.[67] The increase would have been even greater had it not been for the high death rate caused by disease and climate, the late marriage of women (because of their years in service), and the overwhelming preponderance of men. Nevertheless, English migration to the Chesapeake established the largest pocket of European population on the eastern seaboard in the seventeenth century, and by the last two decades of the century the population was beginning to grow through natural increase.[68]

Between the 1620s and 1680s, the expansion of the tobacco economy and the great increase in population pushed English settlement around much of the Chesapeake Bay (figure 3.8). From the core established along the lower James, settlers spread along the banks of the major maritime thoroughfares (the York, Rappahannock, Potomac, Patuxent, Choptank, and Chester rivers), as well as along minor bays, creeks, and inlets. As the land surrounding the Bay comprised part of the coastal plain that sweeps down the eastern seaboard, settlers encountered relatively flat terrain, although there was rolling country on the Northern Neck of Virginia as well as extensive areas of marsh on the Eastern Shore. Apart from occasional clearings made by Indians for growing corn, the landscape of the Bay was covered with deciduous forest; one traveler likened the Virginia shore to "a forest standing in water."[69]

Inevitably, the spread of settlement around the Chesapeake brought the English into conflict with the native peoples.[70] Unlike Newfoundland or the West Indies, the English encountered a large native population (the Powhatan) in the Chesapeake, subsisting on the rich marine resources of the Bay as well as on small-scale farming. At first, the English depended on the Powhatan for food, bartering metal implements for corn and squash. But as tobacco became established and settlement expanded, the English increasingly encroached on native lands. Feeling threatened by this expansion, the Powhatan launched raids on the settlements along the lower James in 1622, killing more than 300 colonists. Revenge attacks on the Powhatan followed; over the course of the seventeenth century, the natives were either pushed back from the tidewater toward the Blue Ridge Mountains or decimated by intermittent warfare. All the while, European diseases made inroads into the population, much as they were doing elsewhere along the eastern edge of the continent. By the end of the colonial period, English aggression and European diseases had largely removed the native population from the Chesapeake. The English assault on native peoples in Virginia marked the beginning of a more general European invasion of native land along the eastern seaboard that would last for more than 250 years and extend across the southern half of the continent.

After removal of the native peoples, English settlers had relatively easy access to land. Compared to the "tight little island" of Barbados, where the small amount of potential agricultural land and the great demand for sugar plantations quickly led to shortages of land and rising prices, the Chesapeake had an enormous amount of uncleared land available throughout the seventeenth century.[71] In such circumstances, the price of land remained low. At the same time, the small European population ensured that the price of free labor was high, allowing immigrants to earn sufficient wages to purchase or rent land.[72] Moreover, a head right system, instituted originally by the Virginia Company and adopted later by the colonial government, allowed each free settler to acquire 50 acres, and a further 50 acres for every member of family or indentured servant brought to Virginia. For those immigrants who survived the rigors of disease and climate, the Chesapeake offered considerable opportunity to acquire freehold land; by 1700, some two-thirds of heads of households owned land.[73] The region had become the first of several "good poor man's countries" along the eastern seaboard.[74]

Over the course of the seventeenth century, a thinly settled countryside of isolated plantations, landing places, stores, and churches, interspersed among extensive stands of deciduous woodland, emerged in the Chesapeake (figure 3.9).[75]

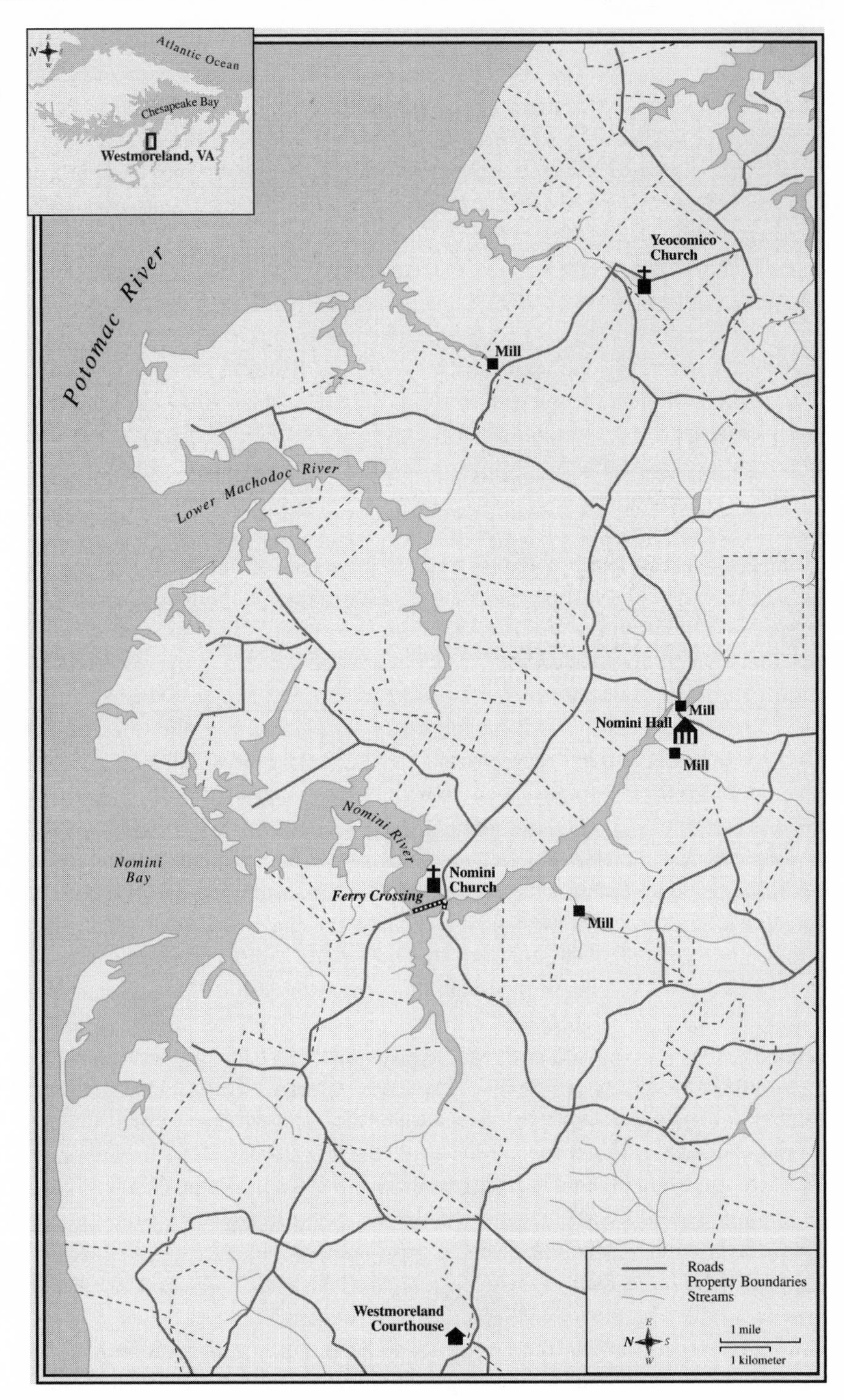

FIGURE 3.9. Dispersed settlement in Westmoreland County, Virginia, 1643–1742. After David W. Eaton, *Historical Atlas of Westmoreland County, Virginia* (Richmond, Va.: The Dietz Press, 1942), 67–73.

FIGURE 3.10. A reconstructed middling tidewater plantation at Historic St. Mary's City, Maryland; planter's house (left), tobacco barn (right). Image courtesy of Historic St. Mary's City.

The availability of land led to the dispersal of settlement along rivers and creeks; plantations usually were distributed between a quarter and one-and-a-half miles apart and commonly fronted tidewater. Such a littoral pattern of settlement—what has been called a "tobacco coast"—had much in common with the settlements of the resident fishery along the coasts of Newfoundland. With easy access to water transportation, planters "deliver[ed] their Commodities at their own Back doors" to English vessels, which served, in effect, as floating stores.[76] Such direct trade with English shippers meant that there was little need for service centers in the Chesapeake, and this discouraged development of towns and villages (see chapter 5). The scattered settlements, patches of cleared land, scrub, and immense stands of woodland in the region stood in marked contrast to the densely populated countryside of England, with its fields, farms, villages, and market towns.

The plantations themselves consisted of irregular parcels of land, laid out by the metes-and-bounds method, and a nucleus of farm buildings (figure 3.10).[77] These buildings consisted of a planter's house, outbuildings, tobacco barn, and corn crib. For much of the seventeenth century, the vast majority of planters, having poured much of their capital into land and labor, lived in simple, one-story dwellings, with one or two rooms on the ground floor and a sleeping space in the attic. Such structures were built of wood; the main structural posts were set in the ground (much like the tilts in Newfoundland) while the walls and roof were covered in split clapboard.[78] Chimneys usually had a frame structure and fire walls filled with wattle and daub. Wealthier planters covered the interior of their houses with plaster made from oyster shells, making the walls and ceilings look "as white as snow."[79] With plentiful supplies of wood and few masons in the Chesapeake, wooden buildings were the cheapest and easiest structures to construct. By the third quarter of the century, some planters had acquired sufficient capital to build more substantial dwellings. These so-called "Virginia houses" were wood-framed, one-and-a-half storied, two-room structures, with chimneys built out of brick at each gable.[80] The position of the chimneys minimized heat from the kitchen fire during the long, hot Chesapeake summers. The rapid appearance of a standard "Virginia house" suggests that regional building traditions imported from England soon gave way in the face an abundance of wood, a shortage of labor, and people from different regional backgrounds. Within the dwellings, furnishings were few and modest, again a reflection of the dearth of craftsmen and the high cost of locally produced goods. Other buildings on a plantation included a tobacco barn, built much like the dwelling, and a corn crib.[81]

Around the plantation buildings were ragged, stump-strewn clearings—what one scholar has called "a transient landscape"—that stood in marked contrast to the closely cropped and manicured agricultural landscapes of lowland England.[82] Unlike the intensive use of land in England or, indeed, the West Indies, where good arable land was in short supply and expensive, the system of farming in the Chesapeake was much more extensive. The availability of land allowed planters to practice a form of shifting cultivation, rasing tobacco or corn in a newly cleared field for three or four years until its fertility declined, and then planting another cleared area. In the meantime, the exhausted field was allowed to revert to brush and recover its fertility. Such a system of farming required a considerable acreage of land, but dispensed with manuring and crop rotation, which minimized the amount of livestock and labor needed on a plantation.[83] First worked out in the Chesapeake, this system of farming became widespread along the eastern seaboard, where land was plentiful and cheap but labor was in short supply and expensive. To prevent depredations from livestock, planters fenced in their few acres of tobacco and corn, and left their cattle to pasture among the marshes and their hogs to forage for mast in the woods.[84] Allowing livestock to range freely was standard practice from Newfoundland to the Carolinas, and reflected the enormous amount of uncleared land available along the eastern edge of the continent during the colonial period.

Compared to the sugar monoculture in Barbados, the agricultural economy that developed in the Chesapeake during the seventeenth century was more diverse. With plentiful land and a good climate, planters practiced mixed farming, growing tobacco and a variety of other crops, as well as raising livestock. The tobacco crop, the principal cash product, required considerable care. Plants were grown from seed and then planted out in the fields in mid-May. Each plant was placed in a hill, about knee-high, spaced three to four feet apart. During the growing period, plants were weeded and grubs picked off. In late July and early August, seed tops were cut off, allowing the plants to channel energy into their leaves. The harvest would also begin about this time and last until September. Leaves were cut, speared onto long "tobacco sticks," and then placed in piles to wilt. After a few days, the piles were collected, making sure that the leaves were not bruised or crushed, and taken to a tobacco barn for curing. Inside the barn, the leaves were hung up and allowed to dry. As with drying fish, the curing process was critical to the price the tobacco would make. The best tobacco lost its green color and turned a golden yellow, making what colonial farmers called "bright tobacco."[85] However, too much humidity, rain, or fog blowing off the tidewater—all common occurrences in the Chesapeake during late summer and fall—slowed the drying process and rotted the leaves. Alternatively, if leaves dried too fast, they would remain green, much like dried hay. Good circulation of air was critical. Seventeenth-century tobacco barns were most likely not tight structures, allowing the free movement of air. After about a month of curing, the tobacco leaves were stripped from the stalks, and then packed into bundles. The bundles, in turn, were pressed by a packer into casks, which were marked with the planter's name, and then rolled to the nearest wharf or store for sale and shipment.[86]

Apart from tobacco, planters raised a variety of crops and livestock. The main food crop was corn, grown in hills, much like tobacco, and used as a substitute for wheat. Planters also had gardens, where they grew beans, peas, sweet potatoes, and turnips, as well as apple orchards for the production of cider.[87] Livestock comprised cattle, hogs, and poultry, producing meat, milk, butter, and eggs for the household. Unlike England, where sheep were an important source of meat and wool, the Chesapeake had few sheep, most likely because of the labor demands in keeping them and fending off predators such as wolves. At least in the early seventeenth century, settlers hunted and fished, perhaps drawing as much as 30 percent of their meat from these sources. Yet as domestic livestock increased, game became less important and by 1680 comprised only 10 percent of all meat consumed.[88] By producing much of their own food, planters reduced their outlays for imported provisions and retained capital for developing their plantations. Although the diet may have been monotonous, planters ate better than most English people, and this was reflected, over the generations, in the greater height of Americans.[89]

The society that emerged in the Chesapeake during the seventeenth century was not marked by great extremes of wealth or poverty. At the top of the social hierarchy was a small and not very wealthy class of merchant-planters and gentry. The merchant-planters served as resident factors for English mercantile houses, and also traded on their own account. By the 1680s, some had invested in land and accumulated substantial holdings. The gentry also had acquired large estates. Along the lower Western Shore of Maryland, a small group of gentry had estates ranging in size from 100 to 5,400 acres, with an average of about 1,000 acres.[90] In England, a thousand-acre estate would have been considered substantial, but in the Chesapeake land was cheap and large estates were not worth much. Lacking great rental or agricultural incomes, the leading Chesapeake planters were unable to follow the example of the sugar planters and retire back to England; instead, they remained tied to the region and its staple economy. As the generations passed and native-born elites emerged in Virginia and Maryland, the tobacco planters viewed the Chesapeake, rather than England, as home; they were not "psychological transients."[91] Below the gentry was a large group of small and middling planters. Along the lower Western Shore, this group had estates ranging from 50 to 1,300 acres, with an average size of 300 acres. Such planters had accumulated enough land to support a farm, but not enough for speculation or rental.[92] Although situated on the agricultural frontier of the Chesapeake, these small planters had much in common with their counterparts on the fishing frontier of Newfoundland: They owned the means of production, employed some indentured servants, suffered from the dearth of women, and were frequently beholden to the merchant class.

At the bottom of the social and economic hierarchy was a great mass of indentured servants. Indentured from three to seven years, these single, young men and women provided planters with a source of cheap labor. With indentures agreed upon in England and usually covering the cost of a trans-Atlantic passage, the terms of work rarely reflected the true cost of labor in Virginia. As a result, planters were insulated against the high cost of labor that so bedeviled the Virginia Company. Like many servants in England, those in the Chesapeake lived under the same roof as their masters, and participated in household chores, but their principal tasks were clearing land and raising tobacco. After they had completed their indentures, servants looked to establish themselves on the land. As they had labor but no capital, many worked for a period as wage-laborers, saving to purchase land; others rented land from a planter on a sharecropping basis. In return for the use of land, tools, and tobacco seed, the tenant paid the owner a share of the crop. During the prosperous years of the 1650s and 1660s, sharecroppers were able to accumulate sufficient capital and purchase cheap, uncleared land on the frontier. But by the 1680s, the depression in the tobacco trade and rising population severely limited opportunities to acquire land. Many ex-servants were faced with laboring, tenancy, or moving elsewhere.[93]

A distinctive vernacular culture also emerged in the Chesapeake during the seventeenth century. With the great mass of settlers drawn from the south of England, the Chesapeake dialect reflected the drawl of southern England rather than the clipped speech of East Anglia or the north.[94] But given the mixing of settlers from many different parts of southern England, the peculiarities and nuances of a Hampshire or a Somerset dialect did not survive long. Instead, a generic Chesapeake dialect emerged, incorporating some speech patterns and words from southern England. Cultural mixing was further reflected in the region's place names. Tidewater place names were drawn from royalty and proprietors, and most especially from the source regions of the early settlers. The first counties in Virginia carried names from across England, including Surry, Isle of Wight, Gloucester, Northampton, Warwick, Stafford, Middlesex, New Kent, Norfolk, Essex, York, Lancaster, Northumberland, and Westmorland.[95] Although all the names were taken from England, the mix of royal, proprietary, and English county names had no parallel in England and reflected the mingling of settlers from different regional backgrounds in the Chesapeake.

Among the most populous English colonies on the eastern seaboard in the seventeenth century, Virginia and Maryland supported basic secular and religious institutions. The Crown took over responsibility for governing Virginia in 1625, and the Calvert family administered Maryland from 1634 until 1689, when, again, the Crown stepped in. While Virginia was headed by a governor and Maryland by a lieutenant-governor, representing the Calverts, both colonies set up executive councils, houses of assembly, and central law courts. These institutions were housed in the tiny colonial capitals of Jamestown in Virginia, and St. Mary's City in Maryland. Both colonies were divided into counties, with Virginia further divided into parishes.[96] Each territorial unit had its own institutions of government: at the county level, there were law courts and justices of the peace, as well as places of record for wills, deeds, and cattle brands; at the parish level, the local vestry oversaw the maintenance of the church and rector, provided for the poor, and set local taxes. Counties and parishes were also responsible for maintaining roads, bridges, and ferries.[97] Compared to the variety of governing institutions in England—sheriff's courts, church courts, manorial courts—the structure of government in the Chesapeake had been simplified to its most basic components.

As English immigrants formed the great majority of the colonial population in the Chesapeake, the Anglican Church was brought over and established in Virginia in 1624. Nevertheless, Maryland was created as a Catholic refuge, and Catholics made up an important minority, particularly in and around St. Mary's City. The Anglican Church was established in Maryland only after the Royal takeover of the colony in 1692. In addition, Quakers were tolerated in Maryland, and formed another significant religious group. Anglican and Catholic churches and chapels of ease were dispersed among the plantations, usually situated at accessible points, such as a crossroads, where taverns or "ordinaries" and stores were also established. In the seventeenth century, church buildings were small, mean structures, frequently built out of wood, using the same forms of construction as in planters' houses. Such unimpressive buildings reflected the high cost of labor, and the difficulties of raising revenue from planters struggling to establish farms.

For much of the seventeenth century, the Chesapeake provided a good deal of opportunity for English immigrants to establish themselves on the land. Unlike the West Indies, where a booming market for sugar and limited land attracted considerable capital investment and soon produced social and economic stratification, the Chesapeake enjoyed both a growing market for tobacco and unlimited land. The combination attracted labor from England but not great mercantile or landed wealth. Despite the rigors of climate, the travails of clearing the forest, and the high death rate, the region provided poor and middling immigrants with the greatest opportunity in seventeenth-century English America to get ahead and establish an independence. In many ways, the seventeenth-century Chesapeake was more like the agricultural frontier of family farmers that developed in eighteenth-century Pennsylvania than the staple frontier of sugar planters in the West Indies. Yet the favorable circumstances were not to last. In the 1680s, the stagnation of the tobacco economy, rather than a shortage of land, constricted opportunity.[98] Faced with the deterioration of the tobacco staple, small and middling planters diversified away from the weed. Like other farmers along the eastern seaboard, they increasingly turned to semisubsistence farming, raising much of the food for their families. For poor immigrants without land, there was little that they could do, except move once more, this time to the newly opened frontiers in the Carolinas and Pennsylvania.

The period from the 1680s to the early 1700s marked a turning point in the development of the Chesapeake. The War of the League of Augsburg (1689–1697) and the War of Spanish Succession (1702–1713) disrupted trans-Atlantic trade and caused stagnation in the tobacco economy. But after the Treaty of Utrecht in 1713, the economy rebounded and the Chesapeake entered its second period of expansion, its so-called Golden Age, which lasted until the Revolution. Production of tobacco increased from about 20 million pounds in 1700 to 50 million pounds in 1730, and reached 100 million pounds by 1770.[99] Even though demand in the British market remained stagnant, new markets developed in continental Europe. More than half of the tobacco exported to England was re-exported in 1669, and this proportion increased enormously in the eighteenth century. By the early 1770s, about 85 percent was being re-exported to the Continent. Although significant amounts were shipped to the Netherlands and Germany, the biggest market was in France. The French market began to increase after the Treaty of Utrecht, became significant in the mid-1720s, and had become the largest by 1730.[100] The massive rise in the French market encouraged big metropolitan firms to deal in the tobacco trade, favored the extension of the tobacco business to the British outports, and underpinned the expansion of the tobacco frontier into Southside Virginia and onto the Piedmont.

The great increase in production and export of tobacco had a major impact on mercantile involvement in the tobacco trade. As exports grew, large merchant houses were attracted to the trade, displacing smaller merchants and retailers. In London, the premier tobacco port for the first part of the eighteenth century, imports rose from about 11 million pounds in 1676 to 44 million pounds in 1775, while the number of importers fell from 573 to 66.[101] Over the course of a century, a few large houses, riding out the periodic depressions in trade, engrossed much of the tobacco business. A similar process of consolidation occurred in Bristol. In 1672, 467 traders handled an average of 9,600 pounds of tobacco; by 1742, only 49 merchants were left, each handling an average of 79,000 pounds. The opening of the continental market and the expansion of the tobacco trade during the early eighteenth century also attracted merchant houses in the northern outports. Merchants in Liverpool, Whitehaven, and Glasgow entered the trade, most of them specializing in the re-export trade to the continent. The development of the French market, in particular, propelled Glasgow and Whitehaven into the front rank of tobacco ports. By 1740, Glasgow was second only to London as the country's major tobacco port; by 1758, it had surpassed the capital to become the premier tobacco port in the land.[102] In the late eighteenth century, a handful of "tobacco lords" in Glasgow dominated the trans-Atlantic tobacco trade.

The structure of the tobacco business also changed. As large planters emerged in the Chesapeake in the late seventeenth and early eighteenth centuries, they increasingly followed the practice of West Indian sugar planters in consigning their tobacco to merchant houses in Britain to sell on their behalf. This benefitted merchants in that it guaranteed a cargo for their vessels, and also provided planters with a secure line of credit in England. The great London firm of Perry & Lane handled some of the largest Chesapeake accounts this way during the first two decades of the eighteenth century.[103] In total, consignments accounted for perhaps two-fifths of the tobacco trade. From about mid-century, British merchant houses also established branch stores in the Chesapeake, much as the West Country merchants were doing in Newfoundland at about the same time. Such practice was fairly common among the big London houses, but became particularly widespread among Liverpool and Glasgow houses.[104] The great Glasgow merchants William Cunninghame & Co. operated seven stores in Maryland and fourteen in Virginia.[105] The stores were situated at key shipping points along the coast, on the fall line, and at county courthouses in the new Piedmont counties. The houses sent out factors to manage the stores, as well as clerks and bookkeepers. The factors extended credit and supplies to planters in return for their tobacco crop. The leaf was then forwarded to merchant houses in Britain. Whatever the method of sale, tidewater planters dealt directly with British merchants, rather than going through colonial intermediaries, and this tied continental agricultural production tightly into the Atlantic commercial system

The rise of large British merchant houses in the tobacco trade had a profound effect on the Chesapeake. After standing aside from the tobacco trade for much of the seventeenth century, the great metropolitan houses entered the region with a vengeance during the following century and established their economic grip over much of the tidewater and the backcountry. In pursuing this strategy, they merely were doing what other metropolitan merchants had already done in the West Indies and in Newfoundland. With no major town in the Chesapeake, no large indigenous merchant class, and an extensive coastline allowing easy waterborne access to producers, the region was particularly vulnerable to the penetration of metropolitan capital. In this regard, the Chesapeake was much like the West Indies and Newfoundland. At first, the massive expansion of British credit had beneficial effects, financing the expansion of the tobacco frontier into Southside Virginia and across the Piedmont to the Shenandoah, as well as funding conspicuous consumption among the tidewater gentry.[106] But credit also meant debt, and as time wore on and debts accumulated, this became a major issue for both British merchants and Chesapeake planters. By the Stamp Act crisis in 1765,

Scottish firms alone had extended £500,000 in credit, and many planters were massively in debt.[107] At the outbreak of the Revolution, British houses had some £1,692,000 outstanding in the region, making the per capita debt of Chesapeake planters nearly twice that of other American colonists.[108] After relative economic independence for much of the seventeenth century, Chesapeake planters found themselves increasingly beholden to their British creditors; as an indebted Thomas Jefferson ruefully remarked: "These debts had become hereditary from father to son for many generations, so that the planters were a species of property annexed to certain mercantile houses in London."[109] The aggressive expansion of British mercantile capital and the loss of economic independence by the planters came to head in the early 1770s, and became a contributing factor in the great planters' support for the Revolution.[110]

The economic stagnation of the late seventeenth and early eighteenth centuries also led to a restructuring of Chesapeake society. Many small planters, especially those on marginal land, found it difficult to produce tobacco profitably during the depressed years and dropped out of the business. Meanwhile, larger planters, better able to ride out the downturn, were well placed to profit from the upturn in the economy after 1713. As the tobacco economy expanded in the 1720s, the larger planters increased their production and profited handsomely. By the 1730s, Chesapeake society was becoming increasingly differentiated between an elite of large planters and a mass of small farmers. In Maryland, a handful of very large planters—no more than 4 percent of total landowners—had estates valued at more than £1,000; in the middle was a substantial group of small planters—about 36 percent of the total—with estates ranging from £100 to £500; while at the bottom were poor whites—about 55 percent of the total—with estates less than £100. The great majority of planters with estates valued at less than £500—nearly 91 percent of the total—were essentially small farmers, scratching a living by raising a cash crop of tobacco and subsistence crops for the family.[111]

The tiny minority of great planters drew their living not only from tobacco but also from other activities such as commerce, the professions, government employment, money lending, and speculation in land. As the frontier moved west, leading tidewater planters exploited their political and economic connections to acquire large tracts of land on the Piedmont, in Southside Virginia, and in the Shenandoah Valley. In the 1720s and 1730s, planter William Beverley acquired land on the Piedmont and then moved on to speculate in the Shenandoah, where he shared with a group of friends title to more than 200,000 acres.[112] Similarly, Robert Carter, one of the leading planters in the Chesapeake and owner of a great house at Nomini Hall on the Northern Neck of Virginia (see figure 3.9), operated several plantations in the vicinity of his house and also held undeveloped tracts in Loudon and Fairfax counties on the Piedmont and Frederick County in the Shenandoah (figure 3.11). Such undeveloped land provided speculative profit through sales to incoming settlers, as well as guaranteed estates for the owner's offspring.[113] Unlike Newfoundland merchants and West Indian planters, who reinvested their capital

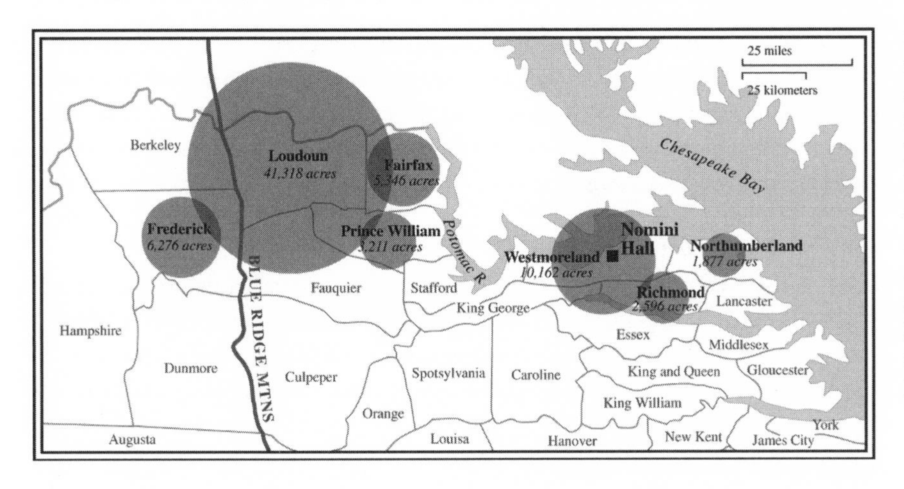

FIGURE 3.11. Land in the counties of northern Virginia owned by Robert Carter of Nomini Hall, 1775. Louis Morton, *Robert Carter of Nomini Hall: A Virginia Tobacco Planter in the Eighteenth Century* (Williamsburg, Va.: Colonial Williamsburg, Inc., 1941), 65, 70.

back in Britain, Chesapeake planters committed themselves and their heirs to the continental settlement frontier.

A further change in the tobacco economy was the supply of labor. During the 1660s, the growth of the English population slowed, and the flow of servants to the Chesapeake leveled out in the 1670s and then declined in the 1680s. Although this hardly affected the small planters, it had a great impact on the larger tobacco growers. Desperate for labor, the large planters, like their counterparts in the West Indies, turned to importing African slaves. During the 1670s and 1680s, larger planters began the changeover from servants to slaves, and by the 1690s slaves were predominant in the work force and remained so for the rest of the colonial period.[114] At first, slaves were imported in small numbers from Barbados and Jamaica, mainly because the Chesapeake was too marginal a market to justify large imports direct from West Africa.[115] Yet by the 1690s, regular shipments of slaves came from Africa into the Chesapeake. Compared to the massive shipments into the British West Indies, the trade to the Chesapeake was relatively minor, comprising less than one-tenth of the West Indies trade. Between 1651 and 1775, approximately 80,000 slaves were imported into Virginia, and about 10,000 into Maryland.[116] Although slave imports into the Chesapeake were of great importance to the development of the region, they were not of major significance to the trans-Atlantic slave trade.[117]

The relative insignificance of the Chesapeake slave trade was reflected in the origins of the slaves. Although tobacco planters, like their counterparts in the West Indies and South Carolina, reckoned that slaves from the Guinea Coast, particularly Coromantine, were the best, they could rarely afford them.[118] While the great sugar planters could pay premium prices for large numbers of slaves from Guinea, tobacco planters bought only one or two slaves at a time and then only from cheaper regions, such as the Senegambia, the Bight of Biafra, and Angola.[119] Tobacco planters also wanted to purchase young adult male slaves, but had to be content with a mix of men, women, and children. Of the slaves imported into the Chesapeake between 1710 and 1760, 52 percent were males, 29 percent females, and 19 percent children (boys equalled 11 percent; girls, 8 percent), a distribution remarkably similar to that for slaves imported into the West Indies.[120] Unlike the slave populations in the sugar islands, however, the slaves in the Chesapeake reproduced. Better conditions on tobacco plantations and encouragement of the planters allowed the reproduction of the work force. From the 1710s, the slave population began to grow through natural increase, and by the 1750s was increasing primarily through reproduction. At the outbreak of the Revolution, the Chesapeake was importing very few slaves and essentially had weaned itself off the slave trade.[121] Compared to the West Indies and the South Carolina low country, this was a singular achievement: The Chesapeake was no longer tied to the slave trade circuit of the British Atlantic but had developed its own internal, demographic momentum.

The natural increase of the slave population in the Chesapeake had important ramifications for the pattern of slave holding. In the early eighteenth century, only a minority of tobacco planters could afford slaves and then only a handful; from the 1690s to the 1740s, the majority of plantations in Virginia had fewer than ten slaves. As slave populations reproduced, however, the holdings grew; by the 1770s, 64 percent of plantations had more than ten slaves, and 29 percent had more than twenty-one slaves.[122] Although the large slave work forces common in the West Indies and South Carolina were rare in the Chesapeake, a small number of very large planters managed to amass several hundred slaves. At his death in 1732, Robert "King" Carter, perhaps the richest man in Virginia, owned 390 slaves of working age.[123] Nevertheless, Carter's slaves, like those belonging to other large planters, were distributed among several plantations or quarters, frequently staffed by no more than an overseer and about ten slaves. Quarters were scattered across the landscape, frequently intermixed with farms of poor whites. Such a pattern ensured that slave work forces rarely dominated the population of a parish, let alone a county, and further differentiated the Chesapeake from the West Indies and the South Carolina low country where slaves made up the majority of the population.

The import and natural increase of the slaves dramatically affected the Chesapeake population. Although the white population increased markedly through the eighteenth century, rising from about 85,000 in 1700 to 398,000 in 1770, the black population also increased from about 13,000 to 251,000 over the same period. Such an increase in the number of blacks changed the racial composition of the total population. From having an overwhelmingly English and white population in the seventeenth century, the Chesapeake became increasingly African and black in the eighteenth century. In 1700, the population was about 13 percent black; by 1750, it had reached 40 percent, a proportion that was maintained until the Revolution.[124] Although the black population became substantial, it never reached the dominant levels found in the West Indies and the South Carolina low country, and this further differentiated the Chesapeake from those other slave-holding regions.

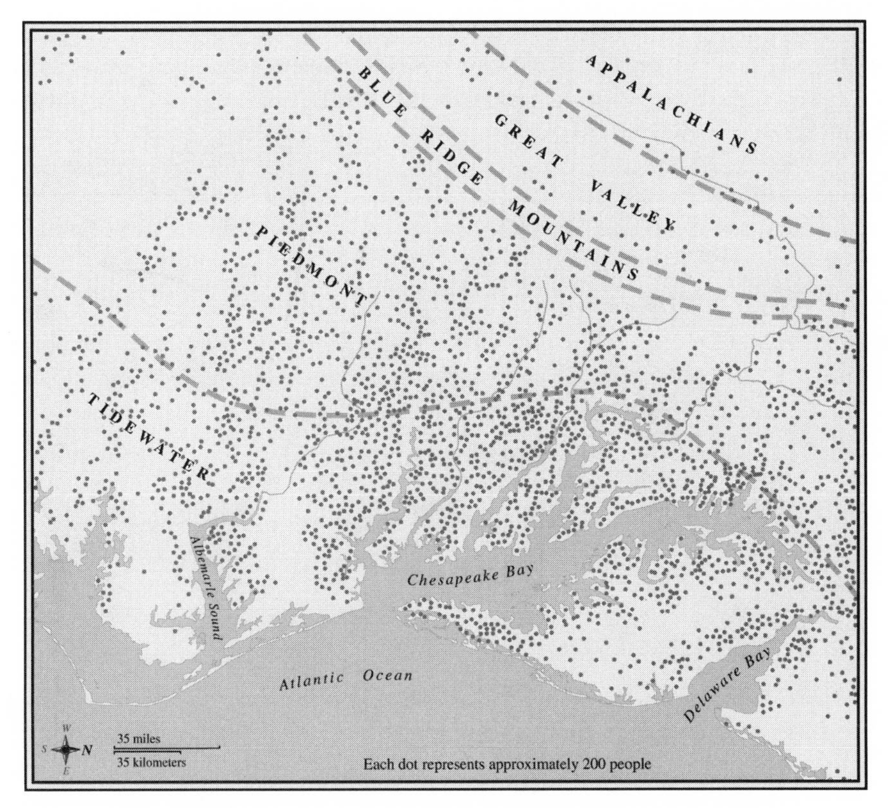

FIGURE 3.12. Settlement in the Chesapeake, 1760. After Herman R. Friis, *A Series of Population Maps of the Colonies and the United States 1625–1790* (New York: American Geographical Society, 1968).

The great expansion of the economy and the increase of population in the Chesapeake during the eighteenth century pushed the settlement frontier far beyond the tidewater. During the 1720s and 1730s, settlement spread onto the Piedmont, into Southside Virginia, and into the Albemarle district of North Carolina (figure 3.12).[125] By the 1740s, settlers had spread over the Piedmont, crossed the Blue Ridge, and started to fill in the Shenandoah Valley. Between 1750 and 1775, the Shenandoah was effectively settled, marking the western edge of continuous American settlement in Virginia. It was also in the Shenandoah that the tidewater plantation complex came up against the southern fringe of a family farming frontier emanating from southeastern Pennsylvania. In the backcountry of Virginia, a different pattern of settlement, economy, society, and culture emerged from that in the tidewater (see chapter 4).

The expansion of the tobacco economy also led to greater investment in the infrastructure of plantations. Although small and middling planters continued to operate plantations consisting of a motley collection of wooden buildings, comprising the ubiquitous "Virginia house," outbuildings, and tobacco barns, the great planters enlarged and embellished their properties.[126] The largest plantations comprised a complex of buildings, comparable, so the Frenchman Durand de Dauphiné thought, to "a fairly large village."[127] At the center of the complex was

FIGURE 3.13. Shirley Plantation, Charles City County, Virginia, 1738, 1772. Overlooking the James River, Shirley was one of the largest and most imposing great houses built by the tidewater gentry. Library of Congress, Prints and Photographs Division, HABS, VA, 19-SHIR, 1-6.

the great house, a building much like the smaller Georgian houses being built by wealthy farmers and gentry in the English countryside during the eighteenth century.[128] At the lower end of elite housing, the one-and-a-half story wooden "Virginia house" was rebuilt or sheathed in brick, sometimes with a gambrel roof to provide more attic space. Houses of the middling elite were commonly two stories with a symmetrical facade, and either a single or double pile plan. Rooms were usually arranged on either side of a central entranceway, hall, and staircase. At the top end of elite housing were a handful of great houses belonging to the richest planters. These were two or two-and-a-half story buildings, with a double pile plan and dependencies on either side (figure 3.13). On the outside, the houses were marked by symmetry, fine brickwork, and classical touches, such as a pediment over the main door; inside, first-floor public rooms were usually embellished with wood paneling and plastered, decorative ceilings.[129] Around these great houses were numerous outbuildings, including kitchens, smokehouses, dairies, slave quarters, stables, carriage houses, schoolhouses, storehouses, workshops, and family cemeteries.[130] On the largest plantations, some of these outbuildings were built out of brick. Interspersed among the buildings were the work areas of the slaves, vegetable gardens, orchards, and the planter's formal garden. Through its large size, architectural splendor, and central position, the great house dominated the complex of buildings, yards, and gardens.

Although the tobacco economy scarcely generated the incomes of the sugar colonies, the greatest Chesapeake planters built considerably more-splendid houses than any sugar planter in the West Indies. Indeed, the mansions along the lower James and on the Northern Neck of Virginia comprised the finest collection of great houses created anywhere in the British American colonies. For the most part, this reflected the difference in commitment by the elites to these colonies. Sugar planters preferred to live in England and transferred much of their income out of the West Indies; tobacco planters remained rooted in Virginia and Maryland and retained their income and capital in the Chesapeake. Great houses were also built in the South Carolina low country, but they were not as common or as grand as those in Virginia and Maryland. Many rice planters were content to have modest rural seats and handsome townhouses in Charleston, whereas tobacco planters, lacking an urban retreat, lavished their attentions on their plantation homes. Unlike the great sugar planters, the tidewater gentry rarely, if ever, visited Britain and had to be content with recreating England in the colonies rather than enjoying England at first hand. Although few in number, great houses such as Stratford, Westover, Shirley, and Corotoman were the ultimate expressions of material success in the Chesapeake, and stood as symbols of the commitment made by the tidewater gentry to the tobacco colonies. Such great houses were the rural equivalents of the grand merchant residences in the North Shore fishing ports.

The diversification in agricultural production that began in the Chesapeake during the seventeenth century became even more pronounced in the following century and further differentiated the region from the West Indies. Booms and busts in the tobacco economy encouraged many planters, whatever their social and economic standing, to lessen their dependence on the metropolitan market. The cultivation of tobacco increasingly was supplemented with the growing of wheat and corn, and raising of livestock, products that were not enumerated under the navigation acts and found ready markets in the West Indies and southern Europe.[131] The great growth of the grain trade, in particular, was a major factor in the development of an indigenous merchant class and port towns in the Chesapeake (see chapter 5). Whereas in the West Indies very little diversification away from sugar took place, the enormous spaces of the Chesapeake and the presence of ready markets for grain and livestock allowed the development of a more diversified agricultural economy, which lessened dependence on the metropole. English and Scottish ideas about improved agriculture also filtered into the tidewater, influencing farming practice and diversification. George Washington saw himself as a gentleman farmer, and experimented with several different crops and new types of farm building on his Mount Vernon plantation.

The economic expansion also produced a stratified, staple society. By the mid-eighteenth century, rural society in the Chesapeake consisted of a tiny planter elite, a large group of small farmers, and an even greater mass of black slaves. As in the West Indies, the planter elite controlled the political levers of power, dominating legislative assemblies, the legal profession, and parish vestries. Planters intermarried among themselves, cementing alliances and preserving family fortunes; by the mid-eighteenth century, the great planters of the tidewater counties had formed "an almost hereditary caste."[132] Like other elites on the colonial periphery, planters looked to Britain for their cultural models, aping the English gentry in their dress, entertainment, and pastimes, and building houses in the prevailing Georgian style.[133] But unlike the leading sugar planters in the West Indies or the fish merchants in Newfoundland, who had homes in England and spent only a few years in the colonies, the great tobacco planters rarely visited Britain. Some had been sent as youths to be educated at Oxford or Cambridge or trained at the Inns of Court in London, and a few crossed the Atlantic again later in life, but, for the most part, the tidewater gentry were domiciled in the Chesapeake. There, their lives revolved around the local contexts of plantation, parish, county, and colony, rather than the trans-Atlantic world of metropole and empire. After those early, first-hand experiences of England, planters living in Virginia and Maryland experienced Britain at second-hand, through the impressions of English visitors and the textual world of letters from tobacco merchants, as well as newspapers, magazines, and books; the leading Chesapeake planters were known for their great libraries. For all the outward signs of English taste and manners, the Chesapeake gentry were rooted in the tidewater: their houses and plantations reflected a deep commitment to the region, and, in the years leading up to the Revolution, they put their own economic and political interests above any cultural attachments to Britain.

If the elite attempted to recreate an English gentry world, small farmers maintained and developed the rural vernacular of the Chesapeake. As the white population increased and European immigration slowed, networks of kin began to emerge in older settled areas, and to thicken over the generations.[134] As time passed, frontier areas became settled neighborhoods of friends and family. Ties with England lessened; links among people in the Chesapeake strengthened. The Anglican Church, as well as new denominations such as the Methodists and Baptists, became increasingly entrenched in people's lives, while hunting, fishing, and horse racing provided popular pastimes. Most likely, small planters in the eighteenth century spoke a stronger Chesapeake dialect than their forefathers, and used a more distinctive vocabulary to describe the natural world.[135] Personal names, too, may have been applied to the landscape more freely than earlier.

The slave culture that developed in the Chesapeake was markedly different to that in the West Indies. The relatively small numbers of slaves on plantations and the preponderance of the white population ensured that black culture would be influenced heavily by the larger white society. While slaves in the Caribbean created their own distinctive creole from English and African languages, slaves in the Chesapeake quickly lost their former languages, probably did not develop a creole, and soon adopted a form of Standard English.[136] Just as a great many immigrant families from the European mainland gave up their own languages after a generation and spoke English, so slaves from Africa used the dominant language. For successful operation of plantations, communication between masters and slaves was essential. Although slaves may have spoken with regional accents and used idioms from Africa, the loss of their indigenous languages and the lack of a synthetic creole represented a considerable erasure of their Old World cultures; a massive amount of oral culture must have been lost. Nevertheless, in other areas of culture, such as dance, music, dress, and personal adornment, slaves maintained fragments from the Old World. A passion for clapping, drumming, and strumming the banjo was common among blacks in both Africa and the Chesapeake. Such African fragments were frequently combined with elements from English culture. Adoption of European musical instruments (notably the fiddle) and the influence of English hymns and song helped create a Chesapeake slave musical culture that was a productive mix of African and English elements.[137]

Although the institutional context in the Chesapeake scarcely changed during the eighteenth century, the material manifestations of political and ideological power became much more obvious.[138] New colonial capitals were built in Williamsburg and Annapolis, symbols of imperial power and the prosperity of Virginia and Maryland (see chapter 5). Beyond these towns, courthouses and Anglican churches were built or rebuilt in brick and the prevailing Georgian style. In a landscape of mainly wooden buildings, these classical brick structures, often standing isolated at a country crossroads, symbolized the permanence of colonial government and the Anglican church in the Chesapeake, as well as the stratification of local society. Leading planters helped fund the construction of churches — Robert "King" Carter built Christ Church on the Northern Neck of Virginia, aligning the building with the driveway leading to his great house at Corotoman — and planters had their own family pews (figure 3.14).[139] In many ways, county courthouses and parish churches were extensions of the power structure so obvious in the great houses and plantations of the tidewater gentry.

By the late eighteenth century, the Chesapeake was the area of longest permanent English settlement in North America. Over the course of 150 years, a distinctive pattern of settlement, economy, society, and culture had evolved, demarcating the region from Britain and from other plantation societies in the West Indies and South Carolina. Historian Jack Greene has argued that "the increasingly hierarchical society of the eighteenth-century Chesapeake probably came closer than that of any other contemporary British-American cultural region to replicating what Harold J. Perkin has referred to as the 'old society' of rural England," and that Virginia and Maryland "looked remarkably English."[140] The deciduous woodlands, the grand brick Georgian houses, and the manners and mores of the tidewater gentry were reminiscent of some aspects of the landscape and society of southern England, but far too much was different to consider the Chesapeake a replica society of England.[141] Although the agricultural economy of the Chesapeake was becoming increasingly diverse in the late eighteenth century, tobacco remained the great staple of the region, and continued to shape much of the rural way of life, particularly in the tidewater. Compared to the varied agriculture and industries of southern England, the Chesapeake remained greatly dependent on a single export crop. Moreover, the stark bifurcation of the Chesapeake population along racial

FIGURE 3.14. Christ Church, Lancaster County, Virginia, 1722–1736. The size and architectural accomplishment of Christ Church owed much to a benefaction from Robert "King" Carter, one of the wealthiest planters in colonial Virginia. Photograph by the author, 2002.

lines had no parallel in England, while the simple division of Chesapeake society into gentry, small farmers, and slaves, with a smattering of merchants, clergy, lawyers, and doctors, scarcely compared to the occupational diversity of English rural society. The racial and social mix in the Chesapeake had no comparison to societies in either the British Isles or in the West Indies.

Just as merchants in the port towns of New England had taken control of the cod fishery, so the tidewater gentry and small farmers had taken hold of the rich land of the Chesapeake and established a viable staple economy. But the English market was too weak during the seventeenth century to support massive capital accumulation, and it was only with the expansion of the European market in the eighteenth century that large amounts of capital began to accumulate in the region. By then, planters had been in the Chesapeake for several generations and saw their future in the colonies. Although a handful had the income to retire to Britain, they preferred to remain in North America, building their fine houses, speculating in frontier land, and holding political office. Big fish in Virginia and Maryland, they would have been minnows in England. Unlike the mobile, trans-Atlantic elites of the Newfoundland fishery and the West Indian sugar islands, the tidewater gentry had put down deep, permanent roots in the Chesapeake. As British mercantile capital increasingly took control of the economy in the early eighteenth century, the stage was set for a clash with the planters. Unlike the elites on the Atlantic islands of Newfoundland and the West Indies, who were integrated into the Atlantic economic and political system, the tidewater gentry, although dependent on the Atlantic world for the tobacco trade, were firmly entrenched in their own developing regional economy, society, and polity. As the economic gale blew in from the Atlantic, threatening to uproot the social and economic structures of the Chesapeake, the tidewater gentry reconsidered their allegiances. By the early 1770s, planters in the Chesapeake were ready to join with merchants in New England in establishing their economic and political independence from Britain.

Rice Plantations in the South Carolina Low Country

Among regions of staple production in early modern British America, the rice plantations of South Carolina stood fourth in economic importance after the sugar plantations in the West Indies, the tobacco plantations in the Chesapeake, and the cod fisheries in Newfoundland and New England. Along the eastern seaboard, however, the Carolina low country was second only to the Chesapeake as a staple-producing region, and by the end of the colonial period generated planter fortunes even greater than those in Virginia and Maryland. As in other areas of staple production, the early development of the low country depended on metropolitan initiative from England, with some colonial support from Barbados. Yet by the early eighteenth century, metropolitan hold over the low country had weakened. The development of rice plantations along the waterways of the low country gave rise to an indigenous planter class, while the rice trade through the port of Charleston led to the emergence of a resident merchant community. By the eve of the American Revolution, low-country planters and Charleston merchants formed a cohesive elite sufficiently prosperous, culturally confident, and politically mature that it was able to join other elites along the eastern seaboard in defying British imperial power.[142]

English colonization of Carolina began with an alliance between English and Barbadian interests.[143] After the accession of Charles II in 1660, a group of courtiers in London acquired the proprietorship of Carolina in 1663 and set about planning its settlement and economic development. Masterminding much of the project were leading proprietor Sir Anthony Ashley Cooper, later the first Earl of Shaftesbury, and his secretary, the philosopher John Locke. They drew up the Fundamental Constitutions of 1669, which envisaged Carolina as an elaborate mix of feudal hierarchy, political liberalism, and religious toleration. The proprietors expected to reap a financial return from the colony through granting land to settlers and collecting quitrents.[144] At the same time that these plans were being devised, the sugar revolution was transforming Barbados. The great expansion of sugar plantations on the island led to rising prices for land and a growing landless population. During the 1650s and 1660s, many lesser gentry and poor whites, unable to get a foothold on the island, looked elsewhere to settle. By the 1660s, significant numbers were emigrating to Jamaica and Virginia, with some looking to Carolina as another potential destination. As several of the proprietors in England had Barbadian connections, West Indian interest in the new colonial venture was encouraged.

After two colonization efforts failed in the mid-1660s, a small colony comprising several hundred Barbadians and English was planted on the Ashley River near present-day Charleston in 1670. Almost from the beginning of settlement, the proprietorial vision developed in the circumscribed world of Carolinian England ran into the colonial reality of settling an expansive continental frontier. Dispensing with the intricacies of the Fundamental Constitutions, the first settlers set about molding the colony to their own needs. Acquiring grants of land on the basis of head right, the colonists took up unsurveyed lots along rivers and creeks, creating a widely dispersed pattern of settlement. Although the proprietors had wanted a prior survey of land before settlement and the creation of compact, contiguous settlements, economic self-interest and "the scent of better land" proved more powerful.[145] The settlers quickly developed an agricultural economy producing basic foodstuffs for domestic consumption, as well as staples for the Caribbean and English markets. Cattle, hogs, lumber, and Indian slaves were sold to Barbados; tar, pitch, and turpentine (naval stores) and deerskins were shipped to England. Barbadians dominated the early colonial government, serving as governors and members of the council and the legislative assembly. As staunch Anglicans, they ignored the principle of religious freedom and established the Church of England in the colony in 1706. Ten parishes were laid out along the coast, six of them named after parishes in Barbados.[146] Such was the sugar island's dominance over early Carolina that John Locke observed: "the Barbadians endeavour to rule all."[147]

If the proprietors struggled to control the colony's internal development, they had even greater problems dealing with external threats.[148] Situated close to Spanish Florida, the pirate-infested Caribbean, and the Yamasee Indians of the interior, Carolina was especially vulnerable. During the War of Spanish Succession (1702–1713), the colonists became embroiled with the Spanish, launching assaults on St. Augustine in 1702 and the interior two years later, as well as beating back an attack by the Spanish on Charleston in 1706.[149] Although the Spanish threat diminished after the Treaty of Utrecht in 1713, the Yamasee Indians began a war against the colonists in 1715. Widespread attacks led to the destruction of plantations and the retreat of European settlement to within thirty miles of Charleston. Military aid from as far away as Virginia and Massachusetts was needed before the Carolinians regained control. Two years later, pirates cleared out of the Caribbean by the British navy took up position along the Carolina coast and harassed local shipping. In 1718, the colonists captured the leading pirates and executed them in Charleston. All these various incidents highlighted the weakness of the proprietary government in controlling South Carolina. As private individuals in England, the proprietors lacked the resources of the state to protect the colony or launch offensives against external threats. With little aid coming from the proprietors, the colonists were forced to take matters into their own hands to protect the colony. This further strained relations between the proprietors and the colonists, as well as strengthening local autonomy. Meanwhile, the English state was becoming increasingly involved in the colonies. A series of measures passed in the 1690s and early 1700s tightened control over the empire, and left little room for proprietary governments. Squeezed between the encroaching power of the metropolitan state and increasing colonial independence, the proprietors bowed to the inevitable and transferred authority over South Carolina to the Crown in 1719 in return for financial compensation. Once again, Old World landed interests had failed to make a success of New World settlement.

While South Carolina was being brought under the wing of the British state, the colony's economy was being integrated into the larger Atlantic world. The emergence of rice cultivation in the 1690s and early 1700s provided the colony with a major commercial staple. By 1720, the export of rice dwarfed the early export trades in livestock, deerskins, and naval stores; by the early 1770s, rice comprised 55 percent by value of South Carolina's total exports.[150] As one commentator remarked in 1761: "The only Commodity of Consequence produced in South Carolina is Rice and they reckon it as much their Staple Commodity, as Sugar is to Barbadoes and Jamaica, or Tobacco to Virginia and Maryland."[151] Like the plantation staples of sugar and tobacco, rice was an enumerated product that had to be exported to Britain before it could be re-exported to the European mainland. After much lobbying by British and Carolina merchants, however, rice was removed from the list of enumerated goods in 1730, and allowed to be shipped direct to markets in the Iberian peninsula and the Mediterranean.[152] Never a staple of the European diet, rice was used as a substitute for grain, particularly after bad harvests. Rice was also exported to the West Indies as cheap food for slaves; in peak years, the sugar islands took about 20 percent of total rice exports.[153] During the 1740s, indigo became a second major staple of South Carolina. Cultivation of indigo was stimulated by wartime disruption of Jamaican and French colonial supplies of indigo to Britain and government bounties on its export. Used as a dye in the textile industry, indigo became increasingly important as the textile industry in Britain expanded during the late eighteenth century. By the early 1770s, indigo comprised 20 percent by value of South Carolina's exports and had become the fifth-most important staple export from the American colonies.[154] Apart from rice and indigo, South Carolina continued to export deerskins and naval stores to Britain, as well as livestock and lumber to the Caribbean.

As the rice economy became established, an indigenous group of planters and merchants emerged.[155] The planters were mostly descended from Barbadian, English, and French Huguenot immigrants who had settled in the low country during the 1670s and 1680s. These second-generation settlers were well placed to take advantage of the new staple, converting family holdings into rice plantations, as well as establishing new plantations along interior rivers and swamps. As rice plantations were enormously expensive to build, South Carolina planters looked for economies of scale by developing large plantations. In St. John's Berkeley Parish on the Cooper River, 38 percent of landowners owned more than 1,000 acres in 1763, a proportion that increased to almost 49 percent by 1793. Moreover, the proportion of plantations with more than thirty slaves rose from 29 percent in the 1720s to 64 percent in the 1770s.[156] Thirty slaves was considered the minimum necessary for operating a rice plantation.[157] By the 1770s, more than half the slaves in the low country lived on plantations with more than fifty slaves.[158] Such labor forces represented enormous investments of capital and help account for the prodigious wealth of Carolina planters on the eve of the American Revolution.[159]

The rice merchants formed the other half of the Carolinian elite. Concentrated in Charleston, the principal port in the low country, colonial merchants served, at least initially, as middlemen between British merchants and low country planters (see chapter 5). Acting as commission agents, they imported manufactured goods from Britain and slaves from Barbados and West Africa; exported the principal staples of rice and indigo, as well as lesser products such as naval stores and deerskins; and extended credit to rice and indigo planters.[160] Although Charleston merchants bore the brunt of the economic relationship with Britain, they also had the opportunity to exploit new trading relationships elsewhere in the Atlantic. The direct export of rice to southern Europe after 1730 and an increasing trade to the West Indies allowed some merchants to bypass British houses and trade on their own account. As they accumulated capital, leading merchants invested in local shipping, urban real estate, and low-country plantations.[161] Charleston merchants also strengthened their ties with low-country planters through intermarriage. By the eve of the American Revolution, the planter-merchant class in South Carolina formed a formidable colonial elite that had more in common with the West India lobby in England than with elites in the Chesapeake or North Shore fishing ports.

As in other New World settlements, the early colonists were faced with a shortage of labor. Although the British Isles had supplied thousands of indentured servants to the plantations in the West Indies and the Chesapeake, the era of white servitude was coming to an end in the 1670s and only a few hundred servants went to South Carolina. Instead, the settlers turned to two other sources of labor: Indians and Africans. In the late seventeenth century, the low country was inhabited by several thousand Indians of various tribes, who made a living from hunting and gathering and growing corn, beans, and squash.[162] Although the early colonists traded with the Indians for deerskins and foodstuffs, they also bartered and captured Indian slaves. Some slaves were sold to planters in the West Indies, others were retained in South Carolina; in 1708, about a third of the colony's 4,300 slaves were Indians. Even so, local Indians were not a long-term solution to the colony's labor problem, especially as large-scale plantations developed. Sporadic warfare, particularly with the Yamasee, depleted local native populations at the same time that European diseases were wreaking their deadly havoc. By the early 1770s, if not long before, a coast that had once been "swarming with tribes of Indians" had been emptied, leaving "nothing of them but their names."[163]

Well before the Indian slave trade ended, settlers in South Carolina had started using African slaves. In the 1670s, Barbadian planters had arrived in the colony with black slaves who were soon employed in the livestock industry, perhaps giving rise to the term "cowboy."[164] As the rice economy developed in the 1690s and early 1700s, the demand for African slaves increased considerably. Between 1700 and 1775, some 98,000 slaves were shipped to the Carolina and Georgia low country, slightly more than were sent to the Chesapeake but far less than the number exported to Barbados, Jamaica, or the Leeward Islands.[165] Unlike the Chesapeake tobacco economy, which gradually weaned itself off imported slaves, the South Carolina rice and indigo economies, desperate for slave labor, continued to require large numbers of imported slaves right up to and beyond the Revolution.

In terms of origin and composition, slaves shipped to South Carolina were much like those sent to Virginia and Maryland. Even though slaves from the Gold Coast were preferred, as they were throughout the British slave colonies, South Carolina drew most of its slaves from the Senegambia, the Bight of Biafra, and Angola.[166] The demographic composition of slaves in South Carolina was also remarkably similar to that in the Chesapeake. Rice planters wanted to buy young, male slaves, "very prime young Negro men" according to one leading Charleston merchant, but they also took women and children; between 1720 and 1770, 56 percent of slaves shipped to South Carolina were adult males, 29 percent adult females, 9 percent boys, and 6 percent girls.[167] Despite the demand for young male slaves, planters had no control over the composition of slave cargoes. Although not perfectly balanced, the slave work force was able to reproduce itself by the 1750s. The ability of planters in the low country, like their counterparts in the Chesapeake, to "breed" their slave work forces marked slave-owning societies on the mainland as fundamentally different from those in the West Indies. In brute demographic terms, slaves in South Carolina, like those in the Chesapeake, could expect to live something close to their natural span of years and reproduce themselves, while those in Jamaica had much shorter life expectancies and did not reproduce.[168]

Yet in terms of size and composition of the low-country population, South Carolina was more like the Caribbean than the Chesapeake. Between 1700 and 1770, the population of the low country increased from about 5,704 to 88,244, a total comparable to Barbados in the 1770s.[169] Of this population, black slaves numbered about 1,200 in 1700, 40,000 in 1750, and 82,000 in 1770.[170] As early as 1710, blacks formed a majority of the population. In 1720, more than 70 percent of the population in four low-country parishes was black; by 1760, the proportion had reached 90 percent in three of them.[171] Such proportions were similar to those in Barbados and Jamaica, and marked the low country as a "negro country." Scottish immigrant George Ogilvie, sent to manage a plantation on the Santee River north of Charleston, reported in 1774 that he was settled "*at least* four miles from any white Person—like the Tyrant of some Asiatick Isle the only free Man in an Island of Slaves—."[172] The low country was the only part of the mainland to have a black majority, and that differentiated the region from the Chesapeake and the South

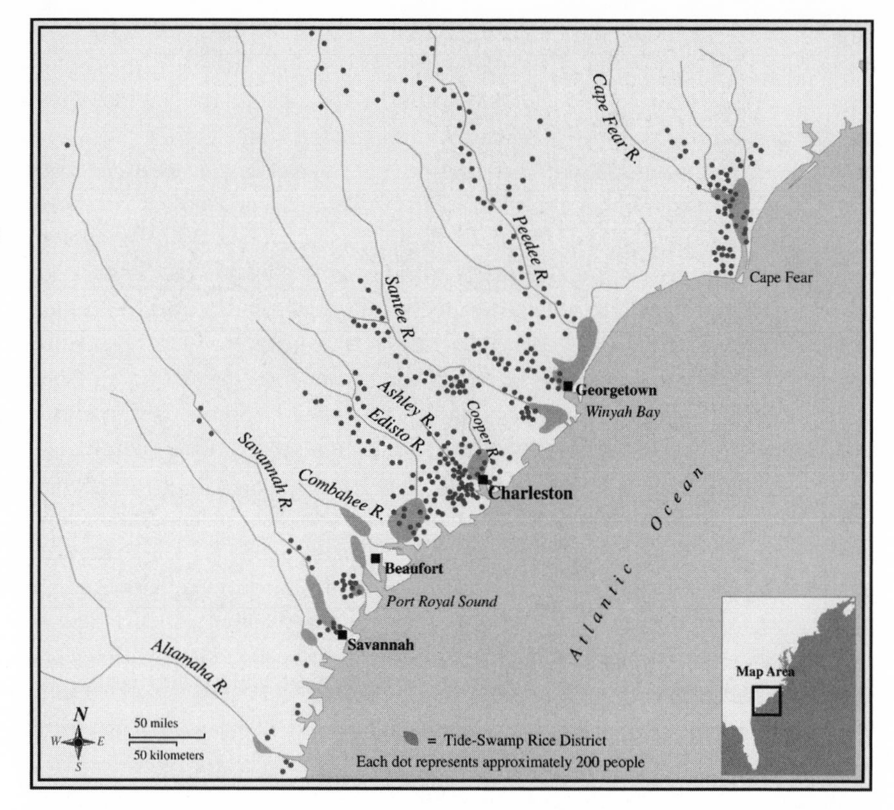

FIGURE 3.15. European settlement in the Carolina low country, 1740. After Herman R. Friis, *A Series of Population Maps of the Colonies and the United States 1625–1790* (New York: American Geographical Society, 1968).

Carolina backcountry, which both had predominantly white populations.[173] The implications of a black majority in the low country were considerable. Whereas slaves in the Chesapeake lived in a largely white world and adopted aspects of Anglo-American culture, those in the low country lived in a mainly black world and developed their own creole culture. Moreover, the white planter elite, like that in the West Indies, was ever conscious of its vulnerability to slave rebellion.

The adoption of rice cultivation led to the spread of settlement through the watery world of rivers, creeks, and swamps that make up the Carolina low country (figure 3.15).[174] In 1680, European settlement was confined to the fledgling colonial capital of Charleston, the adjacent sea islands of James Island and Johns Island, and along the Ashley and Cooper rivers and their tributaries. Twenty years later, colonists had spread south, settling on Edisto Island and along the Stono, Edisto, and Combahee rivers, as well as north, particularly around Winyah Bay. By 1719, settlers had pushed farther south, settling around Port Royal Sound and on Port Royal and St. Helena islands. Although Charleston remained the central hub of the low country, two more towns were founded to service the southern and northern peripheries of the colony: in 1711, Beaufort was established on Port Royal Island; in 1729, George Town was laid out overlooking Winyah Bay. As settlement spread south toward Spanish Florida, Georgia was established in 1732 to serve as a buf-

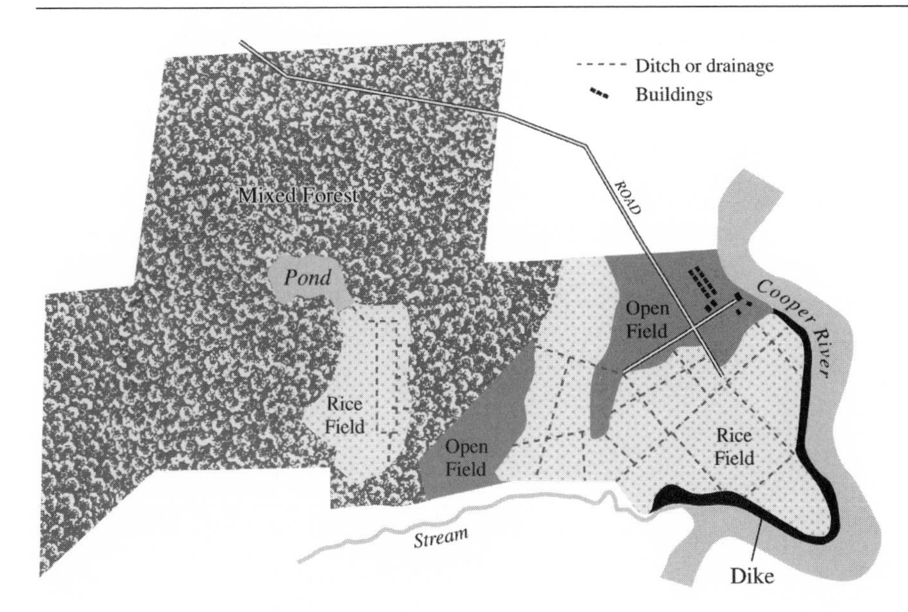

FIGURE 3.16. Bluff Plantation, Cooper River, South Carolina, early nineteenth century. Map shows inland swamp rice fields (left) and tide swamp rice fields (right). After Sam B. Hilliard, "Antebellum Tidewater Rice Culture in South Carolina and Georgia," in *European Settlement and Development in North America*, ed. James R. Gibson, 101 (Toronto: University of Toronto Press, 1978).

fer and as a free-labor, small farming colony.[175] Yet the poor sandy soils of the Savannah area scarcely yielded farmers a living, while the high cost of white labor made extractive industries, such as lumbering, uncompetitive. Nevertheless, the tidewater rivers of the coastal plain were suitable for rice growing. Planters from South Carolina began to move into the colony in the late 1740s, and their numbers increased markedly after the prohibition on slavery was removed in 1752. By the 1760s, Carolinians had incorporated much of the coastal strip from the Savannah River to the Altamaha River into the plantation frontier. To the north, rice cultivation had entered North Carolina (separated from South Carolina in 1691) in the 1730s. Concentrated along the lower Cape Fear Valley, the rice plantations were never as large or as prosperous as those in South Carolina, and were more diversified, usually combining the cultivation of rice with the growing of indigo and the making of naval stores.[176] By the early 1770s, the rice coast stretched more than 400 miles from Cape Fear in the north to the Altamaha River in the south, and up to 50 miles into the low-country interior.

Throughout this low-country landscape, planters and their black slaves adapted the environment to rice cultivation.[177] Although English settlers must have known little about rice, slaves from the coast of West Africa were familiar with rice growing and most likely introduced the specialized techniques to the low country.[178] At first, planters and their slaves used swamps on the flood plains for growing rice. The main areas of swamp rice cultivation are not exactly known but were probably along rivers and on coastal marshes. These "inland swamps" were dammed at their upstream and downstream ends to control water levels, ditches were dug to facilitate flooding and drainage, and fields were planted with rice (figure 3.16).[179] Yet problems with the supply of water, particularly during droughts and freshets, encouraged planters to adapt rice cultivation to "tide swamps" found along the rivers that meander their way across the coastal plain. In 1734, a visitor to the colony crossed the Edisto River about thirty miles south of Charleston and observed: "the land is mostly thereabouts fine cypress swamps, which they count the best for rice."[180] The main areas of tidewater rice cultivation in South Carolina were along the Waccamaw and Santee rivers near Georgetown, the Ashley and Cooper rivers near Charleston, and the Edisto, Ashepoo, and Combahee rivers near Beaufort; the principal areas in Georgia were along the Savannah, Ogeechee, Newport, and Altamaha rivers.[181] The most suitable parts of these rivers for rice cultivation lay between the tidal salt marshes and the freshwater swamps above the tidal zone. Planters relied on high tides to raise the level of fresh water in the rivers to flood their fields, and on low tides to lower the level of water and so drain the fields. Slaves built embankments around the rice fields to protect the crop, and used sluices or "trunks" fitted with doors to control the flow of water (see figure 3.16). In many cases, cultivation of rice in inland swamps existed side by side with cultivation in tide swamps, and in some instances swamp reservoirs were used to supply fresh water to tide swamp fields.

Rice plantations were mere islets amid a largely forested, low-country landscape. George Ogilvie described the low country as "one Continued dead Plain intirely coverd with wood except small spots where Plantations are settled."[182] Travelers moving along the roads on high ground saw mile after mile of pine barren and few signs of settlement; only those traveling by river observed the rice fields set behind their banks and the great houses placed on high ground.[183] On a typical tidewater plantation, rice lands comprised a complex hydraulic landscape of banks, trunks, reservoirs, canals, floodgates, and ditches. Rice fields were usually regular in shape, a quarter or a half an acre in size, enclosed by check-banks, and drained by ditches. Overlooking this landscape was the great house. In terms of size and architectural splendor, the great houses of the low country lay somewhere between the grand plantation homes along the James River in Virginia and the more modest residences in Barbados (figure 3.17). Like tobacco planters in the Chesapeake, rice planters in the low country remained tied to the region, but their commitment was split between their river plantations and their town houses in Charleston. The malarial environment of the low country, as well as the tedium and isolation of plantation life, drove many planters into town during the summer months. During his travels through North America in 1764–1765, Lord Adam Gordon visited Charleston and noted that "almost every family of Note have a Town residence, to which they repair on publick occasions."[184] With a few exceptions, notably Drayton Hall, plantation houses in the low country were not as grand as those along the James, while the town houses in Charleston compared to the finest in the colonies.[185] By the 1730s, some great houses were designed in formal Georgian and Palladian styles, usually adapted from British architectural pattern books. A few great houses (Crowfield, Middleton Place) were also surrounded with formal gardens, comprising avenues of live oaks, magnolias, fish ponds, and

FIGURE 3.17. Middleburg Plantation, Cooper River, East Branch, Berkeley County, South Carolina, 1697 and later. Photograph by the author, 1994.

canals. These gardens were among the most sophisticated designed landscapes in the British American colonies, and a testament to what could be achieved with slave labor. Other buildings on a plantation included an overseers' house, kitchen, stable block, blacksmith's shop, barn, winnowing house, threshing floor, and slave quarter. Slave houses were usually arranged in a street, sometimes parallel to the entrance avenue, and comprised one- or two-unit dwellings.[186] They were commonly constructed out of wattle-and-daub (mud or "tabby" covered the walls) and had palmetto-thatched roofs. Such simple housing may have owed something to African building traditions, and stood in stark contrast to the high-style architecture of the great houses. Beyond the nucleus of plantation buildings lay some cleared land for growing cereals, cultivating indigo, and pasturing livestock; the remainder of a plantation was left in longleaf pine, a source of lumber and naval stores.

The rice plantations of the South Carolina low country represented the most intense application of slave labor in early modern British America. Inland and tidal swamps not only had to be cleared of cypress and gum trees of "prodigious size" growing in the stagnant water, but also drained by ditches and enclosed with enormous earthen banks or dykes. In tidal areas, these banks had to be of "Vast size and Strength" to withstand the daily movement of the water.[187] One embankment laid out on the lower Savannah River in the late 1750s was just over three miles long.[188] Moreover, these banks and ditches had to be maintained. During fall and winter, slaves shored up embankments, cleaned out ditches, and replaced rotted wood in trunks. In spring, they harrowed fields and sowed rice seed, either by hoeing furrows or coating seed with clay to prevent it floating in water. After sowing, the first flooding or "sprout flow" covered the fields to encourage germination. Thereafter, fields were alternatively flooded and drained, allowing plants to grow and weeds to be hoed up. In September, slaves harvested the crop, loaded it on boats or "flats," and transported it to the plantation for threshing. Rice was usually threshed with a flail, and milled with a mortar and pestle: both immensely labor-intensive tasks.[189] During the eighteenth century, wind fans and mechanized mills were introduced on larger plantations to increase productivity.[190]

The organization of work on low-country rice plantations varied according to the tasks at hand. For the massive amount of work needed in clearing swamps, digging ditches, and building embankments, planters usually employed gangs or squads of slaves working in unison under the command of an overseer: a type of labor organization used in the West Indies and the Chesapeake. But for planting and weeding rice, the task system was used. Slaves were set a specified amount of work to be done each day, and then left to get on with it. When the task was completed, the slaves had the rest of the day free. Such an arrangement allowed them time to cultivate their own provisions grounds, which were usually quite extensive. The task system became so common that planters organized their inland swamp fields according to the amount of work a slave could achieve in a day. A common reckoning was that a slave could tend a quarter of an acre each day, which led to many fields being divided into quarter-acre units. With the spread of tide swamp cultivation, less weeding was required, and the task unit was expanded to half an acre.[191] Even more than the sugar plantations in the West Indies or the tobacco plantations in the Chesapeake, the rice plantations of South Carolina were physically defined by their fields of labor.

The planter class in the South Carolina low country, like its counterparts in the West Indies and the Chesapeake, dominated rural society. Indeed, like those in the West Indies, the "opulent and lordly planters" of the low country held sway more completely than planters in the Chesapeake because there was no significant poor white farming class in the region; white farmers in South Carolina were mostly confined to the backcountry (see chapter 4).[192] The rise of the planter class began with the shift to large-scale rice production in the 1730s and was fully formed by the 1750s.[193] By the Revolution, low-country society comprised a tiny white planter class, numbering perhaps two thousand, and an enormous black slave work force.[194] In a brutally materialistic society, planters were "chiefly known by the number of their slaves, the value of their annual produce, or extent of their landed estate."[195] Planters controlled the political life of the region, dominating parish vestries and the legislative assembly. Through marriage with other planters, they formed an hereditary "rice aristocracy"; they also married leading Charleston merchants, creating a powerful economic and political elite.[196]

South Carolina planters also created a distinctive cultural life. Like their counterparts in the Chesapeake and the West Indies, as well as the merchants in North Shore fishing ports, Carolinians looked to Britain for cultural leadership and adopted fashions and tastes of the metropole. This was reflected in the architecture and furnishings of the finer plantation houses in the low country and in the town houses in Charleston. Indeed, two planters on the Cooper River in the mid-1750s went so far as to name their plantations after Kensington and Hyde Park in London, and ordered from England "handsome Views of those two places with the adjacent Woods, Fields, & Buildings & some little addition of Herds, Huntsmen, &ca., &ca, but not too expensive in the Painting."[197] The landscapes were destined for the chimney breasts of the hall and parlor in their respective houses. Yet unlike West Indian planters, absenteeism in England appears to have been rare; low-country planters remained committed to their region and, above all, to their capital city, Charleston.[198] Just as London served as the metropolis of empire, so Charleston acted as a mini-metropolis for the low country. And just as merchants and absentee planters formed the West India lobby in England to protect their interests in the sugar trade, so low-country merchants and planters formed a unified block in Charleston to protect the rice economy.[199] Through common economic and political interests, Carolina merchants and planters were welded together into a formidable colonial elite.

Despite the immense control that South Carolina planters wielded over the settlement, economy, society, and politics of the low country, they had little influence over the development of a black vernacular culture. Slave populations on low-country plantations were sufficiently large that they could support the development of a creole language and culture, a feature common to the slave populations in the West Indies but not those in the Chesapeake. The creole language, known as Gullah, that evolved in the low country appears to have been an amalgam, drawing on vocabulary and grammar from several West African languages, pidgin used along the coast of West Africa and onboard English slaving ships, and Standard English.[200] The resulting hybrid language allowed slaves from different ethnic backgrounds to communicate, and formed the foundation for a distinctive black culture. Although African slaves brought memories and practices of their particular cultures with them, the mixing of people from different ethnic groups on the plantations ensured that elements, rather than entire cultural complexes, would be reproduced in the low country. Like the development of Creole culture in the West Indies, many nuances and details of particular West African cultures were discarded, and only those elements that proved useful or common to several groups survived. Among generic elements of African culture transferred to the low country were a passion for drumming, singing, and dancing; handicrafts of pottery and basketry; simple agricultural implements; and dietary preferences for yams and millet.[201] Such elements were mixed with others drawn from the planter culture and plantation economy. As in the West Indies, low-country slaves adopted European musical instruments, and ate locally raised corn and rice.

From the beginning of English settlement, the Carolina low country had an institutional framework. The proprietors appointed a governor who administered the colony and named an executive council; local property-holders elected representatives to the legislative assembly. This government structure continued after the Crown took over the colony in 1719.[202] The colonial government as well as the principal court were located in Charleston. Circuit courts were established in

FIGURE 3.18. Pompion Hill Anglican Chapel, Cooper River, East Branch, Berkeley County, South Carolina, 1763. Photograph by the author, 1994.

outlying settlements only after backcountry unrest during the 1760s. The principal institution in the rural parts of the low country was the parish church. As in the Chesapeake and the West Indies, Anglican churches served the religious needs of the small white community, while vestries provided local government (figure 3.18). The plantocracy had the most prominent pews and controlled the vestries. Dispersed among the plantations, the small, brick Georgian churches were as much symbols of planter ascendancy as the great houses along the rivers and the town houses in Charleston.

By the late eighteenth century, a distinct plantation economy, society, and culture had emerged in the South Carolina low country. The suitability of the low country for rice cultivation, coupled with European and West Indian demand for rice, created the colony's staple economy. The relatively late settlement of the colony ensured that a fully formed slave plantation system, first worked out in Barbados, was transferred to South Carolina and quickly adapted to rice production. A large slave labor force powered the plantations, creating one of the most specialized agricultural landscapes in the colonies. A simple pattern of settlement consisting of dispersed plantations and isolated Anglican churches soon emerged. While the planter elite that controlled this landscape removed itself to the sophisticated world of Charleston, the black slave population that inhabited the low country created a distinctive creole vernacular culture. With its roots in Barbados, the Carolina low country appeared to have many of the economic, social, and cultural characteristics of the West Indies.[203]

Nevertheless, there were significant differences. First, the economic returns from rice were never as great as from sugar. Although some rice planters were wealthy enough to retire to England, absenteeism never became endemic to the region. Charleston, rather than London or Bath, became the planters' home away from home. As a result, planters in South Carolina, like those in the Chesapeake, remained resident in the region. Second, the opportunities for economic expansion, particularly the extension of the rice economy into Georgia in the 1750s, were greater than in the West Indies and further tied the planters to the seaboard. Third, the slave work forces in the low country were never as large or as brutalized as those in the West Indies. Slaves managed to reproduce, creating their own demographic momentum. Finally, the opening of the Carolina backcountry for settlement in the early eighteenth century created a different economy and society to that in the low country and set up a new dynamic in the colony. Backcountry settlers provided a counterbalance to the planter interest in Charleston, and helped direct the development of the colony away from the Atlantic. Such a dynamic simply did not exist in any West Indian sugar island. Despite the overtones of the Caribbean, the South Carolina low country was not Barbados or Jamaica but a distinct staple region set on the eastern edge of the continent.

Continental Staple Territories

The extension of English economic power to the eastern seaboard of North America during the seventeenth and early eighteenth centuries was much less successful than to the Atlantic staple regions of Newfoundland, the West Indies, and Hudson Bay. Although English merchants and proprietors began the colonization of New England, the Chesapeake, and the Carolina low country, they soon lost control over settlement, defeated by the enormous availability of agricultural land and the difficulty of commanding immigrant labor. As metropolitan capital withdrew, colonial merchants and planters emerged and developed new staple economies. Even though these economies were tied to the markets of the Atlantic world, colonial merchants and planters invested in the expanding towns and agricultural frontiers of the continent. Much more than the elites in the British Atlantic, American elites straddled two worlds: the world of Atlantic trade and the world of continental agricultural expansion.

During the seventeenth and eighteenth centuries, the staple economies along the eastern seaboard drew on enormous amounts of capital, labor, and provisions. At first dependent on England for capital and credit, colonial planters and merchants gradually accumulated capital to finance their own production and trade, although Chesapeake planters remained heavily indebted to metropolitan merchants at the time of the Revolution. Planters and merchants also relied on the metropole for indentured servants during the seventeenth century as well as African slaves in the eighteenth. Nevertheless, the port towns of New England and the plantations of the Chesapeake developed self-sustaining labor forces; only South Carolina and Georgia were economically dependent on slave imports at the end of the colonial period. The continental staple regions were thus much less reliant on the trans-Atlantic trades in labor than the Atlantic regions. Moreover, New England, the Chesapeake, and South Carolina enjoyed a surplus of provisions, some of which were exported to Newfoundland and the West Indies. In terms of capital, labor, and supplies, New England, the Chesapeake, and South Carolina were less reliant on the metropole, which gave them some economic independence within the larger British empire.

Even more important, the staple exports from the American colonies were not completely tied to Britain. To be sure, tobacco was an enumerated article and had to be shipped to British ports, but dried fish and rice were shipped direct to markets in southern Europe. The lesser exports of grain and lumber also found ready markets in the Caribbean. Trade in these various articles usually was handled by local merchants, rather than by British houses, and this further lessened the colonies dependence on the metropole. Although American merchants and planters were heavily indebted to British houses for manufactured goods, which had to be paid for by the sale of the various colonial staples, colonial elites managed to accumulate capital from the staple trades and create their own economic spaces along the eastern seaboard.

Taking profits from the dynamic world of Atlantic commerce, American elites reinvested in the equally expansive world of the continental frontier (see chapter 4). Fish merchants in New England invested in urban real estate and agricultural land; tobacco planters in the Chesapeake speculated in property beyond the tidewater; and rice planters in South Carolina extended the plantation frontier south into Georgia. While the elites of the British Atlantic had no equivalent to the immense agricultural frontier of North America and reinvested in the shires and towns of the British Isles, the elites of the eastern seaboard reinvested in the continental agricultural frontier.

The three staple economies also created settlements, populations, and societies along the eastern seaboard that were different from those in the Atlantic staple regions. Although the plantation landscape of South Carolina came closest to that in the West Indies, the North Shore port towns and the Chesapeake plantations were set in more diversified countrysides than those in Newfoundland or Jamaica. Moreover, the populations supported by the staples were more balanced and diverse. The male worlds of the fishing stations in Newfoundland and the fur posts in Hudson Bay had no counterpart along the American eastern seaboard, while the slave populations of the West Indies were larger and more oppressed than those in the Chesapeake or South Carolina. Societies too were significantly different. Although the fish merchants in New England and the planters in the South were much like their equivalents in Newfoundland and the West Indies, they sat atop more diversified societies. The North Shore port towns had a larger social range than the Newfoundland outports, while the Chesapeake plantations were set amid a landscape of small farms. Only the South Carolina low country came close to replicating the social extremes of the sugar islands, but even in the low country, planters eventually had to deal with a significant backcountry farming population.

Colonial elites also lived in different spatial and temporal contexts than the merchants in Newfoundland or the planters in the West Indies. New England fish merchants may have had ties across the Atlantic and no doubt had spent some years at sea, but they were not as mobile as their counterparts in the West Country, who regularly moved back and forth across the Atlantic. Similarly, the planters were tied more closely to the Chesapeake and South Carolina than their equivalents in Barbados, the Leewards, and Jamaica. For American elites, port towns, plantations, parishes, counties, and colonies provided the spatial contexts of life; there was little direct participation in the spacious, trans-Atlantic world experienced by other elites elsewhere in British America. There was also a different sense of time. Colonial elites were largely cut off from direct experience of the Old World, whereas transient West Country merchants and English sugar planters maintained deep roots in England. Colonial merchants and planters created their own worlds and their own histories, made manifest in their grand homes and plantations, while merchants in the West Country ports or planters resident in London maintained established worlds and a continuum with the past.

In sum, the staple regions along the eastern seaboard were territorial spaces of production that had no parallel elsewhere in the British empire. Whereas the metropolitan-dominated space of the British Atlantic managed to encompass the Atlantic islands and the rim of Hudson Bay, it was never able to gain equivalent purchase over the continental colonies. Instead, the staple regions along the eastern seaboard developed their own economic, political, and cultural integrity. When these spaces began to flex and expand in ways not congruent with the larger imperial space, conflict was bound to emerge. In the decade before the American Revolution, the North Shore ports, the Chesapeake, and the South Carolina low country all became spaces of conflict within the larger British American empire.

Chapter 4

Agricultural Frontiers: New England, the Mid-Atlantic, and the Southern Backcountry

Unlike the Atlantic staple regions of Newfoundland, the West Indies, and Hudson Bay, the staple regions along the American eastern seaboard were situated in larger continental spaces that had enormous agricultural potential. Although New England, the Mid-Atlantic colonies, and the southern backcountry proved unsuitable during the colonial period for plantation crops and thus did not attract substantial English investment, the availability of agricultural land in these areas proved immensely attractive to Europeans with limited capital and an aptitude for hard work. First in New England and then, later, in the Mid-Atlantic colonies and the southern backcountry, agricultural frontiers developed that would set these regions apart from the staple spaces on the seaboard as well as those on the southern and northern margins of the continent. These great agricultural frontiers, offering immense opportunity to immigrant Europeans and generations of Americans, would help turn the seaboard colonies away from the Atlantic toward the interior of North America. This drive westward would, in turn, set up clashes first with the native peoples and then with the French and the British, and contribute powerfully to the breakup of colonial British America.

New England

After the Chesapeake, the second area of permanent English settlement along the eastern seaboard was in New England. Although West Country fishermen had established fishing stations in the Gulf of Maine from about 1610, the first agricultural settlement was not planted until the Pilgrims arrived, somewhat inadvertently, at Plymouth in 1620. Yet this was a tiny toehold on the coast. It was not until 1630, with the creation of the Massachusetts Bay Colony and the start of the Great Migration, that the English gained a firmer foothold. As the English population increased, settlement expanded, native populations were pushed aside, and new colonies were established in Rhode Island, New Haven, Connecticut, and New Hampshire. Over the course of the colonial period, the patterns of settlement, economy, society, and culture that developed in these political units created one of the most well-defined agricultural frontiers on the eastern seaboard.[1]

The English migrations to New England comprised the few hundred Pilgrims, who went to Plymouth Colony in the 1620s, and between 13,000 and 21,000 emigrants who arrived in the Massachusetts Bay Colony between 1630 and 1642.[2] The outbreak of the English Civil War in 1642 brought the migration to a close; for the rest of the colonial period, only a few hundred settlers trickled in, mostly Scots-Irish who settled at Londonderry, New Hampshire. Although the early English migration to New England has been termed the Great Migration, the flow was minor compared to the enormous numbers of indentured servants going to the West Indies and the Chesapeake during the same period, or the migrations of fishermen to Newfoundland each year. One estimate puts the Great Migration at just 30 percent of the total English outflow across the Atlantic during the 1630s.[3] Yet this discrepancy should not be surprising. During the early seventeenth century, massive amounts of English capital were invested in the West Indies, the Chesapeake, and Newfoundland, much of it in migrant labor. In New England, English merchants were interested in exploiting the cod fishery along the coast, but not the agricultural potential of the interior. The agricultural settlements of New England developed in the lee of the fishery, attracting a relatively small number of English immigrants but not large amounts of mercantile capital.

Although the Great Migration to New England drew migrants from virtually every county in England, the greatest flows were from eastern and southern England, particularly the East Anglian counties of Norfolk, Suffolk, and Essex; London and the Home Counties of Hertfordshire and Kent; and the southwestern counties of Wiltshire, Somerset, Dorset, and Devon (figure 4.1).[4] From this considerable area, two major migrant streams can be identified: first, a migration of families, drawn mainly from East Anglia, the Home Counties, and the West Country, who had Puritan sympathies; second, a migration of single, young men, drawn from London and Devon, who were attracted by prospects of employment in agriculture, trades, and the fishery. The migration from East Anglia—approximately 38 percent of total migrants in one study—comprised mainly of families, focused on the Boston area.[5] During the early seventeenth century, East Anglia was a center of religious nonconformism. Many of the migrants from the area were Puritans, who feared religious oppression in England, and wished to join Puritan leader John Winthrop in building a "holy city upon the hill" in the New World. Similar Puritan congregations existed in the Home Counties and the West Country. As the migration got underway, migrants frequently recruited other family members as well as friends to join them, creating a chain of migration across the Atlantic.[6] Particular towns and villages in England became linked to specific townships in New England. Hingham, Massachusetts, drew 40 percent of its families from East Anglia, most of them from the Hingham area in Norfolk.[7] Other family migrations most likely linked eastern Kent to the South Shore of Boston (Scituate, Plymouth, Sandwich), the Wiltshire/Berkshire area to the Merrimack

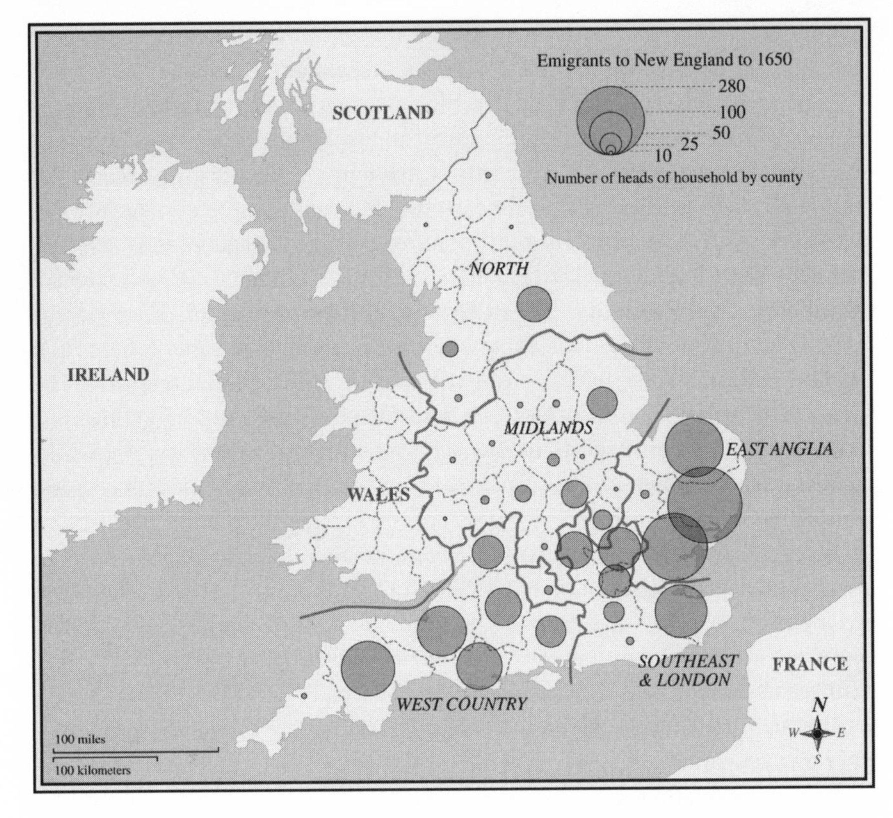

FIGURE 4.1. Regional origins of English migrants to New England. After Martyn J. Bowden, "Culture and Place: English Sub-Cultural Regions in New England in the Seventeenth Century," *Connecticut History* 35 (1994): Figure 1, 74.

Valley (Salisbury, Newbury, Amesbury), and southwest Dorset to the South Shore (Dorchester) and the Connecticut Valley (Windsor).[8]

The migrations from London and Devon were much different. Although both areas sent families to New England, the migrations appear to have been weighted toward single, young men, comprising perhaps a third of total male migrants.[9] As the largest city in England, London had enormous opportunities for employment and attracted migrants from across the country. Out of this pool of labor, thousands of indentured servants were sent to the Chesapeake and the West Indies during the seventeenth century. A small fraction of this labor appears to have fetched up in New England, perhaps contracted to merchants and tradesmen, who themselves had emigrated from London to Boston. In contrast, the migration of single, young men from Devon to New England was part of the larger migration stream associated with the Newfoundland fishery.[10] Virtually all the indentured servants from the county came from the areas of Barnstable-Bideford, Plymouth, Dartmouth-Teignmouth-Newton Abbot, and Exeter-Topsham. Significantly, there was no migration from Poole in east Dorset, a port that became involved in the Newfoundland fishery only later in the seventeenth century. Merchants in Barnstable, Plymouth, and Dartmouth organized the New England fishery, sending out servants each year to the fishing stations along the coasts of southern Maine and New Hampshire, as well as to the capes of Massachusetts (see chapter 3).

The immigration of a large number of families into New England created a new demographic pattern in the English American colonies. Unlike the overwhelmingly male populations that existed in the Chesapeake, the West Indies, and Newfoundland during the early seventeenth century, the rough balance between males and females in New England ensured that the colonial population would reproduce. Moreover, the healthy environment and the comfortable standard of living enjoyed by many colonists meant that people lived longer in New England than in England, with perhaps half the colonial population surviving to age seventy. With couples marrying at a relatively young age—about twenty-two for women and twenty-five for men—women bore numerous children, typically seven or eight, six or seven of whom survived to adulthood.[11] As a result, the population increased rapidly, doubling every generation (approximately twenty-five years).[12] By 1650, the total population of New England was about 22,800. At the turn of the eighteenth century, it was 92,700, and by mid-century had reached 360,000. By 1770, it was 581,000.[13] At that time, approximately 13 percent of the region's population was urban, leaving a rural population of about 505,000.[14] Of this rural population, approximately 205,000 lived in Massachusetts, 160,000 in Connecticut, 54,000 in New Hampshire, 50,000 in Rhode Island, and 36,000 in the territories that later became Maine and Vermont.

From the footholds established around Massachusetts Bay in the 1620s and 1630s, the English population expanded along the easily accessible coastal lowlands and interior river valleys.[15] By 1675, there was a ribbon of settlement along the coast, stretching from Cape Cod as far north as Damariscotta, Maine, and as far west as Greenwich, Connecticut (figure 4.2).[16] Inland, patches of settlement existed along the Merrimack River, the head of Narragansett Bay, and along the Thames and Connecticut rivers. Yet the outbreak of King Philips War in 1675 between the Indians and the English saw the contraction of the frontier of settlement, with major retreats from southern Maine, the upper reaches of the Connecticut Valley, and part of Rhode Island. Even after the end of the conflict in 1677, the frontier was far from secure; the outbreak of war with France and its native allies in 1688 led to further disruption, which lasted more or less continuously until the Treaty of Utrecht in 1713. During the war years, only areas in southern Connecticut, Rhode Island, and southern Massachusetts, which were relatively distant from the conflict with New France, were able to expand. But after the war, the frontier grew rapidly. By the early 1750s, the various settlements in southern New England had coalesced, while those in northern New England had reached up the Connecticut and Merrimack valleys, as well as along the coast of Maine as far as Waldoborough. By then, all the best land in southern New England had been taken. After 1760, further areas were made available (figure 4.3). The British conquest of New France removed the French and Indian threat to New England, allowing expansion to the peripheries of the region, while the clearance of French Acadian settlers from Nova Scotia in

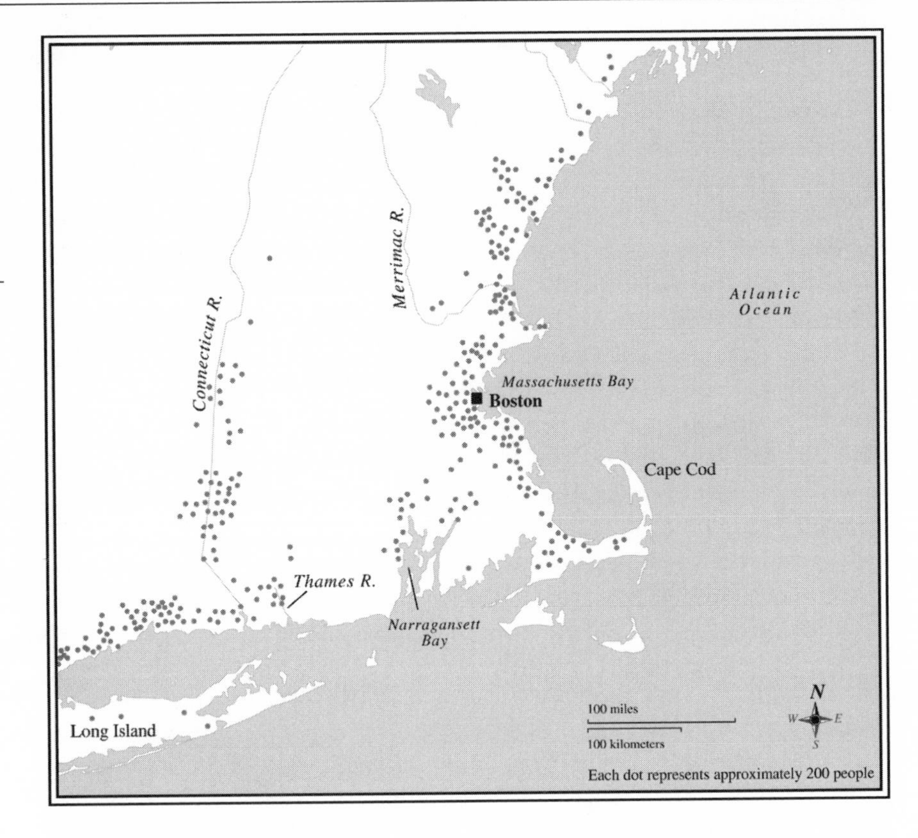

FIGURE 4.2. English settlement in New England, 1675. After Herman R. Friis, *A Series of Population Maps of the Colonies and the United States 1625–1790* (New York: American Geographical Society, 1968).

1755 opened up their lands.[17] During the 1760s and early 1770s, the New England frontier expanded northward into southern Vermont, northern New Hampshire, and central and eastern Maine; at the same time, several thousand New Englanders removed across the Gulf of Maine to Nova Scotia, taking up prime farmland along the Annapolis Valley and on the marshlands around the Bay of Fundy, or establishing fishing stations along the South Shore.[18]

As English settlement spread through New England, it came up against native populations. In the southern part of the region, European diseases had made substantial inroads into the Pequots, the Narragansetts, and the Wampanoags before the first Puritans arrived in the 1620s.[19] In 1622, Thomas Morton observed that the Indians had "died on heapes as they lay in their houses . . . And the bones, and skulls . . . made such a spectacle that . . . it seemed to mee a new Golgotha."[20] Immensely weakened by disease, these native populations soon came under further pressure. As English settlers moved into the Narragansett Country and the Connecticut Valley in the mid-1630s, they encroached on native hunting grounds. In 1637, war broke out between English colonists and the Pequots living near the southern end of the valley; in a brief but bloody battle, the Pequots village at Mystic Fort was attacked and the inhabitants "perished in promiscuous ruin."[21] Further attacks along the coast wiped out most of the rest of the tribe. In 1675–1676, raids

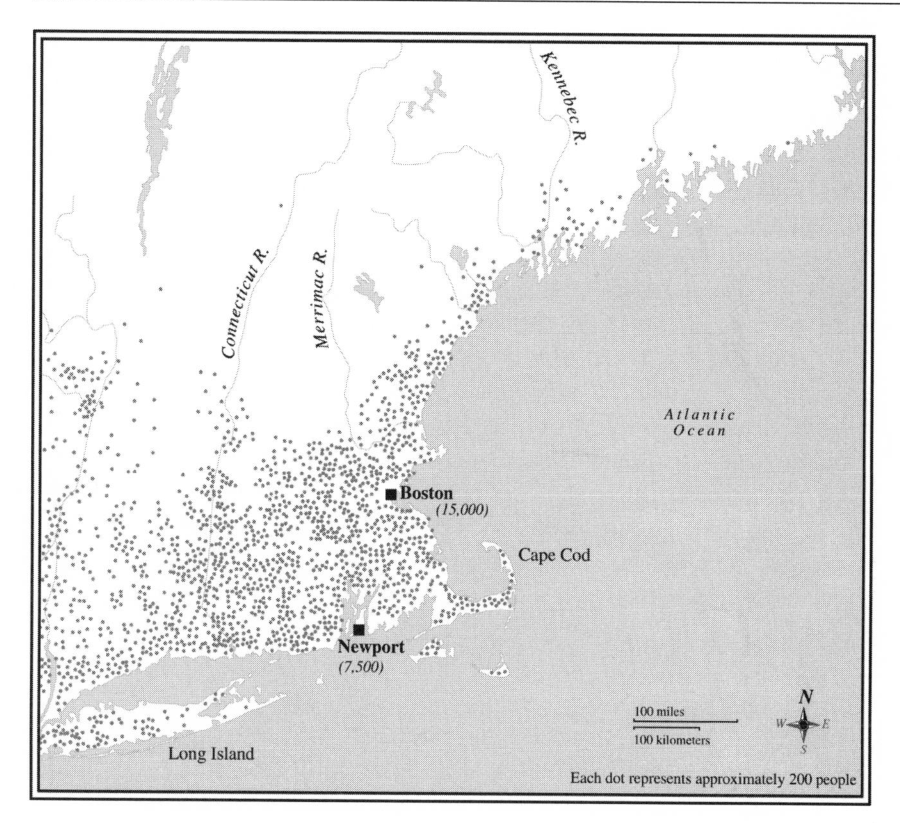

FIGURE 4.3. Settlement in New England, 1760. After Herman R. Friis, *A Series of Population Maps of the Colonies and the United States 1625–1790* (New York: American Geographical Society, 1968).

by the Wampanoags on the Plymouth settlers spread into a more general conflict, known as King Philip's War. More than half of all towns in New England were attacked, and a dozen were destroyed. But the destruction of Indian food supplies by the English, as well as attacks on the Wampanoags by the Mohawks from New York, soon ended the war. By 1700, the shattered remnants of the Indian tribes in southern New England, numbering less than 10 percent of their pre-contact population, were a much reduced threat.

In the northern part of the region, Wabanaki tribes occupied a borderland region between the English in the south and the French settled along the St. Lawrence River and in Acadia. Unlike the Indians in southern New England, who were soon enveloped by English settlers, the Wabanaki had diplomatic room to maneuver between the European powers, as well as physical space in which to retreat. As the Massachusetts Assembly observed in 1708, the Indians "have the Advantage of Retiring for Shelter to the Obscure Recesses of a vast rude Wilderness, full of Woods, Lakes, rivers, ponds, Swamps Rocks and Mountains, whereto they make an Easy and quick passage by means of their Wherries, or Birch Canoes of great swiftness and light of Carriage."[22] Even so, the southern margin of this Indian world was subject to English harassment and attack. Over the course of the late seventeenth and early eighteenth centuries, the frontier between the English and the Wabanaki moved back and forth across southern Maine.[23] English advances of settlement along the coast and up the major river valleys frequently were subject to Indian raids. Nevertheless, the steady encroachment of English settlement led to the retreat of the Wabanaki into the interior, particularly after Dummer's War in the 1720s. Yet British control over the northeast was not fully complete until the conquest of New France in 1760.

From the early seventeenth century, colonial governments in New England encouraged settlement by making grants of land to Puritan congregations and individual entrepreneurs.[24] After receiving a township grant, the congregation set up a town government to oversee development and allocate land to its members. Although the Puritans may have been equal in the sight of God, there was little equality about the allocation of land. Depending on their social and economic standing, some congregational members received large grants of several hundred acres, others much smaller allotments. Individual entrepreneurs, such as William Pynchon in the Connecticut Valley, also received large land grants, which, in turn, were divided into smaller parcels and sold to incoming settlers. By the eighteenth century, speculative development of frontier lands was widespread; in the 1750s, a group of leading Boston merchants formed the Kennebec Proprietors to develop an enormous land grant in Maine.[25] What had started as a Puritan "errand into the wilderness" very quickly turned into a speculative rush for spoils.

In the first years of settlement, some Puritan congregations reproduced a pattern of settlement associated with open field villages common in parts of southern and eastern England. In Sudbury, Massachusetts, settlers created a nucleated village of clustered farmhouses surrounded by open fields and common pastures (figure 4.4).[26] A township council oversaw the cultivation of the arable fields and the stocking of livestock on the commons. Such open field agriculture was widespread not only in the British Isles but also in continental Europe, and reflected the tremendous pressure of population on land; it was an intensive form of farming that made use of scarce resources. In early New England, however, the ratio of population to land had been reversed. The availability of land soon led to the disintegration of nucleated villages and their replacement by dispersed settlement. Instead of living close together, settlers moved onto their own farm lots. In Sudbury, this shift took place in 1658 with the second lotting of land.[27] In many other townships, farms were dispersed from the beginning of settlement. A dispersed pattern allowed farmers to control their own land, free from township oversight, and to use it as they saw fit.

As the handful of open field villages disappeared from the landscape and dispersed farms became widespread, a distinctive pattern of settlement began to take shape in New England (figure 4.5). In almost all areas where dispersed farms were common, regular ranges of rectangular lots were laid out.[28] As early settlement was confined to the coast and valley bottoms (intervales), this landholding pattern gave farmers access to low-lying pasture and hay lands as well as hillside arable and wood lots. It was also a practical and easy system to lay out. The Puritan meeting

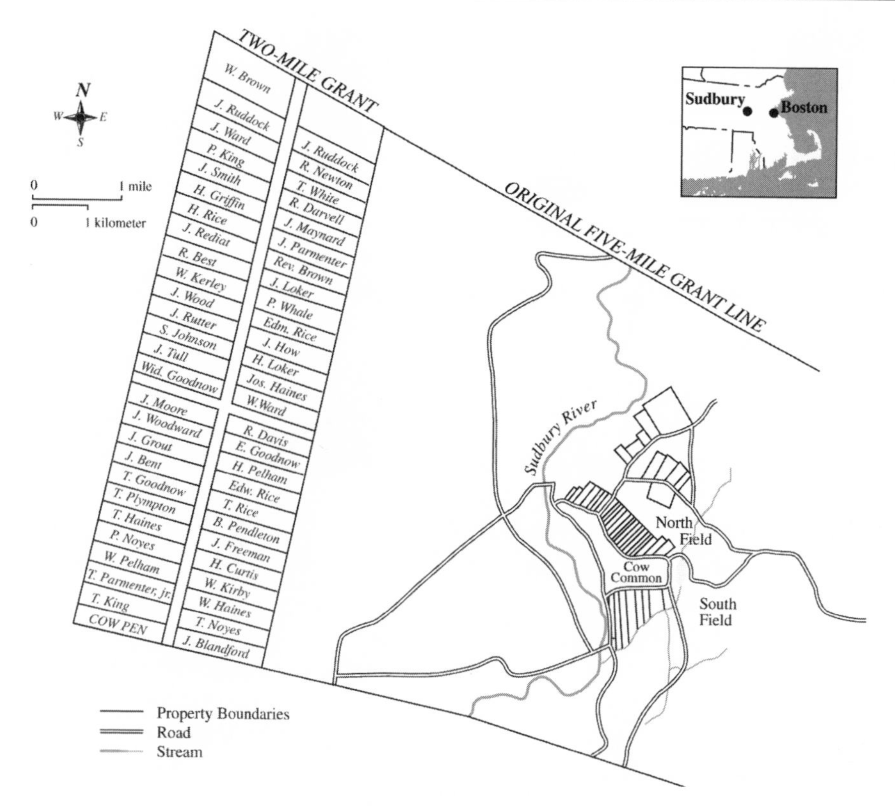

FIGURE 4.4. From nucleated to dispersed settlement: Sudbury, Massachusetts. After Sumner Chilton Powell, *Puritan Village* (Middletown, Conn.: Wesleyan University Press, 1963), Figure 19.

house, which served as both a religious building and the administrative center of a township, was usually located at a central, accessible location.[29] Apart from a parsonage, the town burying ground, and a militia training field, the meeting house usually stood alone.[30] For the colonial period, much of rural New England had a spare pattern of settlement, consisting of dispersed farms, mills, and meeting houses, as well as rectangular lots and fields, which contrasted strikingly to southern and eastern England and its intricate pattern of hamlets, villages, and irregular fields. Only after the Revolution and the expansion of rural trade did the meeting house become the center of a nucleated village of stores and service buildings.

The establishment of a farm in New England required capital and labor. Many of the families who arrived during the Great Migration had some savings, sufficient to secure land and purchase basic implements, livestock, and seed. For those who arrived without resources, the shortage of labor ensured relatively high wages and a chance to accumulate enough capital to rent, lease, or purchase land. Further success depended on the labor that a farmer could command. With large families common, farmers could draw on the labor of sons and daughters, while extra hands could be hired at busy times of the year. In the fertile Narragansett Country of western Rhode Island, a small group of wealthy farmers owned slaves. Acquired from Newport merchants engaged in the slave trade, these slaves made

FIGURE 4.5. Detail from *A Chart of the Harbour of Rhode Island and Narragansett Bay*, by J. F. W. DesBarres, 1777. One of several charts of the New England coast produced by the Royal Navy and published in the *Atlantic Neptune*, the chart of Rhode Island and Narragansett Bay is perhaps the most detailed representation of a New England landscape in the late colonial period. This detail shows the compact city of Newport, dispersed farms and rectilinear fields, and a network of roads and ferries. J. F. W. DesBarres, *Atlantic Neptune*.

up the largest concentration of bound labor in New England; in 1750, there were over 800 slaves in the Narragansett Country, although few farmers owned more than a handful each.[31] Such a pocket of bound labor was almost unique in New England, but it hardly compared to the large slave work forces in the southern plantation colonies.

For much of the colonial period, farming in New England was extensive, reflecting the availability of land and the shortage of labor. As in the Chesapeake, labor was devoted to clearing land and creating a farm rather than to intensive cultivation. A cleared patch of land was farmed for several years until its fertility declined, by which time a new area had been cleared. In this way, settlers gradually pushed back the forest, an especially laborious task in upland areas. The first generation of settlers brought with them a variety of farming practices from the different regions of England, but these soon gave way to the realities of the environment and markets in New England.[32] On most farms, the greatest improved acreage was devoted to pasture and grass, with only a small fraction in arable land and orchard.[33] Arable fields were enclosed, either with fieldstone walls or fencing, while livestock were allowed to forage in the woods, a practice that was opposite to that in England where animals were usually penned. As trade increased, particularly in lowland areas, farmers increased productivity. In the early 1660s, at least some farmers in Massachusetts grew "English grasses" (probably clover) to improve their pastures; by the eighteenth century, the use of red and white clover had become common in the Connecticut Valley.[34] Such intensification reflected the growing importance of the livestock economy and the West Indies market.

Compared to the various staple regions, the agricultural economy of New England was much less connected to the Atlantic world. Whereas the great bulk of the sugar, rice, tobacco, dried fish, and furs produced in the staple areas was exported overseas, much of the agricultural production of New England was consumed locally. Farmers followed a duel strategy, producing sufficient crops and livestock to support their families as well as surpluses to sell or barter in order to acquire manufactured goods, provisions, labor, or land.[35] Farm families consumed milk, butter, cheese, eggs, beef, pork, poultry, corn, fruits, and vegetables raised on their farms, as well as traded these products to neighbors and local merchants.[36] Of course, the balance between consuming goods on the farm and selling them into the market varied from farm to farm and from region to region. Some farm families consumed much of what they produced, others sold more of their produce into the market. Whatever the details of the strategy, the purpose was to support the farm family in a "comfortable subsistence," and provide an inheritance, usually in the form of land, for the next generation.[37]

Yet as markets emerged around the Atlantic rim, particularly in the West Indies, demand increased for New England products (figure 4.6). The New England–West Indies trade first emerged with the expansion of the sugar economy in the late seventeenth century, increased significantly during the first half of the eighteenth century, and, as the general Atlantic economy expanded in the latter part of the century, grew enormously in the two decades before the Revolution. The trade comprised a variety of agricultural and forest products, including salted beef and pork, horses (for driving sugar cane crushers), cattle, flour, grain, and lumber.[38] Between 1768 and 1772, New England shipped 99 percent of its livestock exports, 88 percent of its wood, and 79 percent of its flour and grain to the British West Indies.[39] Large quantities of produce were also smuggled to the French West Indies. Of the agricultural and forest products shipped to the British West Indies, New Hampshire and Massachusetts (including Maine) supplied 75 percent of the wood, Connecticut provided 75 percent of the salted meat and livestock, and Rhode Island yielded 64 percent of the flour and grain, although a good part of the Rhode Island produce came from neighboring Connecticut.[40] Just as the sugar colonies were a cornerstone of the New England fishery, so the islands became a major buttress of the region's agricultural and forest economies.

The variations in the export trades suggest that by the late eighteenth century there was considerable regional specialization in New England's rural economy (figure 4.7).[41] Such factors as location, terrain, soil, climate, vegetation, and time

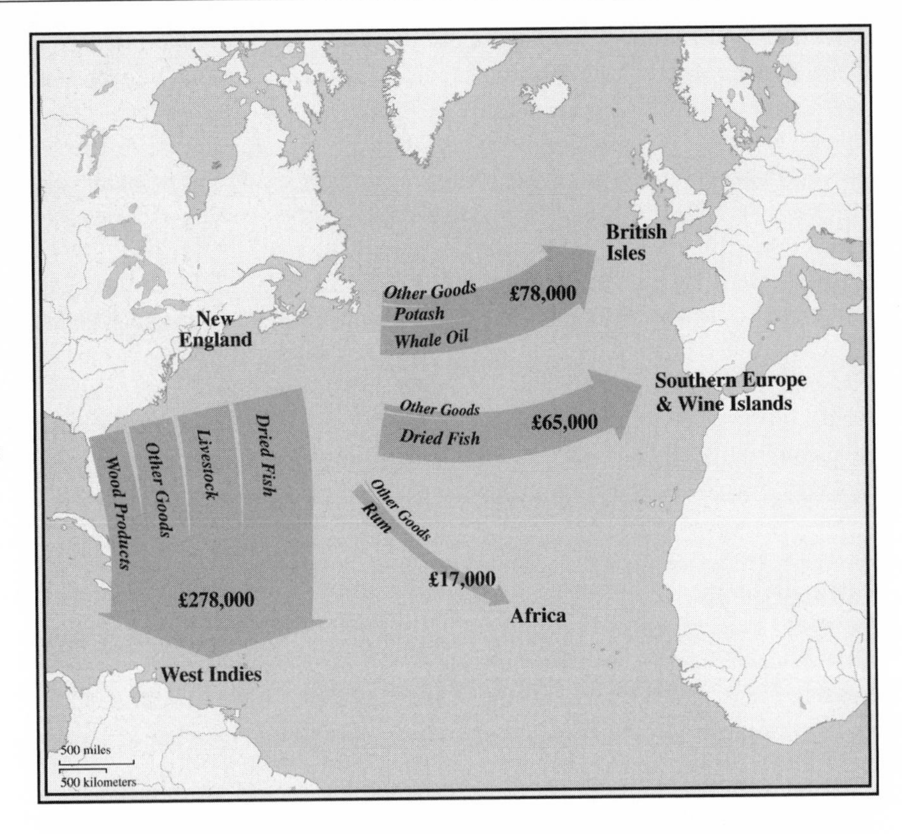

FIGURE 4.6. New England overseas trade, 1768–1772. Average annual export values calculated from James F. Shepherd, "Commodity Exports from the British North American Colonies to Overseas Areas, 1768–1772: Magnitudes and Patterns of Trade," *Explorations in Economic History* 8 (1970–1971).

of settlement all influenced the pattern of development. Along the northern New England coast, northern Massachusetts, southern New Hampshire, and southern Maine quickly emerged as a distinct lumbering zone. Although some farming occurred along the coast, particularly where salt marsh was available, the glaciated, rocky interior was more suitable for lumbering. During the late seventeenth and early eighteenth centuries, the timber industry was centered on the Merrimack and Piscataqua rivers; both areas produced lumber for shipbuilding and export to regional and Atlantic markets.[42] By 1725, lumber was beginning to come from Casco Bay.[43] With the withdrawal of the French and Indian threat to central and eastern Maine in 1760, lumberers moved up the Kennebec River as well as "downeast" to the Penobscot and Machias rivers (figure 4.8). Throughout this extensive lumber zone, the rural population combined farming and lumbering, producing a variety of subsistence crops and livestock, and earning supplementary income from lumber cleared off their farms or from working for merchants engaged in the timber trade.

South of the lumber zone, in the coastal lowlands of eastern Massachusetts, a patchwork pattern of farming prevailed. Across much of the region, glaciated terrain, rocky outcrops, and thin, frequently sandy soils, as well as ponds and swamps, limited agriculture, but the main river valleys, such as the Merrimack, provided

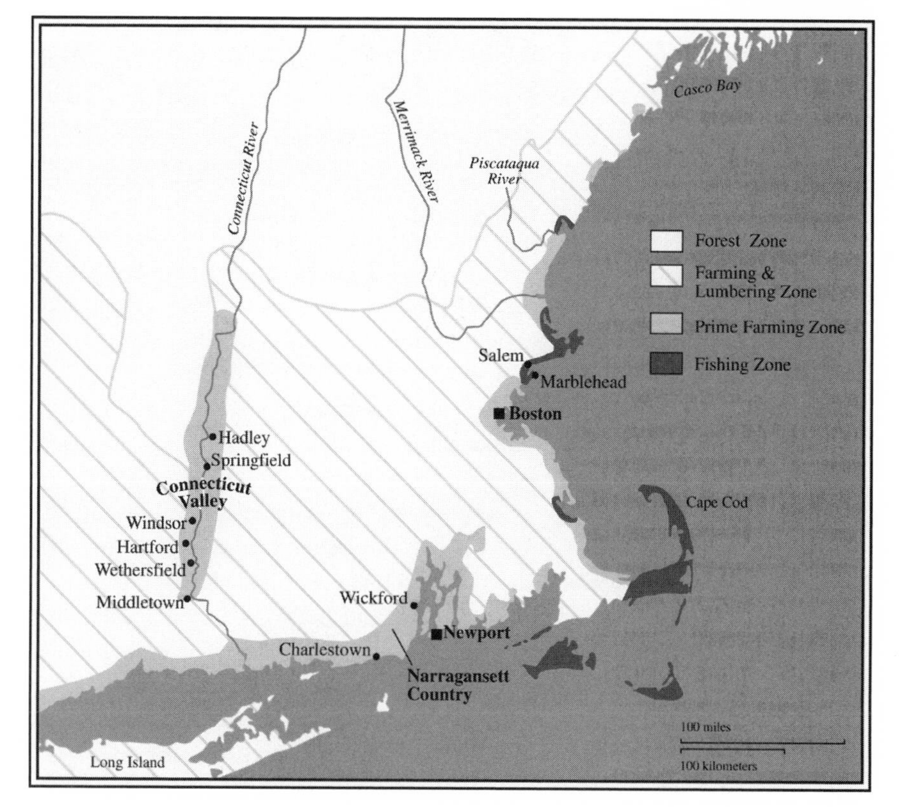

FIGURE 4.7. Generalized land use in New England, late eighteenth century. After Joseph S. Wood, *The New England Village* (Baltimore: Johns Hopkins University Press, 1997), Fig. 1.2, 23.

FIGURE 4.8. Sawmills, Machias, Maine, circa 1770. Lumberers move downeast: two mills have been built on the falls of the Machias River; a third is under construction. Coniferous forest can be seen in the background. J. F. W. DesBarres, *Atlantic Neptune.*

good hay and pasture land, while the coastal salt marshes, such as those around Ipswich Bay, could also be exploited. Moreover, the proximity of urban markets in Boston, Salem, and Marblehead encouraged commercial connections. As a result, the agricultural economy included prosperous farms along the intervales, raising beef, dairy products, wool, corn, rye, and hay for urban markets, while farms in the interior produced basic subsistence crops, although farmers in some towns in Worcester and Middlesex counties raised surpluses of corn and rye in the 1770s.[44] In the immediate vicinity of Boston, there was also some market gardening, with farms producing milk, butter, cheese, fruits, vegetables, and cider; in the mid-eighteenth century, Dorchester, some four miles from the city, was known for its orchards that produced an "abundance of Cyder."[45]

Much better agricultural land was available in the Narragansett Basin. Here, the gentle terrain, rich soils, and milder climate of Aquidneck Island and the Narragansett Country (the area between Wickford and Charlestown), as well as the presence of Newport, led to the development of a commercial farming zone. By the mid-eighteenth century, Aquidneck was "throughout like a Garden from the Industry of the farmers who keep there [sic] ground very Clean."[46] Farms produced a mix of grains, wood products, salted meat, and livestock for the local urban market, as well as for export to the West Indies. During the eighteenth century, the Narragansett Country was famous for the breeding of Narragansett Pacers for the Caribbean market, as well as for its large dairy and beef herds.[47] Local farmers accumulated enough capital from these exports that they took on the trappings of country gentry and became known as the "Narragansett Planters."

Beyond Rhode Island, the commercial farming zone increased in scale: The lowlands of Connecticut and western Massachusetts, particularly those along the Connecticut River, formed the agricultural heart of New England. The rich alluvial terraces of the Connecticut Valley provided fertile soils for cultivation and pasture, while the surrounding uplands provided forest products and fur-bearing animals. From the beginning of English settlement in the 1630s, the region oriented itself to outside markets, first with the fur trade organized from Springfield, and then, increasingly, with the export of grains and livestock.[48] During the late seventeenth century, the valley served as the breadbasket of New England, producing wheat, rye, and corn. As wheat production fell in the 1660s due to outbreaks of rust, livestock raising became more prominent. By the 1750s, fat cattle were being raised on intervale pastures in the mid-valley, particularly around Springfield, and then driven south to Hartford for shipment to the West Indies, or overland to Boston.[49] There was also some specialization in crops: onions were grown in Wethersfield and tobacco in the area around Windsor and Hartford.[50] As the settlement frontier moved north into the upper reaches of the Connecticut River, as well as onto the surrounding uplands, forest products entered valley trade, much of it passing through Hartford or Middletown to the West Indies market.

After 1750, the agricultural frontier in New England shifted from the coastal and interior lowlands to the uplands. Although the terrain was rougher, the soils thin and acidic, and the growing season shorter, upland farmers still made connections to markets. Unable to produce all the foodstuffs that they needed, hill farmers concentrated on raising livestock and dairy products, selling cattle to lowland farmers for fattening, as well as cattle, butter, and cheese to local merchants for shipment to urban markets.[51] Hill farmers also raised a variety of other products, including corn, rye, potatoes, sheep, and pigs, mainly for subsistence. In addition, farm families participated in the local labor market; men worked on lowland farms, cut lumber, or developed an ancillary craft such as blacksmithing or shoemaking; women did spinning and weaving, and also worked in the fields, especially at harvest time.[52]

The agricultural and forest sectors of the New England economy generated considerable employment in manufacturing and trade. Agricultural products such as grain, meat, and hides were processed by millers, butchers, and tanners; raw logs were sawn into deals, shakes, staves, and clapboards at local sawmills (see figure 4.8). All these various products were collected by small country merchants and traders, who then forwarded them to merchant wholesalers in the port towns. As this required transportation, goods were either sent overland by carters and drovers, or by water on river craft or along the coast on schooners and sloops. Imported foodstuffs and manufactured goods were shipped in return. There was also some rural manufacturing of consumer goods, including homespun cloth, shoes, and furniture. By the mid-eighteenth century, prosperous parts of the New England countryside, such as the Connecticut Valley, were also supporting skilled craftsmen in the building trades.[53]

Although many of the activities associated with manufacturing and trade in New England were conducted in port towns along the coast (see chapter 5), some clustering occurred at small urban centers in the interior by the late eighteenth century. Across much of the region, this growth scarcely amounted to more than a handful of country stores at a crossroads or a collection of grist and sawmills at a good waterpower site; but in the thickly settled and prosperous Connecticut Valley, there appears to have been some development of street villages. In 1788, French traveler Brissot de Warville observed that the Connecticut Valley "gives the impression of being one continuous town. As soon as we left Hartford, we entered Wethersfield . . . it extends for a long distance along the road and is filled with well-built houses."[54] The growth of such villages reflected the expansion of local and long-distance trade, as well as the prominence of rural manufacturing.

Compared to England or the staple regions along the eastern seaboard, rural New England was much less economically and socially stratified. The great estates of the English lowlands, the mercantile fortunes of Salem and Marblehead, and the planter prosperity of the Chesapeake and Carolina low country had no equivalent in the New England townships. The availability of land, the absence of a major staple, and the high cost of labor all mitigated against the accumulation of agricultural wealth. The great bulk of the rural population managed to acquire land and attain a minimal independence from farming. In 1765, Scottish traveler Lord

Adam Gordon observed in the vicinity of Boston that "the levelling principle . . . everywhere operates strongly. . . . Everybody has property and everybody knows it."[55] Even so, rural New England was far from being socially or economically undifferentiated; variations across time and space gave rise to differences in wealth and social position. For English emigrants who first settled around Massachusetts Bay in the 1620s and 1630s, there was considerable opportunity to acquire freehold land. In the 1640s and 1650s, a relatively egalitarian rural society emerged. But as the second generation came to maturity, agricultural land in the townships around Massachusetts Bay became increasingly scarce.[56] As the price of land rose, so the opportunity to acquire property diminished and the numbers of adult males without land increased. Second- and third-generation settlers, unlikely to inherit the family farm, had few alternatives but emigration to the frontier.[57] In the late eighteenth century, Brissot de Warville observed in the Connecticut Valley that "with the population growing so rapidly, many people are emigrating from a region where they already feel overcrowded [and] land is expensive."[58] The great expansion of the New England frontier reflected this demographic cycle. Moreover, important subregional variations in rural wealth existed. In areas with fertile land and access to markets, such as the Narragansett Country and the Connecticut Valley, farmers grew prosperous from agricultural trade.

The growth of the agricultural economy also supported a mercantile class.[59] Situated in the port towns at the mouths of the many rivers that drained New England, as well as in interior towns, merchants were well placed to accumulate the small quantities of country produce that came down the valleys and to forward them on to regional or overseas markets. In return, merchants distributed imported manufactured goods and flour to the rural population. As early as the 1640s, the Pynchon family in the Connecticut Valley town of Springfield was growing wealthy from the fur trade; by the eighteenth century, merchants in the main valley settlements as well as in the port towns were accumulating wealth from the sale of agricultural and timber products to the West Indies.[60] A handful of merchants in the Connecticut Valley became so prominent that they earned the sobriquet "River Gods."[61]

The development of a mercantile class gave rural society in New England a different cast from that found in England or the southern colonies. The dominance of merchants meant that they, rather than a rural gentry or a plantocracy, wielded economic power. Instead of the age-old, semi-feudal English relationships between landlords and tenants, which were often encumbered within a web of customary rights and obligations, the connections between merchants and farmers in rural New England were more often defined in terms of credit and debt in the store ledger book.[62] Old World paternalism and custom gave way to New World patron-client relations. Such economic control gave New England country merchants the levers of political power at the local and colonial levels, and created a relatively uniform political, economic, and social elite in the region.

The English settlement in New England also gave rise to a distinctive regional culture. English immigrants brought with them considerable cultural baggage, including several dissenting religious sects, the dialects of eastern and southern England, and a variety of building techniques. Out of these English ways, a New England vernacular culture emerged.[63] From the beginning of settlement, the Congregational Church of the Puritans formed the nucleus of New England religious life. In contrast to the sway of the established Anglican Church in the southern colonies, the West Indies, or Newfoundland, the Congregational Church dominated central Massachusetts and the Connecticut Valley. The westward expansion of Puritan settlers from the townships around Boston created a Congregational wedge across the middle of New England. Nevertheless, doctrinal disputes among the Congregationalists during the mid-seventeenth century led to the splintering of the Church and the creation of dissident enclaves on the periphery. Just as the Puritans had escaped persecution in England by moving overseas, so separatist congregations around Massachusetts Bay used the space of the New World to resolve religious conflict. Expelled from the Massachusetts Colony, Roger Williams founded the Baptist Church at Providence, Rhode Island, in 1636, and established a tradition of religious toleration in the colony. During the seventeenth and eighteenth centuries, Rhode Island became a haven for other separatist and ostracized groups, including Quakers, Anglicans, and Sephardic Jews. Meanwhile, on the western fringe of the Congregational core, Anglicanism was established in southern Connecticut; on the northern fringe, the influx of West Country settlers brought Anglicanism to New Hampshire and southern Maine.[64]

Regional dialects as well as place names from southern and eastern England were transferred to New England. The nasal whine of Norfolk became the Yankee twang of the East Anglian settlements in and around Boston, while West Country speech could be heard in the fishing and lumbering communities along the northern New England coast and in some inland towns.[65] But as the population increased and as settlement expanded, English dialects mixed, particularly in the interior of Massachusetts and Connecticut, and a more generic New England accent emerged.[66] Place names, too, were transferred and recombined. During the early seventeenth century, English immigrants, like their counterparts in the Chesapeake, commemorated the market towns and counties that they had come from. In the Boston area, East Anglian settlers created Suffolk and Essex counties, as well as the towns of Ipswich, Haverhill, Wenham, Toppesfield, and Boxford. Yet these names, all derived from eastern England, were intermixed with names from other regions of England, such as the Home Counties names of Reading, Woburn, and Middlesex, and the West Country names of Amesbury, Salisbury, and Dorchester.[67] Just as the intermingling of settlers from different backgrounds led to a common New England accent, so the mixture of place names from different regions of England created a distinctive geography of New England names. Moreover, the pattern of naming changed over time. By 1700, ties with England had weakened; second- and third-generation settlers, who moved to the frontier, increasingly named places after themselves, land speculators, or prominent New

FIGURE 4.9. Chapman-Hall House, Damariscotta, Lincoln County, Maine, 1754. Photograph by the author, 1989.

FIGURE 4.10. John Perkins House, Castine, Hancock County, Maine, 1765. Photograph by the author, 1987.

Englanders, rather than after dimly remembered market towns and counties in England.

Cultural convergence was also reflected in the early building practices around Massachusetts Bay. Settlers brought a variety of construction techniques and architectural styles from different parts of England, but soon discarded them in favor of a common building tradition in New England.[68] The great variety of materials used in English housing, such as stone, brick, wood, thatch, wattle and daub, was soon pared down in New England to wood, which was used not only for framing the structure but also for siding (clapboards) and roofing (shingles). Brick and field stone were used only for the construction of chimneys and cellar walls.[69] The great variations in plan and elevation of English housing were also reduced to a one or one-and-a-half story, one-room hall house, or a two-story, two-room hall and parlor house, usually extended at the back with a lean-to enclosing a kitchen (figures 4.9 and 4.10). In the one-room dwelling, the chimney was placed at the gable end; in the two-room house, it was placed centrally. The two-story, central chimney house had been appearing in England as the standard house of the yeoman class in the sixteenth century, but it shared the landscape with a great number of older, differently styled and sized houses.[70] When the house-type arrived in New England, it was one of only two basic house-types in the region; English variety had been pared down to Yankee uniformity.

Moreover, the social and economic leveling that occurred in rural New England stripped out the top and upper-middle layers of housing common in England. The aristocratic mansions and the gentry manors found in the English countryside had no comparison in the New England townships. Instead, the great majority of housing in the region was extremely modest. Although the central chimney, two-story yeoman house has become one of the icons of the region, relatively few of these houses were built during the colonial period.[71] Most dwellings were one or one-and-a-half stories, and had only one or two rooms. According to the Federal Direct Tax Census of 1798, one-story buildings outnumbered all two-story buildings by two to one in Worcester County, Massachusetts, while one-story dwellings outnumbered two-story, double-pile houses by as many as seven to one.[72] Across Massachusetts, one-story dwellings formed the majority of houses. Nevertheless, pockets of two-story housing existed in the colony as well as in Connecticut. By the late eighteenth century, two-story houses predominated in long-settled Essex County, Massachusetts, particularly in the agricultural townships along the Merrimack River and around Boston; these houses were also found in the wealthier parts of the Connecticut Valley, such as Hadley, Windsor, and Wethersfield.[73] Most likely, similar concentrations could be found on Aquidneck Island and in the Narragansett Country in Rhode Island.[74] Yet these two-story houses, comparable to the homes of yeomen farmers or lesser gentry in England, usually belonged not to farmers but to merchants and professionals, who had accumulated capital from rural trade and the perquisites of government.

Although a well-developed vernacular building tradition had emerged in New

England by the late seventeenth century, English influence appeared in the form of the Georgian style during the following century and helped shape a polite, formal architecture. From the 1730s, Georgian architectural elements—symmetrical facades, central hallway, chimneys pushed to the gable ends, classical pediments over front doors—and a more hierarchical arrangement of internal space—formal rooms at the front of the building, informal work spaces at the back—began to appear on merchant residences in the port towns; by the next decade, these elements increasingly characterized dwellings in prosperous parts of the countryside. By the 1750s, the Georgian style had a strong hold over the "River Gods" in the Connecticut Valley. Large, two-story, double-pile houses, some with gambrel roofs, and with central doorways embellished with impressive scroll pediments began to appear, cementing the elite's social and economic position in the Connecticut Valley (figure 4.11).[75] And yet, these Georgian-style houses, like those going up in the North Shore port towns, were built out of wood, using construction techniques worked out in the region over the previous century, and were unmistakably part of a developing New England material culture. Wealth from West Indian trade and British ideas about formal architecture intersected with New England materials to produce a distinctive regional building tradition that had no exact counterpart across the Atlantic.

New England culture was also reflected in the religious landscapes of the region. The prevalence of Congregationalism, particularly in rural areas, ensured that the Anglican landscapes of England would not be reproduced. In England, the parish church was usually located in a village, had a close relationship with the local gentry, and was used mostly for religious functions. The building was most often a rectangular, stone structure, with a long principal axis. In New England, the meeting house was usually located at an accessible location, such as a crossroads, was supported and administered by the local population, and was used for both religious and municipal functions. The building was a square or rectangular wood-framed structure, with a short principal axis (figure 4.12). Indeed, the meeting house, as the name implies, owed much to the house, at least in its earliest form; as in seventeenth-century domestic buildings, the principal entranceway was on the long side. Yet by the middle of the eighteenth century, trans-Atlantic Georgian and Anglican influences were beginning to recast traditional New England meeting houses.[76] The influence of Christopher Wren's London churches (1670s) and James Gibbs's St. Martin-in-the-Fields (1721–1726) was reflected not only in colonial urban churches (see chapter 5), but also in rural meeting houses.[77] In imitation of Anglican churches, particularly Christ Church in Boston (1723), bell towers and spires began to be added to meeting houses (figure 4.13). At the same time, the secular function of the meeting house began to be accommodated in a separate building (the town house), leaving the meeting house solely as a religious structure. These changes transformed the meeting house into a church. Nevertheless, the Congregational churches that began appearing in the New England landscape during the late eighteenth century were not mere replicas of their English

FIGURE 4.11. Scroll pedimented doorway, Samuel Porter House, Hadley, Hampshire County, Massachusetts, circa 1761. Library of Congress, Prints and Photographs Division, HABS, MASS, 8-HAD, 3-5.

Anglican counterparts. As with the central chimney house, the Congregational church had married English Georgian style and form with New England materials to create a distinctive New England vernacular building.

The growth and development of the English colonies in New England also generated a complicated institutional structure. A mix of royal and corporate charters created the institutional space of the New England colonies and usually specified the form of government. Massachusetts, Plymouth, Rhode Island, and Connecticut had houses of assembly (General Courts), with the assemblymen elected by either church-based or property-based franchises. In the 1670s and 1680s, the English government, in an attempt to gain greater political control over the colonies, appointed royal governors to New Hampshire and Massachusetts, although the

FIGURE 4.12. Meeting House, Harpswell, Sagadahoc County, Maine, 1757–1759. Photograph by the author, 1989.

freedom of action of the Massachusetts governor was compromised severely by his dependence on funding voted by the legislature. Rhode Island and Connecticut elected their own governors. Within these colonial spaces, lesser institutions administered smaller areas. At the county level, there were law courts, following the English model, while at the township level, there were elected councils, similar to parish councils in England. Towns were also responsible for raising militias, drawn from all able-bodied men in the local area, and these forces were used to defend township and colony from French and Indian attacks. Such a form of defense diffused military power throughout the region, and was fundamentally different from the British practice of a standing army and regiments garrisoned in county towns. The town form of government, as well as the local militia, gave New England communities considerable institutional autonomy.[78]

By the late eighteenth century, rural New England had developed its own distinctive patterns of settlement, population, economy, society, and culture. In

FIGURE 4.13. Congregational Church, Farmington, Hartford County, Connecticut, 1771. A New England icon emerges: This Congregational church illustrates the transition from the traditional, square, side-entrance meetinghouse to the rectangular, Georgian church with tower and spire. Library of Congress, Prints and Photographs Division, HABS, CONN, 2-FARM, 2-8.

a physical and commercial setting that provided little basis for a highly specialized staple trade, an agricultural economy had emerged that supported one of the largest concentrations of people on the eastern seaboard, as well as a society that was more egalitarian than any found in Britain or in the staple regions elsewhere on the continent. For the founding generation of Puritan settlers and for many of their descendants, the rich intervale land of New England provided sufficient basis for a great many families to gain a basic competency and for a few to achieve a measure of prosperity. The fragments of English culture that had come across the Atlantic were also selectively pruned and recombined to produce a unique New England rural culture. Such economic security and cultural self-confidence expressed itself at the political level in the independent attitudes of the several colonial assemblies. The agricultural frontier of New England had created a region much different from either the staple regions of the British Atlantic or those found elsewhere on the eastern seaboard.

The Mid-Atlantic Colonies

English involvement in the Mid-Atlantic region came only after the Dutch and the Swedes had begun European colonization during the early seventeenth century.[79] In 1625, the Dutch West India Company, following up on Henry Hudson's exploration of New York harbor, established the colony of New Netherland and built a fur-trading post at New Amsterdam; in 1638, the Swedes entered Delaware Bay and created the colony of New Sweden. Both colonies led a marginal existence, and were eventually taken over. The Dutch conquered New Sweden in 1655, only to lose all their continental possessions to the English in 1664. The English rechristened New Netherland as New York and then split off New Jersey, which was divided into East and West Jersey and placed under separate proprietary governments; in 1702, the two halves were reunited as a Crown colony. In 1681, Englishman William Penn gained a proprietary grant to Pennsylvania, an area that lay between New Jersey and Maryland; his descendants retained nominal control over the colony until the Revolution.

At the time of the English colonization, the Mid-Atlantic represented the last major area of easily accessible agricultural land available for settlement on the eastern seaboard. To the north, New England was filling up; to the south, the Chesapeake tidewater was settled and the Carolinas were being colonized. As the gentle topography, rich soils, and mild climate of the Mid-Atlantic proved ideal for farming, the region soon became a magnet for immigrants, quickly gaining the sobriquet the "best poor man's country" in America. But unlike New England and the Chesapeake, which were settled primarily from England, the Mid-Atlantic colonies were settled by a variety of ethnic groups from northwestern Europe. The immigration and subsequent intermingling of these various groups in the Mid-Atlantic produced a regional culture that was quite different from those found

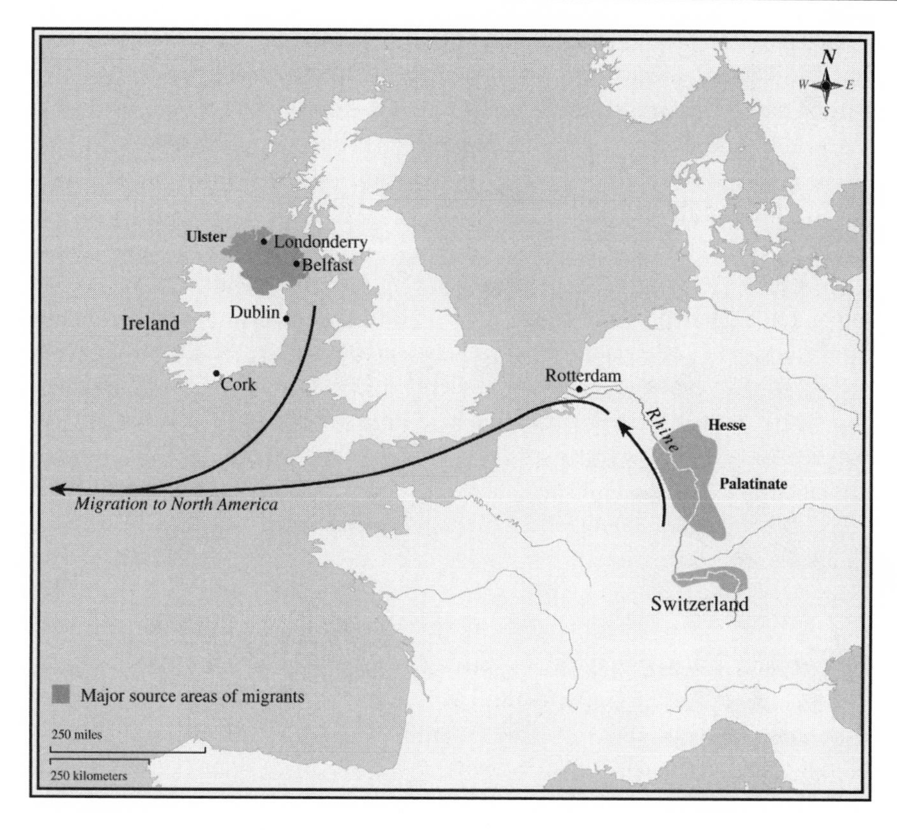

FIGURE 4.14. Regional origins of Scots-Irish and German migrants to North America, 1700–1775. After Marianne S. Wokeck, *Trade in Strangers: The Beginnings of Mass Migration to North America* (University Park: Pennsylvania State University Press, 1999), Maps 1 and 3; Cecil J. Houston and William J. Smyth, *Irish Emigration and Canadian Settlement: Patterns, Links, and Letters* (Toronto: University of Toronto Press, 1990), Figure 2.1.

elsewhere along the seaboard. As the population of the Mid-Atlantic expanded during the late eighteenth century, settlers spilled out of the region, taking their agricultural economy and rural culture south into the backcountry of Virginia, the Carolinas, and Georgia, and west into the upper reaches of the Ohio Valley. After the American Revolution, the Mid-Atlantic, as well as its southern backcountry extension, served as the "seed bed" for American agricultural settlement west of the Appalachians.[80]

The migrants to the Mid-Atlantic colonies were drawn from an extensive area of northwestern Europe. During the early seventeenth century, only a few Europeans trickled into the Mid-Atlantic, mainly Dutch settlers in New Netherland and Swedes and Finns in New Sweden.[81] After the English took over both colonies in the early 1660s, the numbers of immigrants increased, particularly with the opening of settlement in Pennsylvania in the 1680s. Between 1670 and 1700, some 24,000 migrants arrived in the Mid-Atlantic colonies, many of them English and Welsh Quakers.[82] After the disruption of the war years (1689–1713), when few Europeans appear to have emigrated to the region, migration again picked up, with some 126,000 arriving between 1710 and 1770. Most of these emigrants came from two principal sources: the fringes of the British Isles and southern Germany (figure 4.14). Of those from the British Isles, a minority came from Lowland Scotland, attracted to the colonies after the Act of Union between England and Scotland in 1707.[83] Many more came from Ireland, particularly the northern province of Ulster. During the seventeenth century, several thousand Presbyterian Scots moved from the Scottish lowlands to northern Ireland where they created a Protestant enclave; by the early eighteenth century, their descendants looked to the American colonies as a further frontier of settlement. Between 1729 and 1774, some 53,000 Scots-Irish arrived in Pennsylvania, the main place of arrival on the seaboard, with the greatest inflow after 1766.[84] Apart from these immigrations, there were substantial migrations of "foreign Protestants," some of them Huguenots from France but the great majority Germans from the Palatinate, a principality on the River Rhine bordering Switzerland. A small group of German Quakers arrived in Pennsylvania and settled at Germantown, just outside of Philadelphia, in the early 1680s, but the greatest influx of German settlers—some 50,000 of them—arrived between 1745 and 1755. Visiting Philadelphia in 1751, West India merchant James Birket noted that "the Number of German & Irish Passengers who Annually come here and Settle with their wives & familys [is] often 4 or 5000 in a year."[85] In total, as many as 100,000 Germans emigrated to the British American colonies, more than two-thirds of them to the Mid-Atlantic region.[86] By the late eighteenth century, German settlers formed the largest ethnic group in Pennsylvania.

Like the earlier English settlers in New England, the majority of European migrants to the Mid-Atlantic came in family or extended kin groups. Many of the migrants were small tenant farmers or craftsmen who saw the agricultural potential of the Mid-Atlantic as a chance to re-establish themselves and their families on the land. The Scots-Irish were drawn to the region because the amount of land available for their children to settle in northern Ireland was diminishing rapidly; similarly, the Germans in the southern Rhineland were faced with a rising population, shortage of land, and increasing feudal taxation. For both groups, the Mid-Atlantic offered a safety valve, a means to escape the social and economic rigidities of Old World landed society and the Malthusian dilemma of too many people subsisting on too little land. In addition, significant numbers of single, young males were recruited as indentured servants to work in urban trades or on the farms of the region.

The massive immigration of Europeans combined with rapid natural increase led to the growth of a substantial Mid-Atlantic population. In 1650, the non-native population was a mere 4,000 people. By 1700, it had increased to 53,500; fifty years later, it had reached 296,000. In 1770, the population was 556,000. If the urban population comprised approximately 10 percent of the total, the rural population was about 498,000, slightly less than the 505,000 in the New England countryside. Of this rural population, perhaps 248,000 lived in Pennsylvania and Delaware, 146,000 in New York, and 105,000 in New Jersey.[87]

As the European population in the Mid-Atlantic increased, the frontier of agricultural settlement expanded. From the first footholds on the shores of New York harbor and around Delaware Bay, the settlement frontier moved inland,

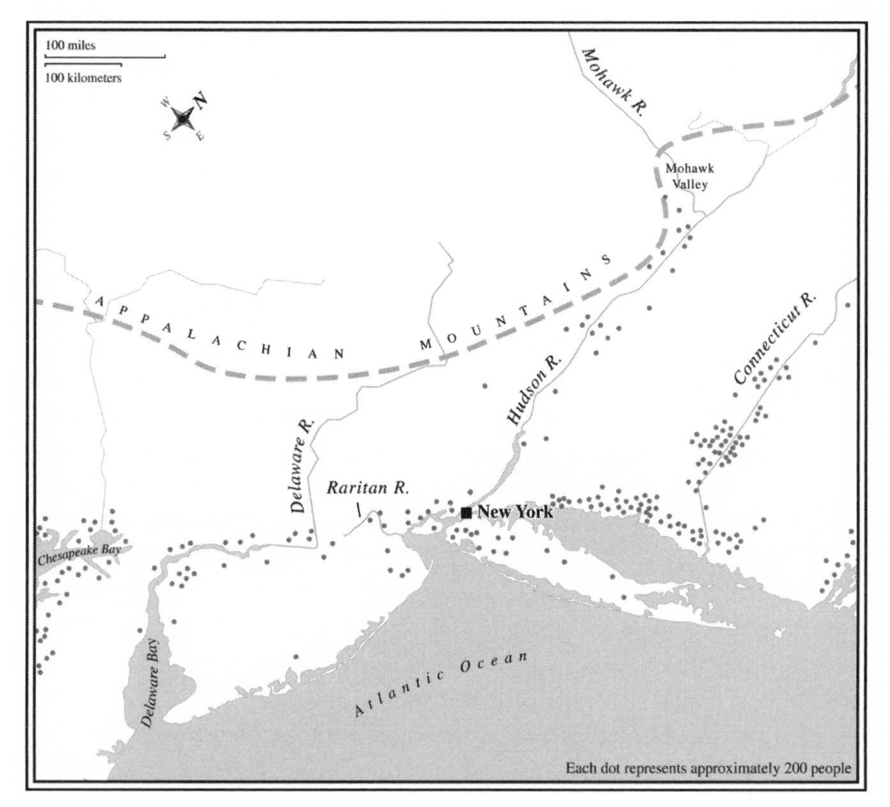

FIGURE 4.15. European settlement in the Mid-Atlantic colonies, 1675. After Herman R. Friis, *A Series of Population Maps of the Colonies and the United States 1625–1790* (New York: American Geographical Society, 1968).

reaching up the Hudson River valley, south across New Jersey, and west across southern Pennsylvania (figure 4.15). As early as 1700, settlers from New York and East Jersey had moved up the Raritan River valley and coalesced with settlers from Pennsylvania moving up the Delaware River valley, creating a wedge of settlement that connected the Mid-Atlantic colonies. Over the next fifty years, settlement expanded up the Hudson River as far as Schenectady as well as along part of the Mohawk River, spread across the Piedmont Plain in New Jersey, and extended across southeastern Pennsylvania to the Appalachian Mountains (figure 4.16). Between 1750 and the mid-1770s, Mid-Atlantic settlement began to press into the upper reaches of the Hudson and Mohawk, worm its way through the Ridge and Valley country into the upper Ohio, and spill into the Great Valley that led south through Maryland and Virginia to the Carolinas and Georgia.[88]

The rapid expansion of European settlement in the Mid-Atlantic had considerable impact on the native peoples. At the beginning of Dutch and Swedish colonization in the 1620s and 1630s, the Mid-Atlantic was occupied by two major native groups: the Iroquois League in what is now upstate New York, and the Delawares in the area that became southeastern Pennsylvania.[89] Like the Wabanaki in northern New England, the Iroquois occupied a border region between the French along the St. Lawrence and Great Lakes, the English in western New England, and the Dutch

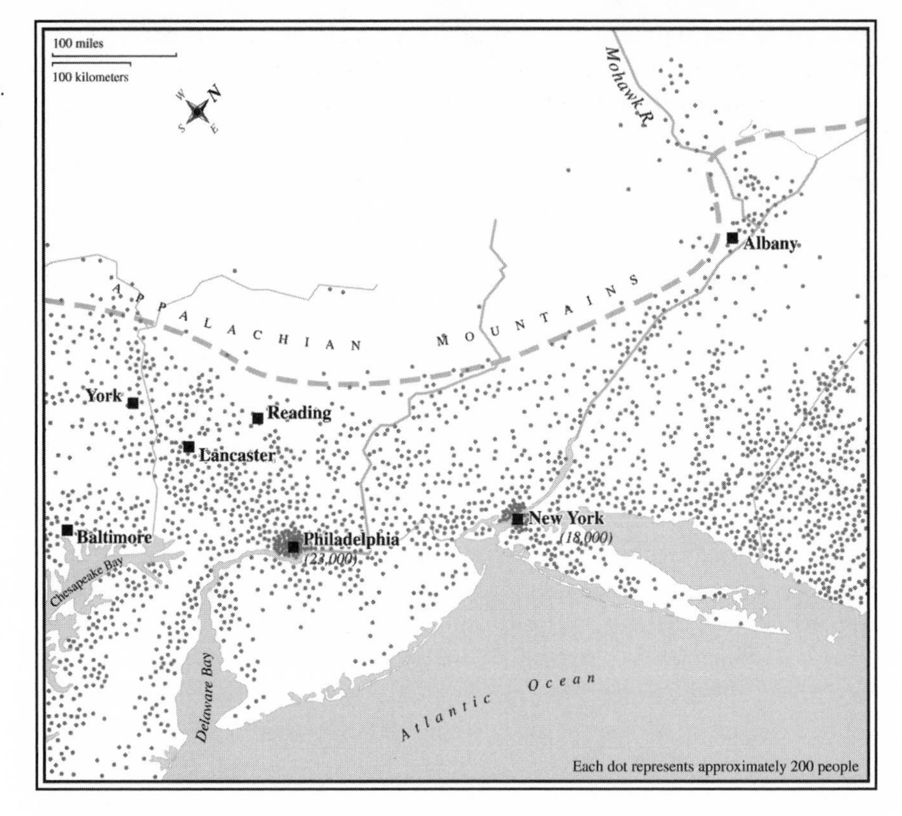

FIGURE 4.16. Settlement in the Mid-Atlantic colonies, 1760. After Herman R. Friis, *A Series of Population Maps of the Colonies and the United States 1625–1790* (New York: American Geographical Society, 1968).

in the Hudson Valley. In this pivotal position, the Iroquois were well placed during the early seventeenth century to play the European nations against each other and maintain a balance of power.[90] Even so, European diseases spread through the Iroquois, taking their deadly toll. By the late seventeenth century, weakened by depopulation, the Iroquois allied themselves with the English, who had taken over New Netherland. Until the American Revolution, the Covenant Chain between the Iroquois and the British created some semblance of stability on the New York frontier and provided the British with an Indian buffer against French expansion. To the south, the Delawares occupied the Delaware Valley and its tributaries.[91] Like other native peoples along the seaboard, they had been weakened by disease, particularly smallpox epidemics, during the seventeenth century. Although William Penn purchased land from the Delaware in the 1680s, the English and the Indians had different conceptions of land use and were frequently in conflict. As European settlement spread through southeastern Pennsylvania, the Delaware's hunting grounds became increasingly constricted. By the 1720s, the Delaware largely had retreated into the interior, toward the Appalachian Mountains, leaving the valley and the plains to the settlers.

As in Virginia, Maryland, and the Carolinas, European commercial companies and landed proprietors made early attempts to control access to land.[92] The Dutch

West India Company made four large land grants along the Hudson River valley to company investors, known as patrons (*patroons*), who were responsible for transporting settlers across the Atlantic and settling them on their estates (*patroonships*) as tenant farmers. Few European peasants wanted to remain as tenants, however, and the scheme soon floundered; only the *patroonship* of Rensselaerswyck, which encompassed more than a million acres on both sides of the Hudson and included Albany, attracted any farmers. By 1650, it had some 200 tenants, but turnover was high and rents remained low.[93] After the English took over New Netherland, further large grants, some with feudal courts (manors), were made in an effort to attract capital and settlers to the colony; but with cheap, freehold land available elsewhere in the Mid-Atlantic region, few immigrants were attracted to the colony. As with Lord Baltimore's feudal experiment in Maryland, attempts to recreate an Old World social order based on feudal landholding had foundered on the New World reality of plentiful land.

By the late seventeenth century, proprietors in New York, New Jersey, and Pennsylvania had become less interested in recreating a feudal hierarchy and more aware of the profits that could be made from selling land to incoming settlers. Although some estates still survived through the eighteenth century, owners of large tracts of land in New York sold off parcels for freehold farms. In neighboring New Jersey, colonial proprietors moved quickly to alienate land, selling land in the form of townships to New Englanders as well as in parcels to other immigrants. In Pennsylvania, proprietor William Penn dispersed the bulk of his land to settlers and speculators, retaining only a small portion for manors. But he held these, too, only for their speculative value, rather than as bases of a new feudal society, and gradually sold them off. Although the process of alienating land in the Mid-Atlantic colonies varied greatly, the end result, with the exception of parts of New York, was the transfer of an enormous amount of land out of the control of proprietors and speculators and into the hands of small family farmers.

The different methods of disposing of land in the Mid-Atlantic colonies greatly influenced the detailed pattern of settlement. Where proprietors and land companies retained some control over the process of alienation, they made attempts at regular surveys. In Pennsylvania, Penn had a large tract running back from the Delaware River surveyed in a ladder pattern of rectangular lots prior to sale, a surveying achievement that was recorded in the Holmes Map of 1687, the only large-scale cadastral map produced in colonial America (figure 4.17). Yet as immigrants poured into the colony after the turn of the eighteenth century, the surveying department quickly became overwhelmed. Settlers took up land in "indiscriminate location[s]," creating a mosaic of irregular lots.[94] A similar mix of regular ranges and irregular parcels developed in New York and New Jersey. Yet whatever the geometric configuration, the pattern of settlement that developed in the Mid-Atlantic, like that in New England and the Chesapeake, comprised dispersed farms. For European travelers, such as the German Gottlieb Mittelberger, who were used to the compact countryside of Europe, the spaciousness of the Pennsylvania

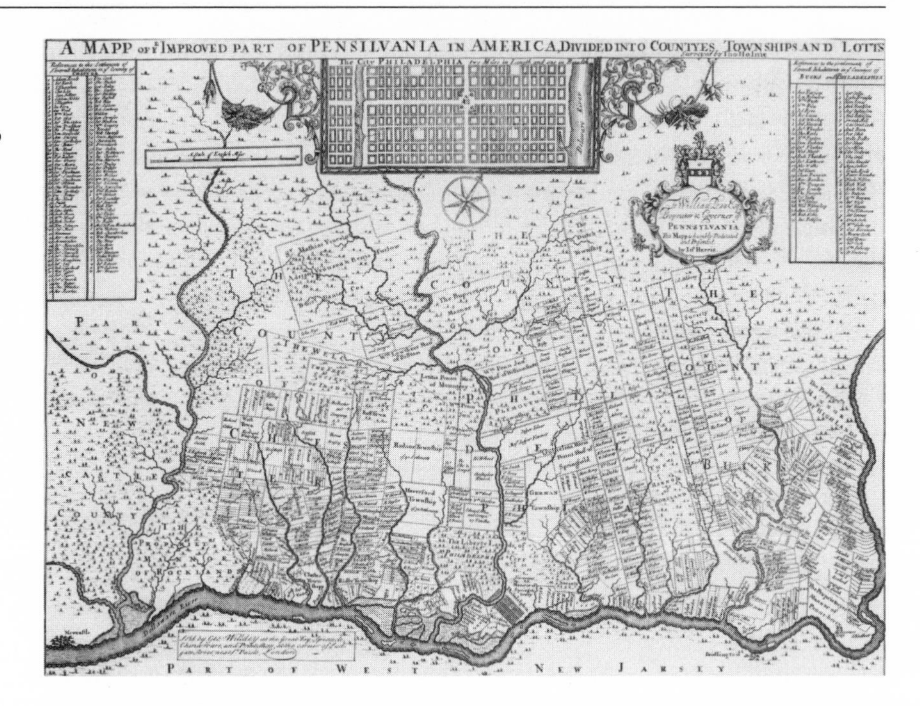

FIGURE 4.17. *A Mapp of Ye Improved Part of Pensilvania in America, Divided into Countyes, Townships and Lotts*, by Thomas Holme, 1687. Proprietor William Penn's vision for the orderly division of land.

landscape was astonishing: "In the country people live so far apart that many have to walk a quarter or a half-hour just to reach their nearest neighbor. The reason for this is that many plantation owners have got fifty or one hundred, even, two, three, up to four hundred morgen [a German measure of land slightly less than an acre] of land, laid out in orchards, meadows, fields, and forest."[95]

As in New England, the establishment of a farm in the Mid-Atlantic colonies required capital and labor. At least some settlers from the British Isles and Germany arrived with capital, perhaps accumulated from the sale of a farm lease or livestock, and were able to buy land; others rented farms for a number of years in the hope of saving enough for their own property; and many took laboring or artisan jobs in the hope of putting money aside for an eventual purchase. Family members provided much of the farm labor, leaving single men at a considerable disadvantage. In Pennsylvania, Brissot de Warville noted that a French farmer who was "without a family" did not have "any poultry or pigeons and makes no cheese; nor does he have any spinning done or collect goose feathers. It is a great disadvantage for him not to be able to profit from these domestic farm industries, which can be carried on well only by women."[96] Farmers also hired indentured laborers, and a few, particularly along the Hudson Valley in New York, held small numbers of slaves.

The agricultural economy of the Mid-Atlantic, like that in New England, was based on mixed farming. With richer soils and a more temperate climate than in most of the northern colonies, the Mid-Atlantic quickly became the breadbasket of British America, producing wheat, rye, oats, corn, and buckwheat.[97] As early as 1700, wheat was recognized as "the farmer's dependence" in Pennsylvania, and wheat bread was standard fare on family tables.[98] It was also the principal crop sold off the farm; between 1725 and 1783, Pennsylvania farmers sold more wheat to gristmill and ironworks stores than all their rye, oats, and corn combined.[99] Apart from grains, farmers grew potatoes, peas, and a great variety of vegetables; they also raised flaxseed, kept apple orchards, and cut hay for their livestock during the winter. Farmers also raised dairy and beef cattle, sheep, and hogs, as well as oxen and horses for haulage. Few farms were without poultry and a few beehives. Although less subregional variation existed in the agriculture of the Mid-Atlantic than in New England, the balance between arable and livestock farming varied according to soils, climate, and markets, with more wheat grown on better soils and a greater emphasis on dairying and horticulture in the shadow of the urban markets of New York and Philadelphia.[100]

Farmers in the Mid-Atlantic, like those in New England, participated in local markets. Even before the development of urban markets and service centers, farmers in Pennsylvania were selling produce to owners of gristmills and ironworks, who not only needed foodstuffs for their families and workers but also operated retail stores to supply dry goods and produce to local consumers.[101] During the eighteenth century, farmers in Pennsylvania visited the local mill store at least ten times a year to sell quantities of foodstuffs—dairy goods, eggs, salted meat, grains, vegetables—as well as lumber and charcoal; farm women, too, sold some of these goods, as well as thread, yarn, and a variety of homespun clothing. In return, store owners paid cash or provided credit, which allowed farmers to purchase provisions and dry goods imported from Philadelphia.

Although the bulk of agricultural produce in the Mid-Atlantic was consumed locally by farm families, tradespeople, and urban populations, significant surpluses filtered into Atlantic markets (figure 4.18). On his trip through the Mid-Atlantic colonies in 1750–1751, James Birket observed that New York had "An Extensive trade to the West Indies . . . For Bread, flour, Pork, Beefe, Horses, Lumber," with "A Great quantity of Bread Flour, and wheat [drawn] yearly from the Jerseys by way of Amboy, and down Rariton river by way of Brunswick" where there was "a very good Corn [i.e. wheat] Country." Similarly, Philadelphia had "a very Extensive trade . . . to all the English Islands in the west Indies for Bread, Flour, Porke, Hams, Indian Corn, Buckwheat[,] Oats, Apples &Ca, Also hogshead & Barrel Staves & heading of white Oak . . . Shingles, Hoops, Bar Iron &C. also live Stock as Sheep, Geese, Turkeys, Ducks & fowles in great Plenty. . . . They have also a good trade for wheat, Staves &ca, to Madeira, Lisbon, And Several parts of Spain, to Say Nothing of that Extensive trade between them & their Mother Country for Black wallnut and other valuable wood of different kinds, Large Qtys of Pig and Bar Iron and that of an Excellent Quality." At Christiana Bridge, a few miles south of Wilmington, Delaware, Birket also noted that he was "in a very good country for Wheat and flour, Abundance of which is brought here for Sale the greatest part of which is sent up to Philada[lphia] in Shallops, and a part Shipt by the Inhabitants for the

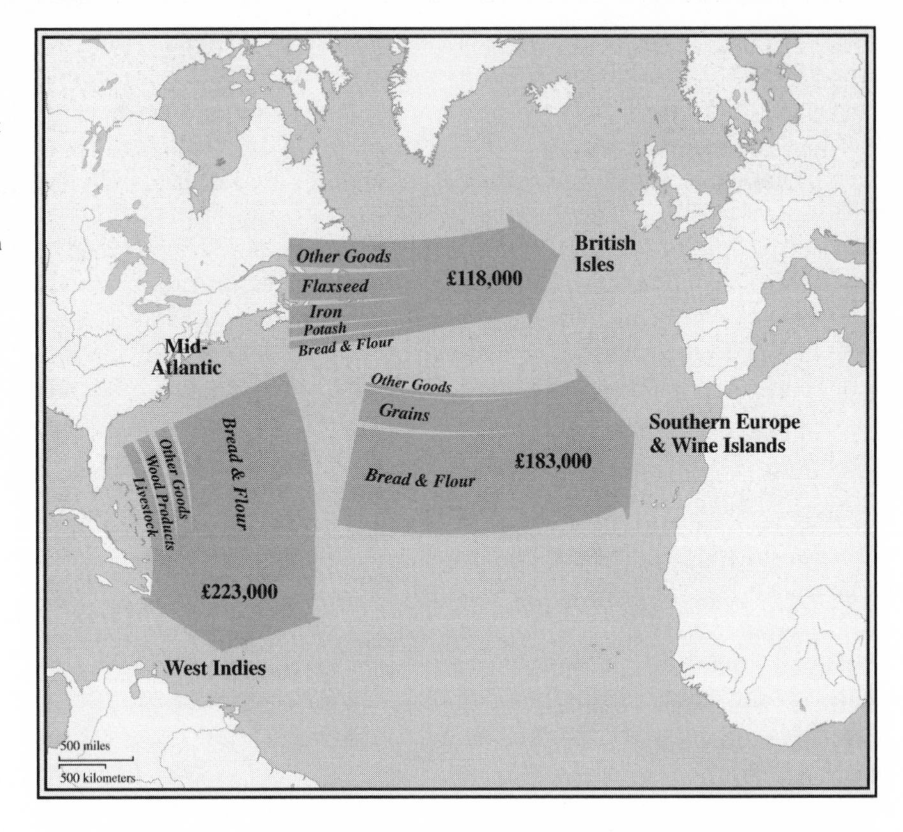

FIGURE 4.18. Mid-Atlantic overseas trade, 1768–1772. Average annual export values calculated from James F. Shepherd, "Commodity Exports from the British North American Colonies to Overseas Areas, 1768–1772: Magnitudes and Patterns of Trade," *Explorations in Economic History* 8 (1970–1971).

Westindies."[102] According to Customs Returns for 1768 to 1772, bread, flour, and grain made up 72 percent of the total average annual value of exports from the Mid-Atlantic region, followed by flaxseed (6.8 percent), wood products (5.6 percent), iron (5.2 percent), and livestock and salted meat (3.8 percent).[103] Compared to the agricultural exports from New England, those from the Mid-Atlantic colonies were much more diverse, and included critically important bread, flour, and grains. Moreover, Mid-Atlantic farmers relied much more on agricultural exports; the per capita value of exports from the Mid-Atlantic was more than twice that from New England.[104] Exports also went to a wider array of markets. Although products with the greatest value were shipped to the West Indies (42.5 percent of total), important markets were found also in Southern Europe and the Wine Islands (34.5 percent), Great Britain (13 percent), and Ireland (10 percent), as well as in the substantial coasting trade in wheat and bread to New England. Significantly, the overwhelming bulk of exports from the Mid-Atlantic region were in nonenumerated products shipped to markets outside the British Isles. As in New England, British merchants had virtually no share of the Mid-Atlantic export trade.

Apart from being the backbone of local and long-distance trade, the wheat economy also had a considerable impact on the development of manufacturing.[105] Unlike the livestock of New England, which could walk to market, the grain har-

FIGURE 4.19. Miller's house (left) and grist mill (right), Millbach, Lebanon County, Pennsylvania, 1752. Built by German settlers, this mill complex reflects the considerable rural prosperity that grew out of the Pennsylvanian wheat economy. Library of Congress, Prints and Photographs Division, HABS, PA, 38-MILB, 1-2.

vest of the Mid-Atlantic had to be carted to granaries and flour mills. As a result, farmers needed wagons; in the 1750s, one observer reckoned that "there may be from 7,000 to 8,000 Dutch Waggons with four Horses each that bring their Produce and Traffick to Philadelphia, from 10 to 100 miles Distance."[106] The demand for wagons gave employment to wagon makers and blacksmiths, and led to the development of the famed Conestoga wagon, a Pennsylvania invention. Moreover, the processing of grain into flour required gristmills, which soon developed at water-power sites. By the mid-eighteenth century, gristmills were widespread across southeastern Pennsylvania (figure 4.19). Although the majority were small or medium-sized operations, employing family members and a few hired hands, a third of the mills were large, commercial operations employing up to seventy-five hands each year.[107] Some of these larger mills were part of mill complexes that included sawmills, tanneries, fulling mills, and foundries. Apart from these industries, craftsmen were employed in blacksmithing, tailoring, and shoemaking, turning out consumer goods for the rural population, and service workers were employed in roadside hostelries and taverns.

Many of these rural industries and services were concentrated at shipping points, crossroads, and water-power sites, which led to the development of local service and market centers. Such towns were most prominent in Pennsylvania, especially beyond Philadelphia's urban shadow, which extended some thirty miles into the interior. As the backcountry filled up and the economy expanded after 1730, county seats were established at Lancaster, York, Reading, Carlisle, and Easton (figure 4.20).[108] Although several of these towns depended for their communications on the major rivers — the Delaware, the Schuylkill, the Susquehanna — which

FIGURE 4.20. *Lancaster County Courthouse, Lancaster, Pennsylvania*, by Benjamin Henry Latrobe, 1801. Although drawn well after the American Revolution, Latrobe's view gives a good sense of the growing prosperity of Lancaster, with its central courthouse square, new Georgian courthouse (built 1787), and row housing. The Maryland Historical Society, Baltimore, Maryland.

facilitated the shipping of agricultural produce to Philadelphia or Baltimore, Lancaster was sited more than a mile from a navigable river and relied, instead, on a thickening network of roads.[109] During the eighteenth century, the town became the focal point of roads in southeastern Pennsylvania and a major market center on the "Great Philadelphia Waggon Road," which ran west from the capital, through the backcountry, and into the Great Valley of Virginia. As the "emporium of the backcountry," Lancaster merchants collected agricultural produce from local and regional farmers, forwarded it to Philadelphia, and received, in return, imported manufactured goods, which were then distributed to settlers across the backcountry. By the outbreak of the Revolution, Lancaster, York, and Reading were flourishing market towns, each with some two to three thousand inhabitants. The largest towns in the interior of British America, they were testament to the commerce and prosperity of rural Pennsylvania.

Aware of the commercial need for well-placed towns, the Penn family founded the first county towns in Pennsylvania as speculative ventures, and most likely shaped their urban plans.[110] The first inland towns had grids of streets and lanes, derived from William Penn's plan for Philadelphia and plans for rebuilding London after the Great Fire, and central market squares or "diamonds," which were most likely inspired by similar spaces created by Quaker settlers in the planned towns of Ulster.[111] A grid of uniform lots was perfect for speculation and development in real estate, while the central market square provided a trading forum for local farmers. Apart from their commercial nature, inland towns were centers of county government; indeed, the county courthouse became a symbol of the Pennsylvania town (see Figure 4.20). Courthouses were usually sited at a prominent location on the square or "diamond," thereby creating the central courthouse square: a morphological, functional, and symbolic feature that would dominate urban planning beyond the Appalachians after the Revolution.[112] Around the "diamond" were relatively spacious town lots, large enough to accommodate dairies and orchards.[113] At least during the colonial era, the development of brick row housing, so characteristic of Pennsylvania towns today, had made little headway and most dwellings were constructed either of wood or stone. Towns also served as religious centers, usually having several denominations and associated church buildings. Nevertheless, the prominence of the courthouse and the variety of religious structures created an urban landscape much different from that found in England. In the English county town, a large, usually medieval, Anglican parish church dominated the landscape; in the Pennsylvania county town, the courthouse, rather than the church, was the symbolic center.

Like New England, the Mid-Atlantic had a relatively egalitarian rural society. The availability of land during the late seventeenth and early eighteenth centuries, the safety valve of an agricultural frontier in the southern backcountry in the late eighteenth century, and the lack of a commercial plantation staple all helped reduce social and economic stratification. The availability of land ensured that the rapid growth of population in the Mid-Atlantic would not outstrip local resources, while the difficulties of raising an agricultural staple meant that land was not engrossed quickly into large plantations. Instead, land was available for family farmers, and the growing of wheat produced a modicum of prosperity. Moreover, the availability of land and numerous laboring and manufacturing jobs provided the poor with economic opportunities. Relatively high wages allowed many eventually to save enough to purchase cheap land. Although few farmers would get rich in the Mid-Atlantic countryside, few would sink to the levels of destitution common in Europe. As a result of this social compression, the Mid-Atlantic, like New England, was a much less deferential society. For Brissot de Warville, used to the social hierarchy of France, the egalitarian attitude of people in the Mid-Atlantic colonies was striking, particularly when they shared a stage-coach together: "a member of Congress sits side by side with the shoemaker who elected him and fraternizes with him; they talk together on familiar terms. No one puts on important airs as people do only too frequently in France."[114]

If rural society in the Mid-Atlantic was similar to that in New England in its level of social and economic equality, it was much different in its cultural composition. Compared to the ethnic homogeneity of New England, the European migrations to the Mid-Atlantic created a heterogeneous mass of ethnic, linguistic, and religious groups. In New York and New Jersey, the English were the largest ethnic group, but there were also significant concentrations of Dutch, Germans, Scots, Irish, and Scots-Irish. In Pennsylvania, the mix was slightly different, with more Germans than English, as well as substantial Scots, Irish, and Scots-Irish populations.[115] Within each colony, clusters of ethnic groups were recognizable. Where land was available, friends and families settled together, creating distinct ethnic neighborhoods. In New York and New Jersey, the Dutch formed a wedge

FIGURE 4.21. Ethnic settlement in the Oley Valley, Pennsylvania, 1750. Note the several religious denominations, and the irregular cadastral pattern. After Philip E. Pendleton, *Oley Valley Heritage, The Colonial Years: 1700–1775* (Birdsboro, Penn.: The Pennsylvania German Society, 1994), 184.

stretching from New York Harbor (including the Jersey Shore) northward up the Hudson to Albany. In Pennsylvania, successive waves of immigrants created bands of ethnic settlement: Groups of English and Welsh close to Philadelphia gave way in the backcountry to an area more heavily settled by Germans, and then to a frontier zone of Scots-Irish (figure 4.21). Although English was the predominant language throughout the Mid-Atlantic, concentrations of Dutch and German settlers provided sufficient critical mass to support their respective languages. The small number of Swedes in the Delaware Valley, however, were insufficient to maintain Swedish, which had virtually disappeared by the time of the Revolution. Further complicating the ethnic and linguistic mix was the religious diversity of the population. While the southern Irish introduced the Catholic Church, the migrants from other parts of northwestern Europe represented a great range of Protestant denominations, including English and Welsh Quakers, Anglicans, Baptists, and Methodists; Scottish and Irish Presbyterians; Dutch Reformed; Swedish Lutherans; Moravian Brethren; and German Lutherans, Seventh Day Baptists, and Mennonites. For English speakers, such as the Scots-Irish, church affiliation was often the primary badge of ethnicity.[116]

Settlers in the Mid-Atlantic drew on elements of this diverse European heritage

FIGURE 4.22. Heinrich Zeller House, Newmanstown, Lebanon County, Pennsylvania, 1745. German peasant housing from the Rhine Valley in eastern Pennsylvania. Library of Congress, Prints and Photographs Division, HABS, PA, 38-NEWM, V, 1-1.

as well as the materials of the New World to fashion a distinctive cultural landscape.[117] In the early Dutch settlements, builders reproduced house types based on Old World prototypes but also took advantage of plentiful local supplies of wood. Many of the one-and-a-half and two-story houses in New Amsterdam had brick facades and crow-stepped gables, and were turned end-on to the street, much like houses in Amsterdam; the colonial houses were also constructed from wood, using Dutch anchor-bent frames.[118] Although many of the Dutch houses in New York City were demolished during the eighteenth century, some Dutch buildings still stood in Albany at the end of the colonial period. In rural areas, Dutch framing techniques lasted longer, but anchor-bent frames increasingly were married with English box framing, creating a hybrid form of construction.[119] Similarly, German settlers introduced other European framing techniques, putting up houses and barns with diagonal bracing and heavily framed roofs.[120] They also built one- or two-story, central chimney, three-room houses, a design derived from peasant housing in the Rhine Valley and Switzerland (figure 4.22).[121] English settlers built one- or two-story hall and parlor houses, one room deep; the two-story variant, known as the "I-house," was widespread throughout the region.[122] Yet by the mid-eighteenth century, these various house types were increasingly being influenced by the English Georgian style. As in New England, Georgian design, reproduced in pattern books, was marked by symmetrical facades and hierarchical interiors. By the late eighteenth century, the two-story, single- or double-pile Georgian farmhouse, built out of wood, brick, or stone, was becoming ubiquitous throughout the Mid-Atlantic region, a symbol of yeomanly success (figure 4.23).

If Georgian order eventually triumphed over the jumble of imported vernacular housing designs, practical considerations led to the adoption, particularly in Pennsylvania, of the Swiss-German forebay barn (figure 4.24).[123] At first, English and Welsh settlers introduced two- or three-story bank barns (a design most likely derived from the upland areas of England and Wales) to settlements around Philadelphia; later, German settlers brought with them the Swiss-German bank barn and distributed it throughout southeastern Pennsylvania. Both barn types were ideal for mixed farms, which needed storage for hay, grains, and farm equipment, as well as housing for livestock during the winter. Hay and grains were put up in the second story and loft space, while livestock were housed on the ground floor. In the Swiss-German barn, an extended forebay allowed hay to be thrown out to cattle in the yard during winter. By the late eighteenth century, the combination of a Georgian farmhouse and a Swiss-German barn was becoming common in parts of southeastern Pennsylvania, and represented a distinctly Mid-Atlantic cultural landscape.

Despite the enormous range of church denominations and the symbolic importance of religion to individual ethnic groups, the religious landscapes of the Mid-Atlantic colonies mostly reflected New World circumstances and the impress of English cultural ideas. With few towns and villages in the region, the great majority of churches and meeting houses were dispersed across the landscape, frequently located at accessible points, such as a crossroads or a ferry landing. As in rural New England, the Chesapeake, and the South Carolina low country, the isolated meeting house set beside the road, with a graveyard close by, was a common feature of the Mid-Atlantic landscape. Although many farmhouses and barns reflected Old World building traditions, few religious buildings were ethnically distinct.[124] A monastic complex called the *Kloster*, built by German Seventh Day Baptists known as the Dunkards at Ephrata, Lancaster County, Pennsylvania, in 1741, had Germanic-looking buildings with steep roofs and small dormer windows, but the *Kloster* was an isolated example. More commonly, religious groups built relatively simple structures marked by a certain Georgian formality and detail. Anglican, Presbyterian, and Lutheran churches looked much like contemporary meeting houses in New England, although they were more often built out of stone or brick rather than wood. Even the Quakers and Mennonites, who stressed simple design in their meetinghouses, were not averse to the barest classical form. Only the arrangements of the interiors betrayed denominational differences.

The institutional framework that overlay the Mid-Atlantic comprised colonial and county governments and religious congregations. At the colonial level, the region was split between proprietary governors in Pennsylvania (including Delaware), and royal governors in New York and New Jersey (after 1702). At the local level, there were counties, parishes, and townships, but only counties served as an effective means of secular government. Unlike England, New England, and the Chesapeake, where parishes and townships served as territorial units for both religious and secular government, the Mid-Atlantic had such religious diversity

FIGURE 4.23. Warrenpoint, Knauertown, Chester County, Pennsylvania, 1756. Library of Congress, Prints and Photographs Division, HABS, PA, 15-KNATO, V, 1-3.

FIGURE 4.24. Swiss-German forebay barn, Berks County, Pennsylvania, 1770. This massive stone and wood-framed structure housed livestock and wagons on the first floor, and grain and hay on the second floor. Library of Congress, Prints and Photographs Division, HABS, PA, 6-CENPO. V, 1A-1.

that no denominationally defined unit possibly could serve as the basis for secular government. Essentially, the intertwined relationship between church and state that had served well enough in areas where one church was dominant or established dissolved in the Mid-Atlantic colonies. As a result, secular government had its own local territory—the county—while the various churches defined their own individual parishes. Such a distinction, first worked out in the Mid-Atlantic region, soon became standard practice in the southern backcountry and then, after the Revolution, in American settlements west of the Appalachians.

The second great agricultural frontier on the eastern seaboard, the Mid-Atlantic region was marked by a more diverse population, more commercial agricultural economy, more prosperous rural society, and more plural culture than existed in New England. On the evidence of the landscape alone, a Mid-Atlantic subregion such as southeastern Pennsylvania looked much different from a New England subregion such as the Connecticut Valley. Nevertheless, the Mid-Atlantic had much more in common with New England than with the Chesapeake or the Carolina low country. The agricultural frontier of the Mid-Atlantic, like that in New England, provided immigrant Europeans with considerable opportunity to establish themselves on the land. Across the rich valleys and plains of the Mid-Atlantic region, the family farm, rather than the plantation, was the preeminent unit of settlement. And, as in New England, the availability of land produced social and economic leveling. Although New York and New Jersey had elements of a gentry class, they were much less prominent than the planters in the tidewater, let alone the aristocracies of Europe. The combination of heavy European immigration and cheap land also engendered considerable cultural change. From a diverse northwestern European cultural heritage, fragments had been recombined in the Mid-Atlantic to produce a regional culture that was much different from any in Europe and more diverse than those in New England and the Chesapeake, but one that ultimately would be more influential for American development.[125]

The Southern Backcountry

As population increased in the Mid-Atlantic and southern colonies during the early eighteenth century, settlers spilled into the backcountry.[126] The last major area of agricultural land available for settlement east of the Appalachians, the backcountry encompassed the Piedmont regions of Maryland, Virginia, the Carolinas, and Georgia, as well as the Great Valley that stretched from southeastern Pennsylvania through western Virginia to the Carolinas and Georgia. For historian Frederick Jackson Turner, this backcountry was the "Old West," the first internal American settlement frontier. By the middle decades of the eighteenth century, it was the most dynamic agricultural frontier in colonial British America, with a major nucleus of population, a significant agricultural economy, and an important rural culture. By the time of the Revolution, the "quintessentially American landscape of

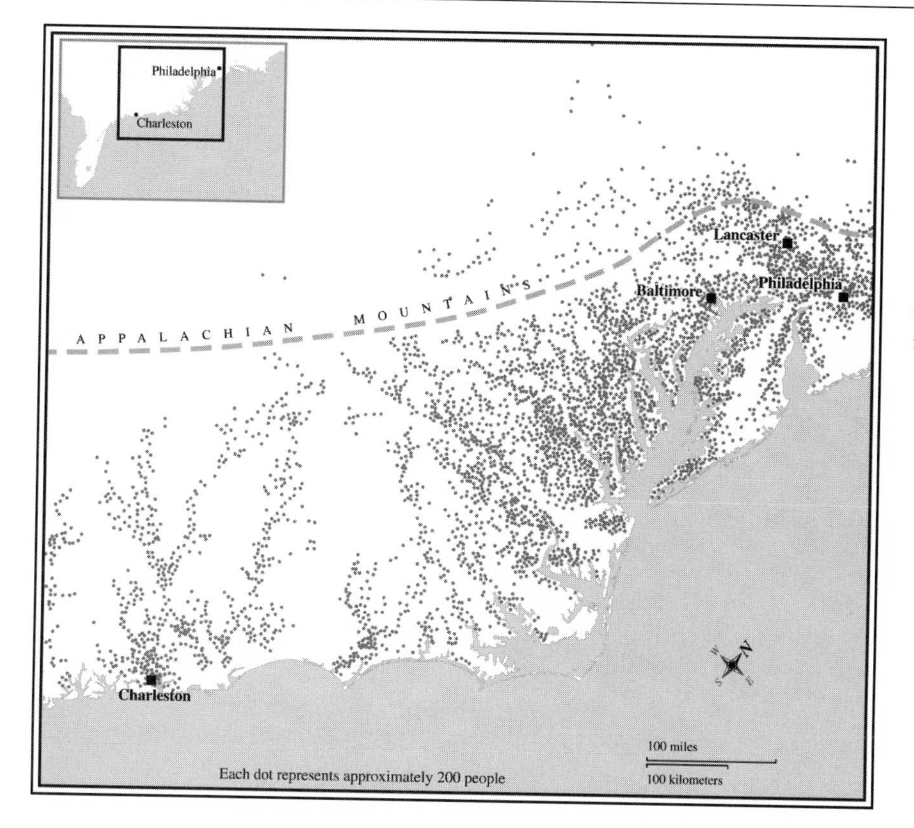

FIGURE 4.25. European settlement in the southern backcountry, 1760. After Herman R. Friis, *A Series of Population Maps of the Colonies and the United States 1625–1790* (New York: American Geographical Society, 1968).

rural farms and small towns" that had appeared first in southeastern Pennsylvania had spread south through the backcountry settlements to western Carolina.[127]

From the beginning of European settlement in the 1720s to the outbreak of the Revolution in the 1770s, the population of the backcountry increased rapidly (figure 4.25). By 1780, the population was about 380,000, roughly equivalent to three-quarters of the rural populations of New England or the Mid-Atlantic colonies.[128] The great bulk of the backcountry population occupied the Great Valley (140,500), while lesser concentrations were spread over the Piedmont and valley regions of Maryland (62,000) and the Piedmont regions of the Carolinas (110,500 in North Carolina, 55,000 in South Carolina) and Georgia (12,000). Given the rapid growth of population, the pace of immigration must have been considerable, with at least 100,000 migrants entering the backcountry during the colonial period. Although many migrants filtered in from the Virginia tidewater or the Carolina low country, the majority of settlers entered the region from Pennsylvania, working their way down the Great Wagon Road, which stretched from Philadelphia through Lancaster to Roanoke, Salisbury, Camden, and Charleston (figure 4.26). In 1753, the Governor of North Carolina reported to the Board of Trade in London that settlers were arriving from the northern colonies in hundreds of wagons. Two years later, a minister in Virginia noted that some 5,000 settlers had crossed the James River

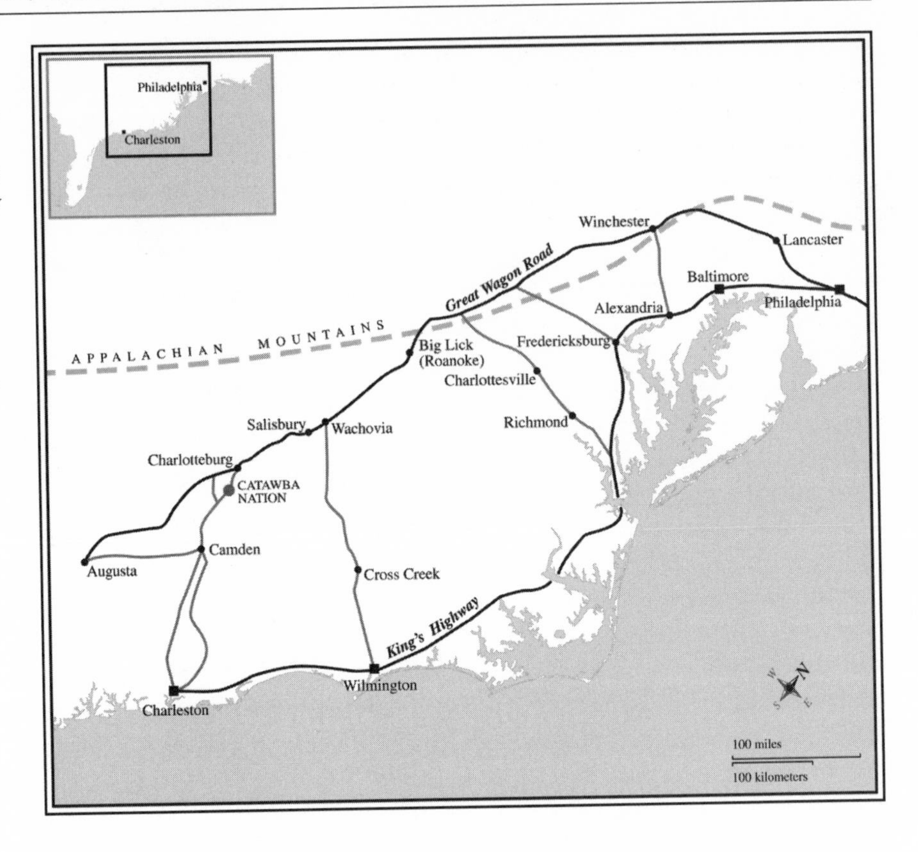

FIGURE 4.26. Principal roads in the southern backcountry, circa 1775. After *An Accurate Map of North and South Carolina*, by Henry Mouzon, 1775; William Dollarhide, *Map Guide to American Migration Routes, 1735–1815* (Bountiful, Utah: Heritage Quest, 2001), 7.

between the months of January and October bound for North Carolina.[129] Most of the settlers from the tidewater and the low country were of English origin, although some French Huguenots, Germans, and Highland Scots filtered in through the Carolinas; those who came from Pennsylvania were mostly Germans and Scots-Irish, with a sprinkling of English.[130] Essentially, two great migrant streams, one from the tidewater and the other from the Mid-Atlantic, merged in the southern backcountry to create an ethnically diverse settler population.

As in New England and the Mid-Atlantic, the migrations into the southern backcountry were composed mainly of families. Many immigrants arrived from Europe or the Mid-Atlantic in family groups, and attempted to settle close to kith and kin. Where land was available, the clustering of friends and family from similar backgrounds created distinctive ethnic enclaves. Such chain migration and group settlement marked the Germans who settled in Frederick County, Maryland, during the 1730s and 1740s.[131] Arriving directly from the Palatinate as well as from Pennsylvania, German immigrants purchased land in the county and settled on dispersed farms close to friends and relatives; at Tasker's Chance, many of the German families had come from the village of Klein Schifferstadt on the west side of the Rhine River. Although the ethnic composition of Frederick County was mixed, there were sufficient Germans to sustain some social cohesion, particularly through the provision of marriage partners and loans within the farming community. Farther south, in the backcountry of North Carolina, Scottish Highlanders created a similar ethnic settlement.[132] Drawn from Campbell-country in western Argyleshire, the first group of Highlanders arrived in North Carolina in the late 1730s, moved more than one hundred miles inland, and settled in the vicinity of Cross Creek and Campbellton (now Fayetteville) in the sand hills region of the upper Cape Fear River, an area that the governor of the colony described as "the fag end of the provin[ce]." After the initial settlement, further Highlanders were attracted to the region, dribbling in during the 1740s and 1750s. Although the Highlanders were never numerous enough to command all the backcountry around Campbellton and had to rub shoulders with English, Germans, French Huguenots, and Lowland Scots, they had sufficient numbers, like the Germans in Frederick County, to preserve vestiges of their culture; Gaelic was still being spoken in the area in the late nineteenth century.

As European and American settlers spilled down the Great Valley and crossed the Piedmont, they encountered remnants of Indian tribes, survivors and refugees from earlier conflicts along the seaboard. Although no Indians appear to have been resident in the Shenandoah Valley by the eighteenth century, groups were still scattered across the Piedmont.[133] Among these was the Catawba settled at the junction of the Catawba River and Sugar Creek in the South Carolina backcountry.[134] Like other native groups, the Catawba had been ravaged by disease but they still mustered between 300 and 500 warriors in the mid-eighteenth century. Yet after a small group had served with the British against the French in 1759 and been exposed to smallpox, the Catawba Nation was decimated.[135] In the space of a few weeks, six villages had been reduced to one. Hardly a threat to the incoming European settlers, the few hundred Catawba were placed on a reservation in the early 1760s.

During the colonial period, settlers could acquire land in the backcountry from the Crown under the head right, treasury right, or military right system, or by purchase from speculators who owned large patents.[136] In the Shenandoah Valley, considerable tracts were held by tidewater planters, such as George Washington, or the proprietor Lord Fairfax, who owned lands that extended from the Northern Neck of Virginia to the Appalachians.[137] Whatever the method of acquisition, land was usually cheap, certainly in comparison to that in Europe or in settled areas of the seaboard, and was thus accessible to settlers with some capital. As population increased, however, the price of land rose and property became more difficult to acquire. In older settled areas, such as the northern Shenandoah, tenancy had become common by the time of the Revolution.[138]

As in the Chesapeake and the backcountry of Pennsylvania, farms in the southern backcountry were usually demarcated by the metes-and-bounds method, creating a patchwork cadastral pattern.[139] Only on parts of the South Carolina Piedmont, where townships were laid out, were there more regularly surveyed lots. Throughout the backcountry, lots were mostly between 100 and 400 acres, although

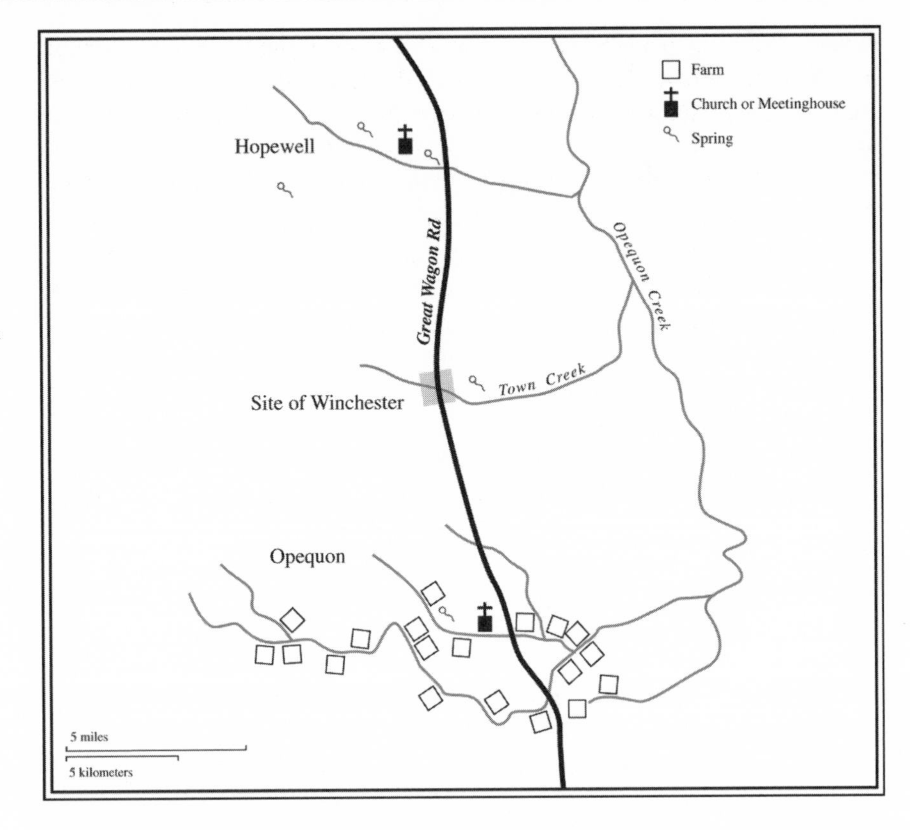

FIGURE 4.27. Open country neighborhoods, Frederick County, Virginia, 1730–1745. After Robert D. Mitchell and Warren R. Hofstra, "How Do Settlement Systems Evolve? The Virginia Backcountry during the Eighteenth Century," *Journal of Historical Geography* 21, no. 2 (1995): Figure 3.

larger lots were common. In the southern part of the Shenandoah Valley, over 93 percent of land patents granted before 1780 were for 400 acres or fewer; in the northern part, the average size of land holdings in the early 1760s was 473 acres.[140] The relatively large size of holdings, as well as the prevalence of undeveloped land held by speculators, created a loosely settled landscape of dispersed farms; in the Shenandoah Valley in the late eighteenth century, farms along Opequon Creek and its tributaries were between one-half mile and one mile apart (figure 4.27).[141] Such a distribution created open-country neighborhoods, a distinctive feature of the southern backcountry landscape.

Like the earlier migrations of families into New England and the Mid-Atlantic, many of the families who settled the southern backcountry arrived with some capital. This allowed them to purchase or rent land, buy seed and stock, and begin establishing a farm. Family members provided much of the labor, with men clearing the bush and laboring in the fields, while women did domestic work. In the fall of 1752, a group of Moravian Brethren, trekking south from Bethlehem, Pennsylvania, to Wachovia, North Carolina, occasionally encountered backcountry settlers, almost always men, who gave them turnips, hay, and other field produce.[142] Most likely the wives and daughters of these backcountry farmers were working in the household or farm yard, scarcely visible to passersby. Slaves were also employed on some farms and plantations. As tidewater planters moved across the Virginia Piedmont and into the northern Shenandoah, they established tobacco plantations and introduced slave labor. By 1780, the northern counties of Berkeley and Frederick contained more than 3,500 slaves, some 16 percent of the two counties' total population. In the southern Shenandoah, where family farming prevailed, about 25 percent of settlers who left wills between 1770 and 1780 held slaves, although usually only one or two per inventory.[143] In general, slavery was more common in the backcountry than in New England and the Mid-Atlantic, which reflected the strength of the institution in the southern colonies.

The migration of settlers from southeastern Pennsylvania as well as from the Chesapeake introduced mixed farming to the backcountry.[144] This agricultural economy included the growing of grains, such as wheat, rye, oats, and corn, as well as the raising of livestock, particularly cattle and hogs. In addition, farmers raised flax, hemp, tobacco, vegetables, peas, beans, and squash, and kept poultry and bees. The cultivation of wheat was introduced by the streams of settlers coming from southeastern Pennsylvania as well as from the Chesapeake, and the crop became the great staple of the Shenandoah Valley and the Carolina backcountry.[145] In 1770, Governor William Bull of South Carolina reported that "proper mills are now daily erected in the back parts as wheat encreases."[146] The growing of tobacco, on the other hand, was introduced from the tidewater. At least until the Revolution, the cultivation of tobacco was concentrated on the Virginia Piedmont (see chapter 3) and in parts of the Shenandoah. By the 1740s, tobacco was being grown in the southern part of the valley, and by the late 1760s substantial amounts were being exported.[147] Nevertheless, the total area devoted to tobacco comprised no more than a few hundred acres.

In contrast to these introduced crops, the raising of corn-fed cattle appears to have been a backcountry initiative, which was later to have a great impact on the development of midwestern agriculture. By the 1750s, farmers on the western edge of the Shenandoah, particularly along the South Branch of the Potomac River, were grazing cattle on the open range as well as fattening them on corn.[148] "The forests afforded abundance of rich food for all stock almost throughout the year with a little corn fodder and straw in hard winter weather," observed one Shenandoah Valley minister in the late eighteenth century, "but many cattle in the spring were weakened and needed help to rise or get out of swamps in which they ventured seeking early grass."[149] The South Branch was particularly suited to corn-feeding; the alluvial soils of the river's bottom land were fertile but the valley was too physically isolated from the neighboring Shenandoah to be competitive in the wheat trade. Cattle, rather than wheat, became the local staple, a product that could be walked easily to market in Philadelphia. From the South Branch, the technique of fattening cattle on corn diffused south into Kentucky after the Revolution and from there into Ohio in the early nineteenth century to form the basis of midwestern agriculture: corn-fed livestock.[150]

Backcountry farms produced food for the family and small surpluses for sale

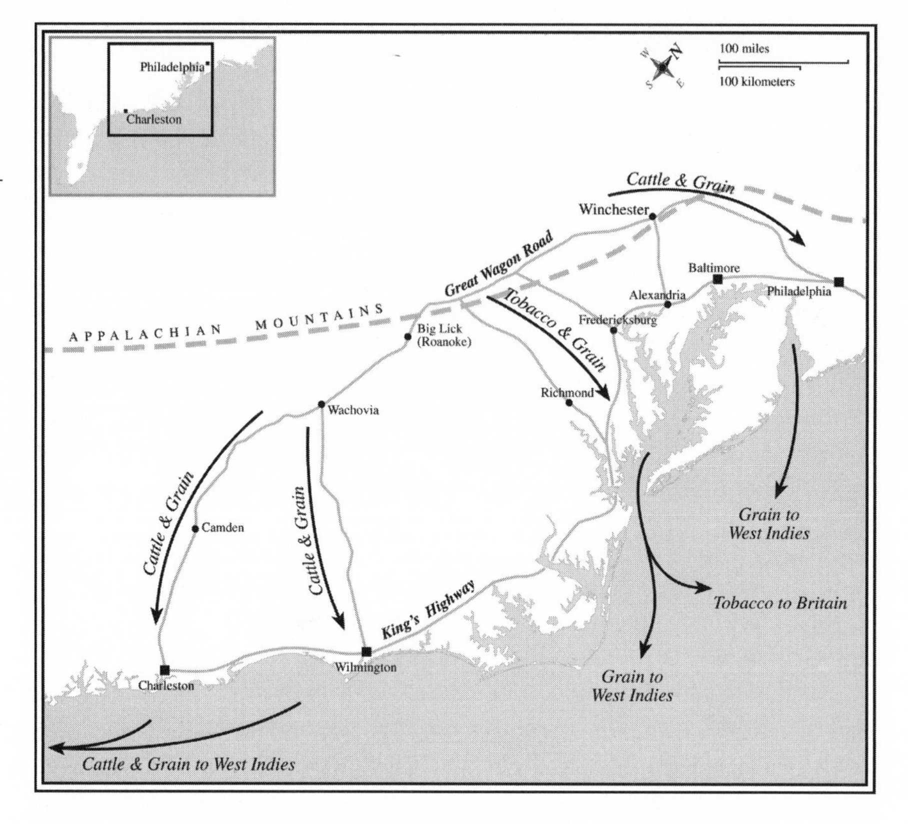

FIGURE 4.28. Backcountry trade, circa 1770. Based on data in Robert D. Mitchell, "The Southern Backcountry: A Geographical House Divided," in *The Southern Colonial Backcountry: Interdisciplinary Perspectives on Frontier Communities*, ed. David Colin Crass et al., 1–35, especially Map 1.2 (Knoxville: University of Tennessee Press, 1998).

in local and long-distance markets (figure 4.28). As in New England and the Mid-Atlantic, there was an active local economy, with farmers exchanging goods and services with their neighbors. Wheat, flour, tobacco, and cattle also were sold to merchants or drovers, who then forwarded the products to urban and export markets. From the northern part of North Carolina and the Shenandoah, wheat and flour were exported via the Great Wagon Road to Philadelphia, and also to the Chesapeake port of Alexandria; tobacco was sent to Alexandria or Fredericksburg where agents of British firms arranged for its export.[151] From the southern part of North Carolina and from the backcountry of South Carolina, flour and tobacco were shipped through Charleston. Cattle, too, were collected from farmers in the Carolinas and Virginia and driven along the Great Wagon Road to butchers in Philadelphia, or through the mountains to troops stationed at Fort Pitt. As early as the 1740s, settlers in the Shenandoah complained about drovers taking herds "down" the valley to Philadelphia, while farmers in Pennsylvania objected to cattle drives "from Virginia and the Southern Provinces."[152] Most likely, much of the wheat and cattle sent to Philadelphia or Alexandria was exported: wheat to markets in southern Europe, salted beef to the West Indies.

Unlike the staple regions along the coast and the agricultural frontiers in New England and the Mid-Atlantic, where coastal shipping and river boats played an

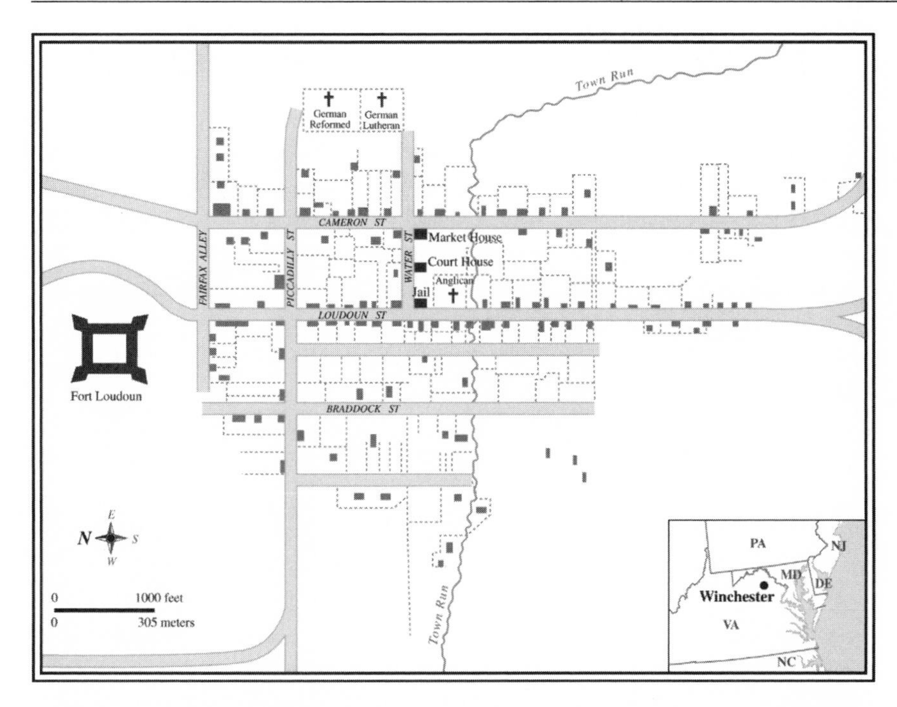

FIGURE 4.29. Winchester, Virginia, 1777. After map by Andreas Wiederhold redrawn in Warren R. Hofstra and Robert D. Mitchell, "Town and Country in Backcountry Virginia: Winchester and the Shenandoah Valley, 1730–1800," *Journal of Southern History* 59, no. 4 (1993): 633.

important part in the transportation network, the southern backcountry depended on the great north-south trunk of the Great Wagon Road and a proliferating number of east-west branches. By the early 1750s, the Great Wagon Road stretched from Philadelphia, through the Great Valley, to the Yadkin River in North Carolina, a distance of 435 miles. After 1760, the road was extended down the North Carolina Piedmont to Salisbury, and then along the Catawba River to Camden, South Carolina, where it joined the road from Charleston.[153] Over the same period, numerous lesser roads were laid out. At the time of the Revolution, the thickest networks were in the Piedmont regions of Maryland and Virginia, with three roads running through gaps in the Blue Ridge to connect the Shenandoah to the tidewater.[154] A thickening network also developed in the northern and western parts of North Carolina, as well as several connections to the coast, particularly down the Cape Fear Valley to Wilmington.[155] By the 1770s, backcountry farmers could send produce north through the Great Valley to Philadelphia, or east across the Piedmont to Wilmington and Charleston. The traffic was so great that Governor William Bull of South Carolina observed that "as many as three thousand wagons per year come in [to Charleston] from the Back Country."[156]

As the rural economy developed and a transportation network emerged, small urban centers were established.[157] In the Shenandoah Valley, the most advanced agricultural area in the backcountry at the time of the Revolution, service centers, such as Winchester, began to appear in the 1740s (figure 4.29). In the following decade, Winchester was created a county seat and a garrison town; by the 1760s, the town was benefitting from the growing commercialization of the regional economy. Although Winchester did not serve as a collecting point for the valley's staples of cattle and wheat, which were forwarded directly from larger farms and mills to Philadelphia or Alexandria, it did act as a retail and distribution center for imported dry goods, which came mostly from Philadelphia. Situated at the crossroads of the Great Wagon Road, which ran north to south, and the road linking Forts Pitt and Cumberland to the tidewater, which ran east to west, Winchester was well placed to serve as the leading central place in the Shenandoah. By the mid-1770s, Winchester was atop an emerging regional hierarchy of central places; from the town, merchants distributed imported dry goods to retail merchants in smaller settlements, country storekeepers, and settlers. Such a pattern of local trade and urban development already had evolved in southeastern Pennsylvania, and further distinguished these agrarian regions from the plantation economies of the tidewater.

Compared to the stratified societies of the Chesapeake or the Carolina low country, the society of the southern backcountry was much more egalitarian. Like the rural worlds of New England and the Mid-Atlantic region, the society of the backcountry was based on small family farms that were in various stages of commercial development. Just as farmers in the Shenandoah were making increased connections to market in the 1760s, so those in the backcountry of the Carolinas or Georgia were struggling to clear the forest and make ends meet. In areas with relatively weak connections to local and long-distance markets, few settlers had accumulated much capital by the time of the Revolution. Moreover, the availability of land slowed social and economic stratification, ensuring that backcountry society remained relatively open Although the population of the Shenandoah Valley was marked by increasing tenancy and landlessness by the 1770s, the availability of land farther south provided a safety valve and helped prevent the extreme stratification that had occurred in the tidewater.[158] In 1770, Governor William Bull of South Carolina observed that the available land in the colony's backcountry was "divided into many hands, and it will be always improved with more industry by the freeholder than the under tenant, as the fruits of his labor are wholly his own." Bull further reflected that such widespread property ownership "nurses a spirit of liberty independency and democracy."[159] For the thousands of migrants from older settled areas of the seaboard and immigrants from northwestern Europe, the southern backcountry held out the greatest opportunity of acquiring an independence on the land in late colonial British America.

The influx of English, Scots, Scots-Irish, Germans, and French Huguenots into the backcountry created a considerable religious and ethnic mix. Although the Anglican Church, the established church in Maryland, Virginia, and the Carolinas, was brought into the backcountry by English settlers from the tidewater, the Church had little purchase on backcountry society. There were relatively few English settlers, and some of them converted to evangelical denominations such as the Baptists and the Methodists. In the Shenandoah, there was only one Anglican church and two chapels.[160] Instead, there was "every denomination of

Christians but the Roman Catholic," according to South Carolina governor William Bull, "and these are subdivided ad infinitum in the back parts."[161] Of these many denominations, perhaps the largest was Presbyterianism. By the end of the colonial period, the Scots-Irish and Lowland Scots had established numerous Presbyterian churches throughout the backcountry, including more than twenty in the northern Shenandoah.[162] German settlers also introduced several denominations common in southeastern Pennsylvania, including Lutheran, Reformed, and Moravian Brethren.

Compared to the established, hierarchical Anglican church of the tidewater, the dissenting churches of the backcountry were community supported, egalitarian, and evangelical. One Anglican minister, who visited the Carolina backcountry in the 1760s, found the "extravagancies" of the Baptists beyond the pale: "One on his knees in a Posture of Prayer—Others singing—some howling—These ranting—Those Crying—Others dancing, Skipping, Laughing and rejoycing. Here two or 3 Women falling on their Backs, kicking up their Heels, exposing their Nakedness to all Bystanders." Such evangelical churches played an important role in providing a sense of community and moral order to fluid backcountry societies.[163]

As in other frontier areas along the seaboard, ethnicity had only a fleeting effect on vernacular housing; most ethnic groups quickly adopted one or other of the standard house types common to the backcountry. The movement of German settlers from southeastern Pennsylvania into the Great Valley introduced the Rhenish house to the northern Shenandoah.[164] Mennonite settlers, who took up land at Massanutten on the South Fork of the Shenandoah in the 1720s, constructed several one-and-a-half- and two-story houses over the course of the eighteenth century. Like similar houses in Pennsylvania, these dwellings usually comprised three rooms dispersed around a central chimney; massive, exposed structural members; and a cellar. As the German migration moved farther south, more Rhenish houses were built in the Yadkin River Valley of North Carolina.[165] In the larger context of the southern backcountry, however, such housing was rare. The great majority of settlers lived in single-story, one- or two-room cabins, a simple, utilitarian house type common throughout the seaboard colonies (figure 4.30).[166] In the backcountry, these dwellings were usually built from squared logs, notched at the corners, a form of construction that derived either from the old Swedish settlements along the Delaware River or from German areas in central Europe.[167] As settlers accumulated capital, they invested in more substantial, two-story dwellings, such as the "I-house," but such houses must have been relatively scarce in the backcountry before the Revolution. Although evidence for colonial-era farm outbuildings is scant, settlers most likely built utilitarian log structures, such as corn-cribs, for storing corn and grain, and, possibly, in the northern Shenandoah, Pennsylvania bank barns to accommodate grain and livestock.[168]

The institutional structure of the Mid-Atlantic also permeated the southern backcountry. Although the parish served as the secular and religious institution of local government in the tidewater and the low country, the parish was not adopted

FIGURE 4.30. The MacIntyre log house, Mecklenburg County, North Carolina, before 1780. A one-story, two-room cabin built out of squared pine logs notched at the corners. Library of Congress, Prints and Photographs Division, HABS, NC, 60-, 3-1.

as a civil unit in the backcountry. Instead, the county, the unit of administration common in Pennsylvania, became standard. The county was responsible for secular government, but left religious oversight to individual churches. As in Pennsylvania, the county courthouse, rather than the parish church, became the symbolic center of the small service towns that developed in the backcountry in the late eighteenth century.[169]

During the mid- to late eighteenth century, European settlement of the southern backcountry created the third great agricultural frontier in colonial British America. Although the backcountry owed much to the Mid-Atlantic, particularly southeastern Pennsylvania, it developed its own unique mix of ethnic groups, agricultural innovations, and vernacular architecture.[170] Yet for all the differences in detail, the region shared some general characteristics with the preexisting agricultural frontiers in the Mid-Atlantic and New England. First, the southern backcountry provided enormous opportunity for Europeans and Americans to establish themselves on the land. Settlers were drawn by the availability of cheap land, and the chance to create an independent livelihood based on mixed farming. Second, the absence of a highly commercial agricultural staple, as well as the availability of land, ensured that there would be no rapid stratification of backcountry society. Instead, rural society was relatively egalitarian with no great landed wealth or landless poor. Third, although ethnicity was important in structuring immigration and settlement, its influence quickly waned. Even though ethnic groups maintained their Old World religious affiliations and certain other cultural traits, they quickly adapted to the material and economic conditions existing in the New World. Whatever their background, immigrants settled on dispersed lots, created mixed farms, learned English, and participated in local institutions. The distinctively American rural world that had been taking shape in the Mid-Atlantic was further refined and enhanced in the backcountry, ready to spread westward after the Revolution.

An American Continental Frontier

Over the course of 150 years, three great agricultural frontiers developed along the eastern seaboard of North America. The first emerged in the Puritan settlements around Massachusetts Bay in the early 1600s, the second in the Mid-Atlantic colonies in the late 1600s, and the third in the southern backcountry in the mid-1700s. By the time of the American Revolution, these three frontiers had expanded and coalesced to form a continuous frontier stretching from Maine to Georgia, a frontier that was poised to spread west, once the British and their Indian allies had been pushed aside (figure 4.31). Compared to the staple regions along the coast and those on the margins of the continent, this great agricultural frontier was a singular American achievement. Whereas the staple regions were hitched to an external dynamic of Atlantic trade and commerce, the American agricultural frontier was enmeshed in an internal dynamic of population growth and geographic expansion. The American frontier had no real counterpart anywhere else in British America during the early modern period.

Although the details varied considerably, the agricultural frontiers in New England, the Mid-Atlantic, and the southern backcountry had certain features in common.[171] Throughout the area that lay between the coast and the Appalachians, potential agricultural land was available in massive quantities. To be sure, native peoples had to be dispossessed and the stony uplands of Massachusetts, New Hampshire, and Maine, the steep slopes of the Blue Ridge Mountains in Virginia, and the pine barrens of the Carolinas and Georgia offered scant agricultural reward, but overall there was enormous potential, far more than existed in the West Indies, Newfoundland, or Nova Scotia. At the same time, there was a marked shortage of labor, as well as a lack of ready markets. In such circumstances, land was cheap, almost worthless in its uncleared state, and wages were high. In Europe, the situation was the reverse: land was in short supply and expensive, labor was plentiful and cheap. As a result, Europeans found it almost impossible to purchase land in the Old World, but relatively easy in the New. The three agricultural frontiers along the seaboard thus offered immense opportunities for immigrant Europeans and generations of Americans to establish themselves on the land.

This American frontier was marked by a large, expanding, and relatively homogeneous population. The total rural population of New England, the Mid-Atlantic, and the southern backcountry was approximately 1,350,000 in 1770, more than

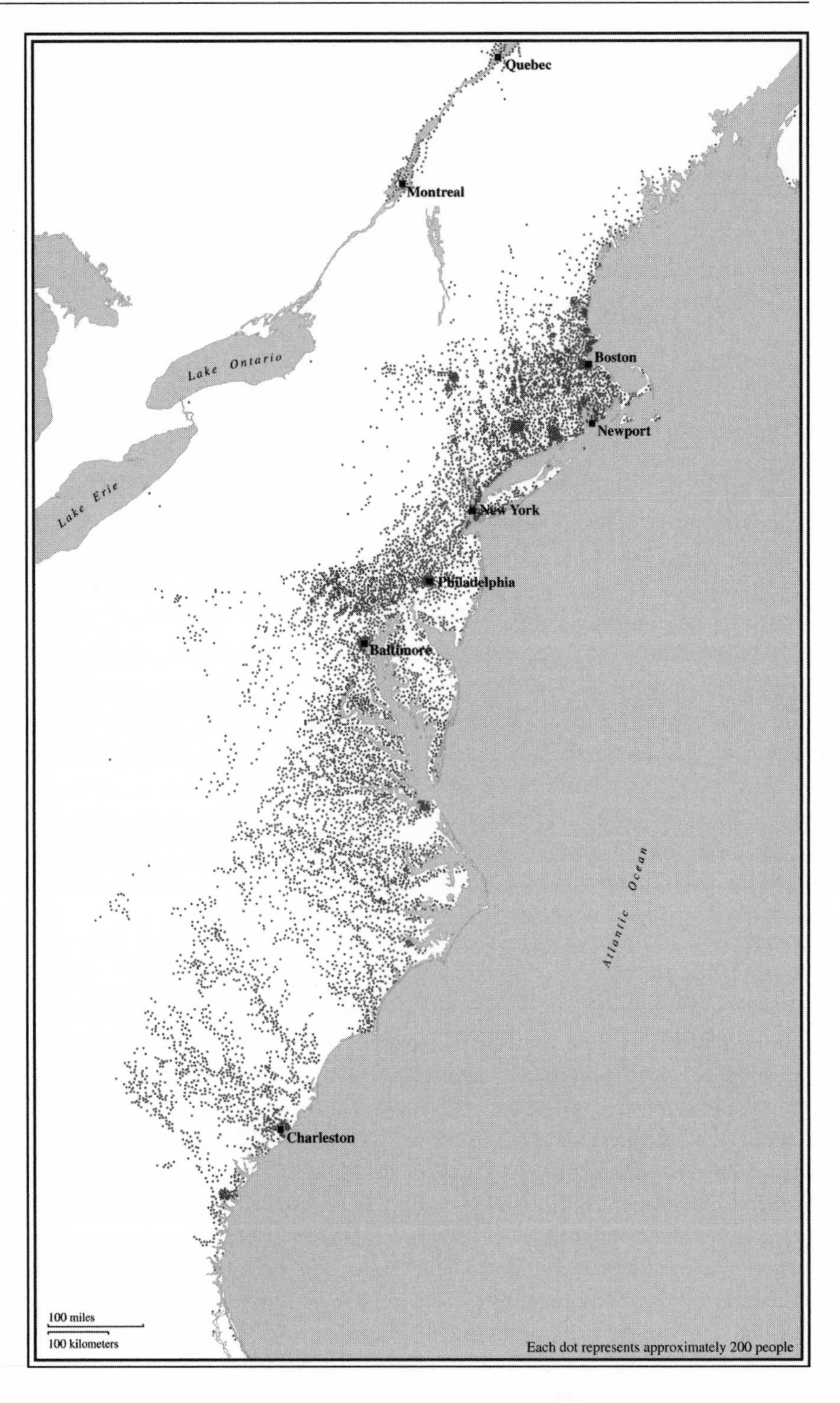

FIGURE 4.31. Continental agricultural frontier, 1770. After Herman R. Friis, *A Series of Population Maps of the Colonies and the United States 1625–1790* (New York: American Geographical Society, 1968). The population data for the settlements along the St. Lawrence are for 1760 and aggregated from Richard Colebrook Harris, *The Seigneurial System in Early Canada* (Madison: University of Wisconsin Press, 1968), Figure 6-5; the data for western Nova Scotia are for 1767 and from Graeme Wynn, "A Province Too Much Dependent on New England," *Canadian Geographer* 31, no. 2 (1987): 98–113, Figure 6.

three times the population of the British Atlantic staple regions. Demographically balanced by the early eighteenth century, this rural population increased rapidly, doubling every generation. As a result, growth of population, rather than circulation of labor and goods, was the great dynamic of the American frontier. Moreover, this settler population was racially uniform. In contrast to the enormous Afro-American populations on the plantations in the Chesapeake, the Carolina low country, and the British West Indies, the populations of New England, the Mid-Atlantic colonies, and the southern backcountry were overwhelmingly of European origin. With only a weakly developed plantation economy in the backcountry, there was scarcely any need for African slave labor. As a consequence, expansion into the interior of the continent, particularly from the Mid-Atlantic and New England regions, created a relatively homogeneous settlement frontier. The implications of having a largely free white population in the northern continental colonies and a largely enslaved black population in the southern colonies would become clearer only after the Revolution.

The growth of the frontier population also stimulated the expansion of the agricultural economy. At one level, this expansion was marked by the taking up of land by thousands of individual settlers and the development of family farms dependent on mixed agriculture. For European immigrants and American settlers, the availability of cheap land and the relatively low cost of creating a farm represented an unparallel opportunity to get ahead and establish an independence in a still largely preindustrial world. At the same time, the growth of a continental market for agricultural produce lessened dependence on Atlantic markets, helping to detach these agricultural territories from the British Atlantic.[172] At a second level, the growth of an internal agricultural economy and the development of agricultural trade led to urban development along the coast (see chapter 5) and the creation of significant mercantile elites. Because agricultural products such as grain and livestock were not enumerated under the navigation acts, merchants in the port towns were free to export produce to markets in the West Indies and southern Europe. The growth of these overseas trades led to urban growth and development, and further detached the continental economy from direct metropolitan control.

Compared to the enormous range and fine gradations of European society and the hierarchical world of the British Atlantic, rural society on the American frontier was much less stratified. The availability of land stripped out the upper and lower levels of the European social hierarchy, leaving a middle class of family farmers on the American frontier. Moreover, the lack of large agricultural markets ensured that the frontier developed along different lines than the staple regions. Without large markets, highly capitalized forms of agricultural production, such as plantation slavery, could not develop. As a result, the social hierarchy in rural New England, the Mid-Atlantic, and the southern backcountry was much more compressed than in the Chesapeake, the low country, or the West Indies. Farmers and country merchants accumulated less wealth than tobacco, sugar, or rice planters;

farm labor comprising family members and perhaps a servant or two replaced the gang labor of the slave plantations. Essentially, the feudal relationships in Europe that depended on the control of land and the commercial relationships of the British Atlantic that depended on the control of capital had been subverted on the American frontier by the availability of land and the weakness of markets.

The magnet of cheap land further influenced the formation of American rural culture. The immigration of hundreds of thousands of people from many different regional backgrounds in northwestern Europe to the completely different circumstances of the American eastern seaboard created a new cultural complex. During the process of settlement, some European ways were enhanced, others were discarded, and new American forms were developed. The availability of land strengthened the economic independence of the family, while the practical demands of making a farm diminished the importance of social background or education. The prevalence of wood, as well as the high cost of skilled labor, led to the rapid replacement of stone and brick building with wood construction. Distinctive forms of vernacular building developed on all three frontiers. Moreover, the mixing of people from different backgrounds weakened established institutions and customary modes of behavior. The majority of settlers were Protestants, many drawn from the dissenting sects on the margins of British society. As a result, the Church of England made little headway in rural New England, the Mid-Atlantic, or the southern backcountry. Unlike the established religious hierarchy of England and the British Atlantic, the American frontier was marked by denominational diversity and congregational control. Other cultural traditions, whether from continental Europe or the British Isles, had little support among the new social configurations on the American frontier. European languages gave way to English, and English regional dialects were eventually replaced by dialects generic to New England, the Mid-Atlantic, or the South. The preservation of West Country and Irish dialects that so characterized the isolated outports of Newfoundland would not be replicated on the American frontier.

The emergence of this new American rural economy, society, and culture strengthened ties within the continent while weakening links back across the Atlantic. To be sure, British merchants traded with colonial merchants along the seaboard and some speculated in land in the interior, but metropolitan capital was involved much less directly in the American frontier than in the Chesapeake, the West Indies, Newfoundland, or Hudson Bay. Instead, the economy that developed in the interior of the continent created its own momentum, primarily serving the local needs of the growing rural population rather than markets around the Atlantic rim. Moreover, connections that immigrant generations had with kith and kin in Europe were increasingly replaced by ties with younger generations on the frontier. Unlike fishermen in Newfoundland, fur traders in Hudson Bay, or merchants and planters in the West Indies, settlers along the eastern seaboard rarely traveled back to Europe. The populations of rural New England, the Mid-Atlantic colonies, and the southern backcountry put down their own roots, developing commitments and relationships within the local worlds of farm, neighborhood, township, county, and colony, rather than with far-off, scarcely remembered people and places in northwestern Europe.

The increase of population, the expansion of internal markets, the growth of trade in nonenumerated goods, and the weak links back to the metropole had considerable significance for the development of early modern British America. While Britain could monitor and control the settlements, populations, and maritime commerce of the Atlantic islands, it had little or no sway over the growth of the American population, the spread of backcountry settlement, the expansion of the agricultural economy, or the development of Caribbean trades. The dynamic of a land-based, agricultural frontier along the eastern seaboard was fundamentally different from that of a seaborne, commercial empire in the Atlantic. The European settlement of the eastern seaboard of North America created a new geographic space that lay beyond the reach of effective British metropolitan control.

Chapter 5

British American Towns

OVER THE COURSE of the seventeenth and eighteenth centuries, a string of urban centers stretching from the fur posts on Hudson Bay to the sugar ports of the West Indies developed in colonial British America.[1] Almost all these centers were port towns, points of contact between the commercial circuits of the Atlantic and the staple and agricultural regions of North America and the Caribbean. The towns served as gateways for the export of staples and the import of immigrants, slaves, and manufactured goods; as markets for local and overseas produce; as centers of manufacturing; as social and cultural hubs; and, in some cases, as centers of colonial administration and military power. Yet for all their similarities, differences in their growth and development were marked. A handful of towns flourished with relatively large populations and a wide array of functions, while others languished with meager populations and few activities. In historian Jacob Price's terms, some urban centers became "general entrepôts," while others remained "mere shipping points."[2] To some extent, differences in urban growth and development reflected the size of each town's hinterland and trade, the degree of urban competition, and the linkage effects of particular staples. Yet the reach of metropolitan power also played a significant role in shaping urban destiny. Just as British merchants had great influence over the growth and development of particular staples, so they helped shape urban expansion. It was no coincidence that the smallest and least-developed towns in British America were found on the islands of the British Atlantic, while the largest and most advanced lay in the American colonies. Such differences had important implications for the development of mercantile elites and political, economic, and cultural leadership in the colonies.

Planting Colonial Towns

In the opening years of English colonization, the sway of metropolitan power along the eastern flank of North America was considerable. English chartered companies and proprietors established the first towns in order to facilitate administration and trade (figures 5.1 and 5.2).[3] The Virginia Company began the process with the founding of Jamestown (1607), and was followed by the Earl of Carlisle at Bridgetown (1628), the Massachusetts Bay Company at Boston (1630), Lord Baltimore at St. Mary's City (1634), the Hudson Bay Company at Fort Albany (1670), the Carolina proprietors at Charleston (1680), and William Penn at Philadelphia (1682). As the English state increasingly became involved in the colonies in the late seventeenth and early eighteenth centuries, it, too, planted towns. In the 1690s, the colonial government encouraged the founding of Annapolis as the capital of Maryland and Williamsburg as the capital of Virginia; in 1749, the British government established Halifax in Nova Scotia as a military counterpoint to the great French fortress at Louisbourg on Île Royale (Cape Breton Island). Through conquest, the English also acquired New Amsterdam (New York) from the Dutch and Port-Royal (Annapolis Royal) from the French.

In the early seventeenth century, English proprietors and chartered companies appear to have given little direction to the laying out of towns. The settlers themselves seem to have laid out Jamestown, Bridgetown, Boston, and Newport, creating towns with haphazard arrangements of streets, blocks, and lots. The only exception was St. Mary's City, the first capital of Maryland, where the proprietor Lord Baltimore appears to have planned the city along grandiose Baroque lines, with radiating avenues and public buildings at prominent intersections.[4] But the weakness of urban growth in the Chesapeake and the transfer of the Maryland capital to Annapolis put paid to the proprietor's civic dream. Yet by the late seventeenth century, the English were more involved in town planning. The growing influence of classical urban design emanating from France and Italy, as well as the need to rebuild the City of London after the Great Fire of 1666, led to a new interest in the laying out of towns.[5] The various plans drawn up for rebuilding London, as well as the "Grand Modell" devised by Carolina proprietor and leading advocate of American colonization Lord Shaftesbury, led to the production of more formal plans for colonial towns.[6] These designs were distinguished by a rigid geometry consisting of a gridiron of wide, straight streets; rectangular, standard-sized lots; and public squares (figure 5.3). Charleston was the first colonial town to be laid out on a grid in 1680, and it was followed by Philadelphia in 1682, Kingston in 1692, Nassau in 1729, Savannah in 1732, Halifax in 1749, and Charlottetown in 1768.[7] Touring through the British American colonies in the mid-1760s, Lord Adam Gordon observed that in Kingston "the Streets [were] spacious, and regularly laid out, cutting one another at right Angles"; in Charleston "the Streets [were] Straight, broad, and Airy"; and in Philadelphia "the regularity of its Streets, their great breadth and length, their cutting one another all at right Angles" helped make "the Noble City . . . perhaps one of the wonders of the World."[8]

The clearest demonstration of state involvement in town planning was in the Chesapeake. After the Glorious Revolution of 1688, the English government tightened control over the American colonies and pursued a deliberate policy of

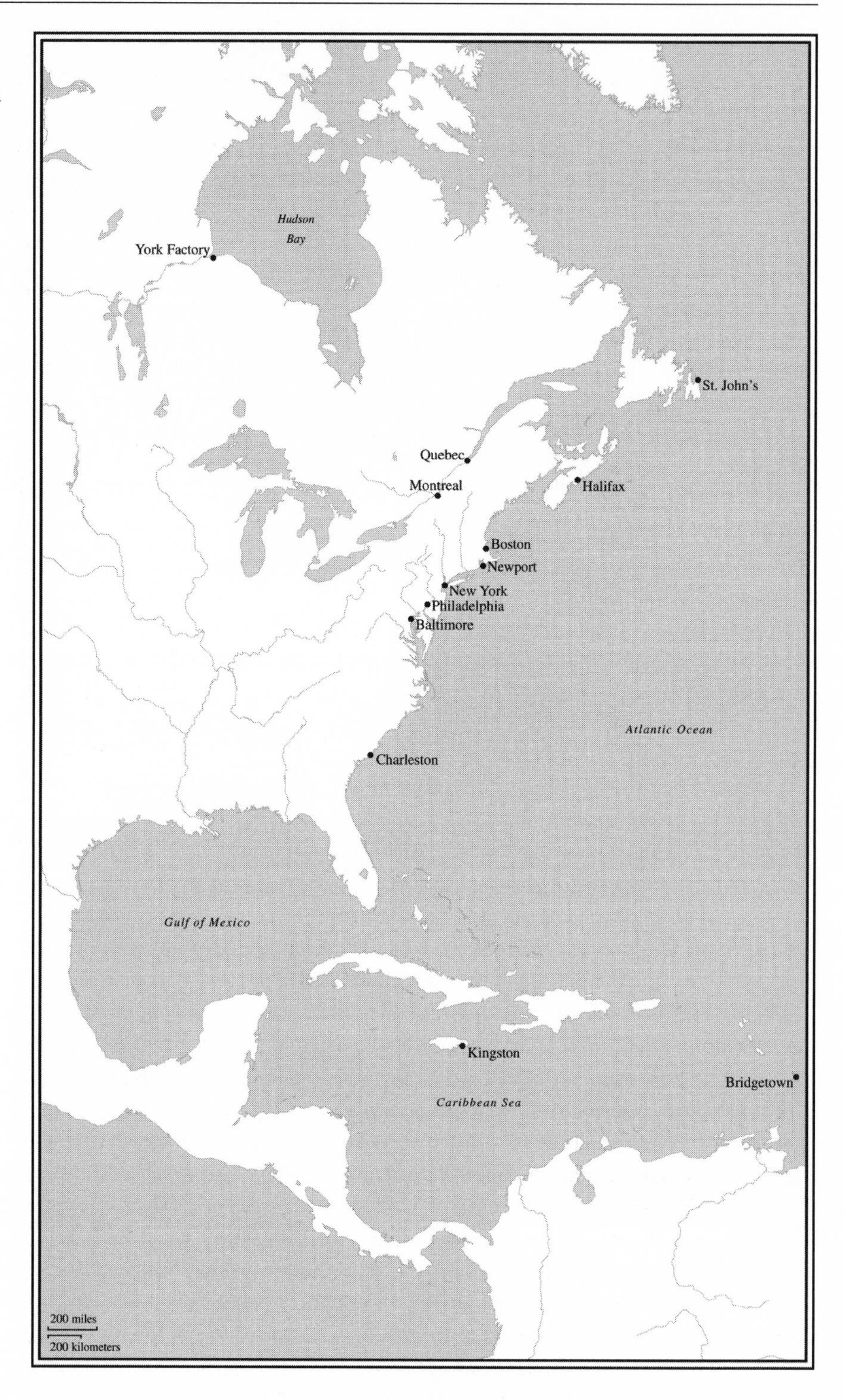

FIGURE 5.1. Distribution of leading urban places in British America, circa 1770.

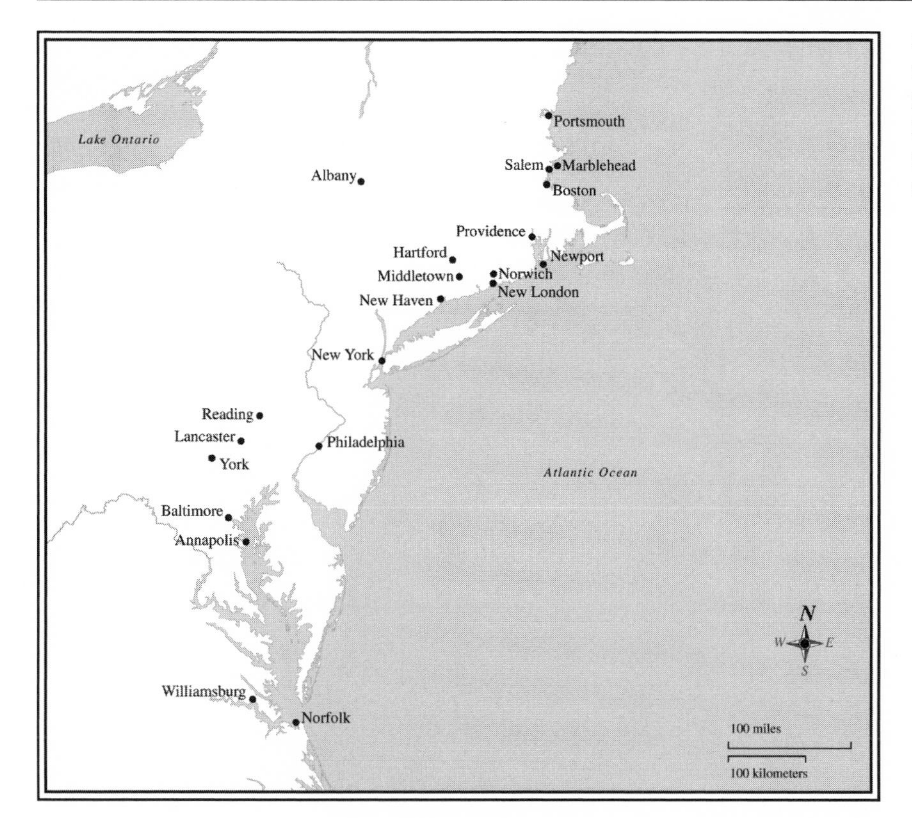

FIGURE 5.2. Distribution of urban places in New England, the Mid-Atlantic colonies, and the Chesapeake, circa 1770.

anglicization. The appointment of Francis Nicholson, first to the governorship of Maryland and then to that of Virginia, had a considerable impact on town founding and planning in the tidewater. A devout Anglican, ardent royalist, and career soldier, Nicholson aspired to "the ideal of officer-governor" imposing royal authority in the colonies.[9] Among his first acts in Maryland was to relocate the capital from St. Mary's City, the stronghold of the Catholic Calvert family, to Annapolis in 1694.[10] Apart from naming the new capital after Princess Anne, the heir apparent after the death of Queen Mary in 1695, Nicholson laid out the town on baroque principles.[11] At the center of the town were two circles, both set on a hill overlooking the Severn River, with avenues radiating outward. One circle was assigned to the colonial capitol, the other to the Anglican church. As the capitol housed the council chamber, colonial court, and administrative offices, as well as the legislative assembly, it was a clear symbol of English authority. So, too, was the Anglican church. In 1692, the Anglican Church became the established church in the colony.[12] On a prominent site overlooking the lower, mercantile town and harbor, Nicholson set aside land for the governor's mansion.[13] If the house had been completed, the institutional apex of Annapolis would have symbolized the new royal order. But the house does not appear to have been started, and in 1698 Nicholson was appointed lieutenant-governor of Virginia. Significantly, later

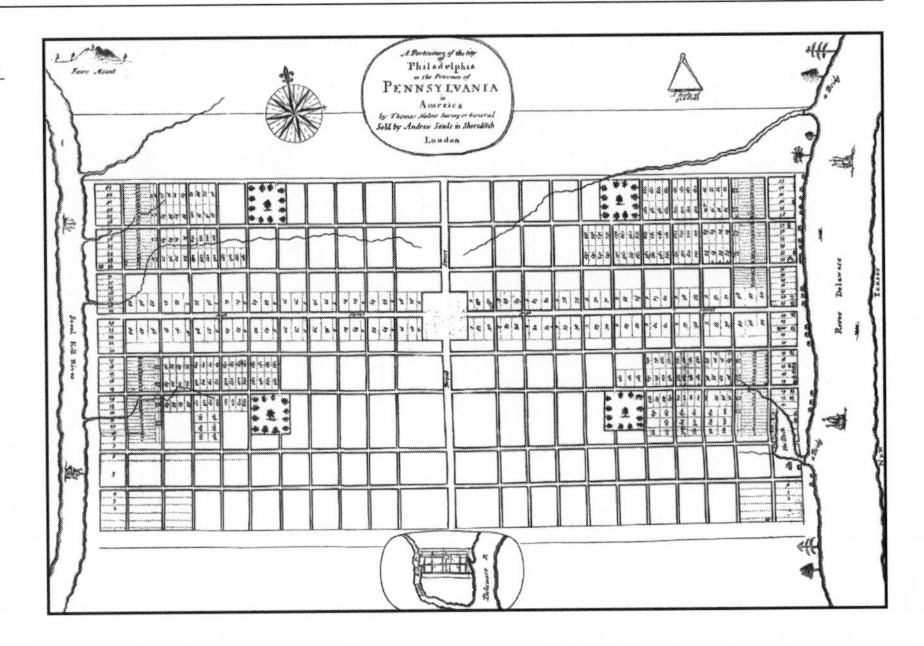

FIGURE 5.3. *Portraiture of the City of Philadelphia*, by Thomas Holme, 1683.

governors of Maryland built residences as close as they could to the capitol and church circles, the symbolic heart of Annapolis.[14]

Nicholson's ambition to express the authority of the Crown and Anglican Church in bricks and mortar had greater success in Virginia. As in Maryland, one of his first decisions as governor was to relocate the capital; in 1699, it was moved from Jamestown to Middle Plantation, which was renamed Williamsburg, again in honor of the reigning monarch. Although several reasons have been put forward for the choice of Middle Plantation as the new capital city, Nicholson made his preference clear in his recommendation to the house of burgesses. "I do cordially recommend to you placeing Your Publick Building . . . somewhere at Middle Plantation nigh his Majesties Royall Colledge of William and Mary which I think will tend to Gods Glory, his Majesties Service, and the welfare and Prosperity of your Country in Generall and of the Colledge in particular."[15] The College of William and Mary was a royal and Anglican foundation. The design of Williamsburg appears to have been the work of Nicholson, and reflected the power structure in the colony (figure 5.4). With the college and the Anglican Bruton parish church already in situ, Nicholson designed his new capital around these two buildings, locating the legislative building or capitol on the opposite end of the main street from the college, and the governor's mansion on a cross axis aligned with the parish church. Although Nicholson never saw the Governor's Palace completed, successive governors embellished it, making it the finest house in Virginia and the prototype for many of the great tidewater plantation houses built in the mid-eighteenth century.[16] In terms of layout and official buildings, Williamsburg was without parallel in British America, and symbolized metropolitan power in the seaboard colonies.[17]

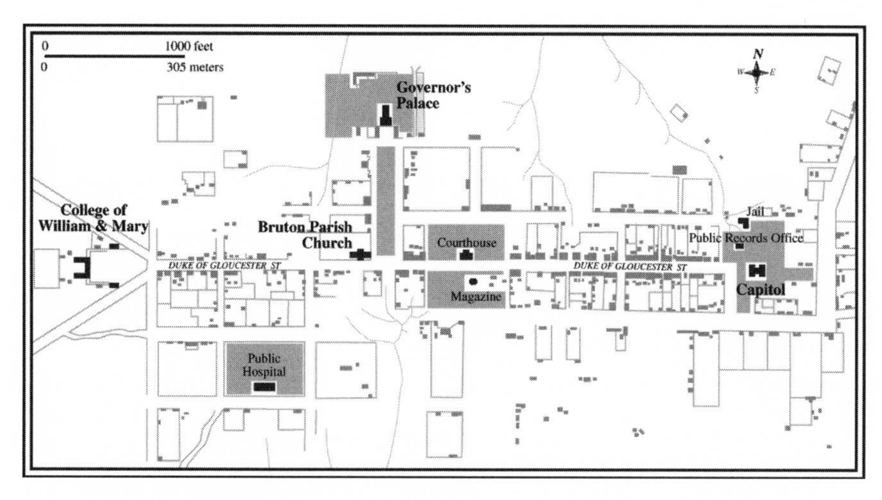

FIGURE 5.4. Plan of Williamsburg, 1782. Tidewater town as imperial city: The layout of Williamsburg owes much to the urban vision of Governor Francis Nicholson. After Lester J. Capon, *Atlas of Early American History: The Revolutionary Era, 1760–1790* (Princeton: Princeton University Press, 1976), 11, which was based on "Plan de la ville et environs de Williamsburg en virginie america 11 mai 1782."

Yet even more striking than the role of the English state in shaping the colonial town was the invisible hand of the free market. At the time that the first towns were planted in North America, the urban landscape of northwestern Europe was being remade.[18] The old feudal allocation of land was giving way to a capitalist market in property; the Crown, church, and guilds were being replaced by merchants as the principal arbiters of land use. As geographer James Vance has noted, "To secure the greatest accumulation of wealth it was necessary always to seek the 'highest economic' use, which . . . encouraged the segregation of uses."[19] This resulted, as another geographer, Martyn Bowden, has observed, in the "concentration in the area of high accessibility of establishments linked by competition and complementarity."[20] Under pressure from commercial rents, distinct zones of land use emerged. Such patterns first appeared in the principal port towns around the North Sea during the early sixteenth century. In Amsterdam and London, medieval feudal land uses were replaced by new mercantile land uses. Yet the new commercial landscape was greatly complicated by the feudal heritage of the medieval period. This was clearly visible in the City of London, where the recasting of land use along commercial lines was made more difficult by the prominence of the great medieval cathedral of St. Paul's, the numerous parish churches, and the many halls belonging to the guilds.[21] In the English colonies in North America, however, new mercantile towns could be established without any hindrance from the past. As a result, port towns founded along the eastern seaboard were extremely pure examples of the free market in land and the dedication of property to commercial use.[22]

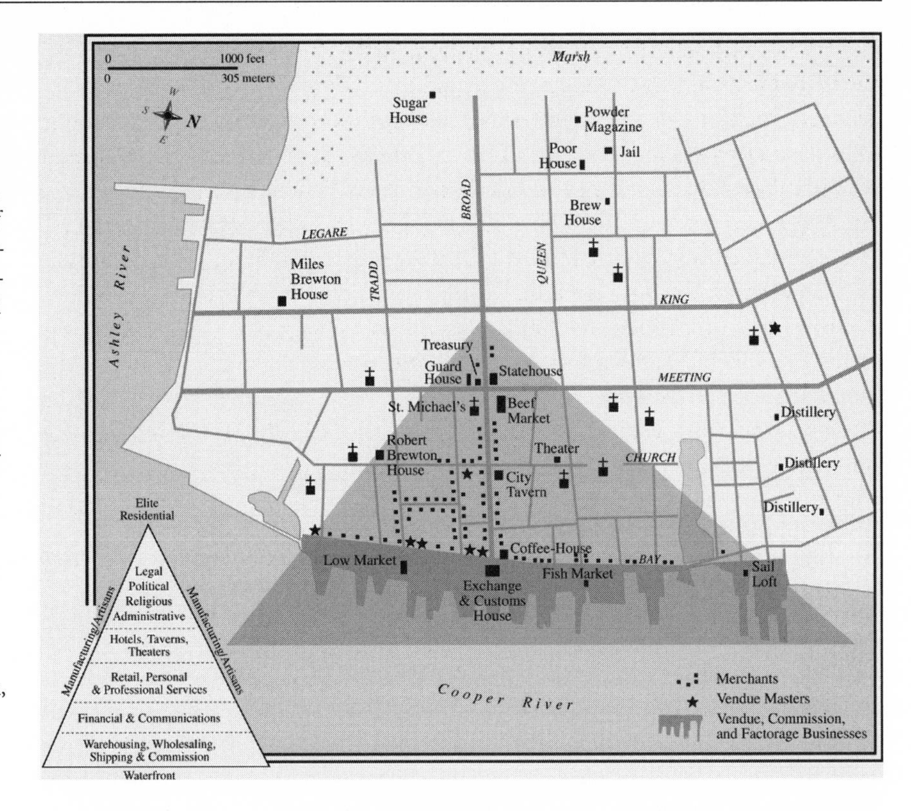

FIGURE 5.5. Mercantile triangle in Charleston, 1780s. Inset shows Bowden's model of the mercantile triangle. Based on data in *Ichnography of Charleston, South Carolina*, by Edmund Petrie, 1788; business and real estate advertisements in *South-Carolina Weekly Gazette* March 1783-February 1784; and the Charlestown Directory for 1782 reprinted in *The Charlestown Directory for 1782 and The Charleston Directory for 1785* (Charleston, S.C.: Historical Commission of Charleston, S.C., 1951).

According to Martyn Bowden, the port towns of colonial North America developed a distinctive mercantile landscape. The free market in land led to the sorting of land use, the segregation of functions, and the creation of a triangular pattern of land use (figure 5.5).[23] At the vertex of the "mercantile triangle" lay the waterfront, the point of contact between the resource hinterland and the trading circuits of the Atlantic. Along the waterfront were wharves and warehouses for loading and storing goods, and wholesale markets (vendues) for auctioning goods and slaves. Inland from the wharves lay the offices of those engaged in trade, such as merchants, factors, and commission agents, as well as financial and commercial intelligence services, such as coffee houses, exchanges, customs houses, merchant banks, and marine insurers. Behind these functions lay a zone of retailing, sufficiently close to the wharves to allow easy transfer and cheap freighting of imported manufactured goods but also near enough to prime residential neighborhoods at the top of the town to cater to the carriage trade.[24] Adjacent to the retailers were taverns and lodging houses, providing hospitality and accommodation for immigrants, mariners, merchants, and tradesmen. At the apex of the mercantile triangle lay the institutional core, comprising the town house, colony house, law courts, and principal churches. All these noncommercial functions were pushed to the back of the town by their inability to pay the high rents of the prime commercial districts: Church and state were content to leave the waterfront to commerce. Aside from zones of commercial and institutional land use, some residential specialization also developed, usually along class and ethnic lines. Mercantile elites generally were found close to the institutional apex of the town, the center of political and ideological power. Lower-class housing was scattered along either side of the mercantile triangle. In these marginal areas, ethnic ghettoes developed. Small manufacturing workshops were distributed throughout the town (with artisans living above the shop) in order to be close to consumers. However, noxious trades, such as butchering and tanning, and those with a risk of fire, such as soap making, tended to be in peripheral areas.[25] Larger industries, such as shipbuilding, ropewalks, distilleries, breweries, and foundries, also were found on the margins of town, attracted by extensive cheap land.[26] Although Bowden outlines the ideal mercantile landscape, the size and specialization of the triangle depended to a great extent on the growth and development of the colonial port town, which, in turn, reflected the structure of the larger colonial economy.

British Atlantic Towns

Because the economic structure of early modern British America was divided between the British Atlantic and the American continental colonies, the port towns in these two areas developed along strikingly different lines. In the British Atlantic, the sway of metropolitan merchants and a charted company over the fish, fur, and sugar trades appears to have significantly stunted urban growth in Newfoundland, Hudson Bay, and the West Indies.[27] Moreover, metropolitan control of the tobacco trade also limited urban growth in the Chesapeake. Under the navigation acts, furs, sugar, and tobacco had to be shipped direct to Britain before being reexported to other markets. As a result, trades in these commodities were simple bilateral exchanges between the colonies and Britain. British merchants organized the trades from home, sending out representatives to organize the export of staples from the colonies and the import of manufactured goods from Britain. In mid-eighteenth century Bristol, and perhaps also in London, Liverpool, and Glasgow, the great majority of shipping engaged in the sugar and tobacco trades shuttled back and forth across the Atlantic (figure 5.6).[28] This was also the pattern in the small fur trade between Hudson Bay and London. The fish trade from Newfoundland, however, was more complicated. West Country merchants had won the right to ship dried fish direct to markets in southern Europe, and this created a trilateral trade around the North Atlantic. According to Price, such multilateral trade encouraged British merchants to relocate to the colonies to oversee their transactions, but this was not the case in the Newfoundland fishery. The importance of the migratory fishery during the seventeenth and eighteenth centuries ensured that fish merchants remained based in the West Country, where they were better able to arrange the shipment of labor, supplies, and provisions to the fishery. West

FIGURE 5.6. *Bristol Harbour with the Cathedral and Quay*, by Nicholas Pocock, 1785. Possibly commissioned by a prominent West Indian merchant and shipowner, this painting shows Bristol in its commercial heyday. In the seventeenth and eighteenth centuries, Bristol was one of the leading outports in England, dealing mainly in the sugar, tobacco, and slave trades. Bristol Museums and Art Gallery.

Country fish merchants moved their operations to Newfoundland only when the resident fishery had grown sufficiently and the Napoleonic Wars had disrupted the migratory system.

Metropolitan control over these various staple trades led to urban growth and development in Britain, and urban underdevelopment in the colonies. In Britain, the "entrepreneurial headquarters effect" generated financial services in London and leading outports, as well as stimulating manufacturing industry in many parts of the country; in the colonies, the trading posts, fishing stations, and landing places remained "mere shipping points." At best, these were proto-urban places rather than port towns. The fur posts on Hudson Bay accommodated only Company personnel, furs, trade goods, and provisions.[29] The migratory fishery to Newfoundland and, later, to the Gaspé and Cape Breton Island, spawned numerous settlements. Yet leading outports, such as Trinity, had few functions apart from storing, processing, and shipping of fish, while the largest port, St. John's, had from 800 to 1,200 year-round residents between the 1750s and 1780s, and was considered a "mere fishing station."[30] Significantly, St. John's began to grow and develop as the principal port on the island only after West Country merchants relocated their businesses to the town during the 1790s. By the early nineteenth century, St. John's was the "emporium and depôt for the whole Island"; the town's merchants imported supplies from Britain, distributed them to fishermen in the outports, received dried fish in return, and then exported the fish to markets overseas.[31] The rise of St. John's led to a corresponding decline in the fortunes of Dartmouth, Teignmouth, and Poole.[32] In the Chesapeake, tobacco was shipped from landing places scattered along the tidewater creeks. These "inconsiderable villages," as one Englishman in late eighteenth century Maryland called them, contained little more than wharves, stores, and warehouses, a complex of buildings similar in function to those at the fur posts and the fishing stations.[33]

The one exception to this pattern of urban underdevelopment was in the West Indies. The two major port towns—Bridgetown, Barbados, and Kingston, Jamaica—were much larger and more elaborate urban places than the fur posts on Hudson Bay, the fishing stations in Newfoundland, or the landing places in the Chesapeake. During the seventeenth century, Bridgetown served as the capital of the richest colony in English America, the principal port in Barbados, and a mini-entrepôt in the Caribbean.[34] With an upwind location outside the Lesser Antillean Arc, Bridgetown was well placed to service the expanding sugar frontier, distributing slaves, supplies, and provisions to the other English islands and collecting sugar for export back to England. By 1680, the town had a population of 2,900, the third largest in English America. Yet in the eighteenth century, the growth of Bridgetown was checked by the rise of other port towns in the West Indies—particularly Port Royal and Kingston—each with their own shipping and trading connections to Britain and West Africa. Furthermore, metropolitan control over the long-distance trades in sugar and slaves discouraged any "entrepreneurial headquarters effect" in Bridgetown. Even so, the town's merchants, particularly those in the Sephardic Jewish community, carved out an economic niche for themselves through their control of regional and local trades. Although Sephardic Jews were on the margins of English society, they were well connected to other Jewish congregations in the Dutch West Indies and American port towns, particularly Newport, New York, Philadelphia, Charleston, and Savannah, and used these links to further regional and interregional trade.[35] Bridgetown also served as the central place or market town in Barbados. Merchants retailed goods and services to the planters, while slaves were allowed a "Negro market" to sell produce to the local population. By the 1790s, the population of Bridgetown had reached 10,000, comparable to that of Newport or Charleston.

After an earthquake destroyed Port Royal in 1692, the principal commercial town in Jamaica was Kingston. During the eighteenth century, Kingston commandeered most of the sugar trade of the island and grew rapidly; in 1775, the population was approximately 11,200, with some 5,000 whites, 5,000 blacks, mostly slaves, and 1,200 coloreds.[36] Yet, like Bridgetown, Kingston remained dominated by British firms and did not develop a sophisticated financial infrastructure. Kingston merchants—a mix of colonials and Sephardic Jews—controlled only regional and local trades. The regional trades included the *asiento*, the reexport of slaves and manufactured goods to the Spanish colonies in the Americas, as well as the shipping of rum, sugar, and molasses to the American colonies. By 1770, these two trades were earning substantial sums between £300,000 and £400,000 annually.[37] Kingston also served as the principal central place in Jamaica, providing goods, services, and a "Negro market." Much of the local retail trade was in the hands

FIGURE 5.7. *The Town of St. John's and Fort Townshend, Newfoundland*, by James S. Meres, 1786. Wharves, warehouses, and fish flakes along the waterfront and the Anglican church and fort at the top of the town delimit the developing mercantile triangle. National Archives of Canada, Ottawa/ C-002545.

of Jewish merchants.[38] The combination of local and regional trades, as well as the long-distance trades, ensured that Bridgetown and Kingston were larger than the proto-urban settlements in Newfoundland, Hudson Bay, and the Chesapeake. Nevertheless, these "mini-entrepôts" of the Caribbean, despite being gateways for the sugar and slave trades, the most valuable of all colonial commerce, never became "general entrepôts" of the Atlantic world.

The relatively meager economic growth and development of port towns in the British Atlantic was reflected in their landscapes. The fur posts on Hudson Bay and the tobacco shipping points in the Chesapeake were so small that the land use associated with the mercantile town scarcely existed; but in the leading port towns of Newfoundland and the West Indies the triangular pattern can be discerned. As a "mere fishing station," St. John's had a long waterfront of wharves, warehouses, and fish stages (nearly fifty in 1728), but an extremely shallow triangle comprising a discontinuous row of taverns and stores back from the wharves and an Anglican church and fort at the apex of the town (Figure 5.7).[39] The financial, communications, retailing, manufacturing, and elite residential functions were almost completely absent, situated back in the West Country. Although a much larger port, Bridgetown also had a shallow triangle. Again, the financial and communications functions were stunted, and the retail zone showed remarkable stability during the seventeenth and eighteenth centuries, a reflection of the relatively small demand from the plantocracy for goods and services.[40] Kingston was much the same. In the early eighteenth century, wholesaling activities were concentrated along Port Royal Street, running along the waterfront, while retailing was thinly spread on Harbour Street, one block back.[41] The town does not appear to have had a well-defined financial and communications area. A small manufacturing sector, comprising a mixture of artisans and craftsmen, occupied side streets around the mercantile core.[42] The institutional apex lay on the Square, and included the Anglican church.[43] Residential segregation was well defined according to ethnicity and class,

FIGURE 5.8. *Court House and Guard House in the Town of St. John's, Antigua,* by Thomas Hearne, 1775–1776. The pomp of imperial power: The governor of the Leeward Islands arrives in his coach before an honor guard drawn up in front of the Guard House (left) and the Courthouse (right). The town's Anglican church is in the background. V&A Picture Library.

with the important Jewish community concentrated in an enclave behind Harbour Street and between King and Orange streets, while the wealthy white mercantile class, having moved out of the center of town, lived in the suburban community of the Penns.[44] Although Bridgetown and Kingston had recognizable mercantile triangles, they were much less developed than those in the leading American port towns.

The lack of capital investment in the port towns of the British Atlantic was further reflected in their buildings. The fur posts on Hudson Bay were no more than walled factories enclosing a small nucleus of utilitarian buildings accommodating company servants, trade goods, and provisions. The fishing stations in Newfoundland were ramshackle collections of wharves, stages, flakes, bunkhouses, stores, and merchant quarters. Today, no building from the seventeenth or eighteenth centuries survives in Newfoundland. Only in the port towns of the West Indies were there more substantial buildings, housing government officials, merchants, planters, professionals, and tradesmen. Yet the transfer of wealth back to Britain by absentee planters and metropolitan merchants ensured that the grand town houses built by the slave and sugar trades were to be found in London, Bristol, Liverpool, and Glasgow, rather than in Bridgetown or Kingston. The most substantial structures in the West Indies were built not by commerce or the plantocracy but by the state. Government and religious buildings, such as houses of assembly, law courts, and Anglican churches, as well as military structures, were the most prominent buildings in Bridgetown, Kingston, Spanish Town, and St. John's (figure 5.8).

The military presence in the British Atlantic was particularly important. Because Hudson Bay, Newfoundland, and the West Indies were easily accessible by sea and bordered other European empires, the principal urban centers had defenses and military personnel. The fur posts on Hudson Bay were as much forts as trading factories, with enclosed walls, swivel guns, and canon. Although no regular army detachments were stationed on the Bay, Company personnel were trained to defend themselves from native or French attack. The leading fishing stations in Newfoundland, such as St. John's and Trinity, had fortifications. St. John's had gun batteries covering the entrance to the harbor, as well as forts on the high ground commanding the town's landward approaches, while Trinity also had a fort at the mouth of the harbor.[45] The sugar ports in the West Indies were defended well, not only against French and Spanish attack, but also against slave rebellion.[46] The harbor mouths of Bridgetown and Kingston were protected by forts, and both towns had barracks.[47] In Kingston, the barracks—housing 200 troops—were located at the top of the town beside the square, which was used as a parade ground.[48] There were also fortifications and barracks in Antigua and St. Kitts; the fortress on Brimstone Hill in St. Kitts was known as the "Gibralter of the West Indies." Given the importance of commanding the Caribbean Sea, the British stationed a squadron of warships in the West Indies during the eighteenth century, and built naval dockyards at Bridgetown, English Harbour in Antigua, and Port Royal in Jamaica.[49] The availability of local dockyards meant that naval vessels did not have to return to Britain for refits. Anglican churches in the leading port towns served not only as parish churches for the local inhabitants, but also as garrison churches for the military. Over time, they became guardians of laid-up regimental colors, and final resting places for many military personnel and colonial officials; in essence, shrines of colonial commemoration. Such ensembles of forts, barracks, parade grounds, and garrison churches comprised theaters of power. Church and military parades were demonstrations of military might that aimed to reassure the local population and, in the West Indies, overawe black slaves (see figure 5.8). Compared to the port towns along the American eastern seaboard, the fur posts on Hudson Bay and the urban centers in Newfoundland and the West Indies were significant parts of the militarized space of the British Atlantic world.

The societies of the various port towns reflected the staple economies and the metropolitan presence. Fur posts on Hudson Bay and fishing stations in Newfoundland were dominated completely by work forces associated with staple industries. In the fur trade, Company posts were staffed by factors and servants; in the fishery, St. John's and leading outports had small groups of government officials, merchants and agents, clergy and doctors, publicans and shopkeepers; and much larger groups of artisans, fishermen, and shoremen.[50] In the fur trade and the migratory fishery, labor forces were indentured for varying lengths of time, and master-servant relations were well defined. In the sugar ports of the West Indies, the urban economy was more diverse and the social range a bit greater. Although merchants and planters controlled the port towns (in Jamaica, merchants dominated Kingston and planters the interior capital of Spanish Town), there was a tiny class of professionals (lawyers, doctors, clergy) as well as small groups of publicans, shopkeepers, and artisans. The bulk of the laboring population comprised free blacks and slaves; in Kingston, more than half the total population comprised slaves in the late eighteenth century.[51] Such bonded labor was controlled through the slave codes enforced on the islands, as well as by the frequent declarations of martial law. As the Jamaican House of Assembly observed in 1760, "the public security required [martial law] and to that every other consideration gave place."[52] Although the methods of controlling labor were far more stringent in the West Indies than in Newfoundland or Hudson Bay, the bulk of the laboring populations in all three areas was under tight discipline. With no agricultural alternative, shipping controlled by merchants, and wilderness, often close at hand, urban labor forces lived and worked in extremely circumscribed worlds.

A significant imperial element overlay these various social groups and helped ensure public order. During the eighteenth century, the British army and navy were stationed in Newfoundland and the West Indies, not only contributing a military presence to the port towns but also commanding the government and the religious life of the colonies. Military governors administered Newfoundland and controlled many of the governorships in the West Indies; the Church of England was the established church in both Newfoundland and the West Indies. Such an imperial presence gave the port towns a different social cast than towns on the American eastern seaboard, and reinforced metropolitan authority, order, and hierarchy in the British Atlantic world.

Continental Towns

Port towns along the American eastern seaboard fared much better in terms of growth and development. The continental colonies included not only the largest cities in British America but also significant urban concentrations in coastal New England and the Mid-Atlantic colonies. By the mid-eighteenth century, the leading American port towns—Boston, New York, Philadelphia, Charleston, and Newport—had relatively complex economies, diverse societies, and sophisticated cultures. The towns were also centers of political leadership, and during the third quarter of the eighteenth century became points of friction between the British Atlantic and the American colonies.

For much of the seventeenth century, Boston dominated the urban hierarchy of the eastern seaboard. As the gateway town for the early English colonies, Boston was "an extensive and elaborate general entrepôt . . . a miniature Amsterdam or London," exporting dried fish, lumber, and agricultural produce to southern Europe and the West Indies, and importing dry goods from England and sugar, rum, and molasses from the Caribbean.[53] By 1680, when several censuses provide useful estimates of population, Boston was the largest town in English America with some 4,500 inhabitants, followed by New York with 3,200, Bridgetown 2,900, Port Royal 2,900, Newport 2,500, and Charleston 700.[54] Yet in the eighteenth century, the urban hierarchy changed dramatically. Along the eastern seaboard, New York and Philadelphia rose to urban dominance while numerous other towns developed. The greatest urban growth was in New England, where a string of port towns emerged (see figure 5.2). These included Portsmouth, Salem, Marblehead, Providence, New London, and New Haven, which grew wealthy on the export of dried fish, whale oil, lumber, and agricultural produce; and Hartford, Middletown, and Norwich, which flourished on the export of agricultural produce from the Connecticut and Thames river valleys. The growth of these various towns cut into the trade of Boston and left the town languishing by the 1740s. The second area of urban expansion was in the Mid-Atlantic. The rapid settlement of the rich farmland of the Jerseys and southeastern Pennsylvania and the development of the grain and flour trades led to the rise of New York and Philadelphia. By the mid-eighteenth century, these two port towns had leapfrogged past Boston and established their commercial supremacy over much of the eastern seaboard.

By the end of the colonial period, American port towns can be sorted on the basis of population and trading range into a distinct urban hierarchy.[55] The first group of towns, comprising Boston, New York, and Philadelphia, were "general entrepôts"; they had populations between 15,000 and 25,000, and traded up and down the eastern seaboard, as well as to the Caribbean, southern Europe, and Britain.[56] The second group, consisting of Newport, Baltimore, and Charleston, were "mini-entrepôts"; they had populations in the 6,000 to 12,000 range, and also traded to the Caribbean, southern Europe, and Britain. A final group, comprising Portsmouth, Salem, Marblehead, Providence, New London, Norwich, New Haven, and Norfolk, had populations in the 3,000 to 8,000 range, and had significant trade with the West Indies.

The principal port towns—Boston, New York, Philadelphia, Charleston, and Newport—rose to prominence through long-distance trades in nonenumerated goods, such as wheat, corn, flour, dried fish, salted meat, lumber, whale oil, and rice. During the eighteenth century, Newport was also heavily involved in the slave trade.[57] Such trades frequently involved multilateral movements and an exchange of a wide variety of goods; New England captains were known "'to trye all ports' and to risk all freights."[58] This pattern of trading encouraged local mercantile control. In 1770, merchants in New England and the Mid-Atlantic colonies owned between 62 and 87 percent of the shipping operating between their regions and Britain.[59] Local control produced "entrepreneurial headquarters effects" in Boston, Newport, New York, and Philadelphia, which generated a constellation of urban activities, such as shipbuilding, outfitting, finance, and insurance. In the Chesapeake and the Carolinas, however, metropolitan merchants owned between 62 and 75 percent of all shipping operating between the colonies and Britain, and urban growth was comparatively weak. Nevertheless, Chesapeake and Carolina merchants were beginning to garner significant trade by the mid-eighteenth century, and this had an effect on urban growth. Although few merchants in the Chesapeake participated in the tobacco trade to Britain, merchants in Baltimore and Norfolk controlled the growing trades in wheat, corn, and lumber to the West Indies, and this was reflected in the growth of these two port towns.[60] In South Carolina, the rice trade through Charleston also was dominated by British merchants and shipping, but, here too, local merchants were establishing some economic independence. Unlike the simple shipment and marketing of sugar and tobacco in Britain, the rice trade was relatively complicated, involving exports to markets in Britain, southern Europe, and the West Indies.[61] British merchants controlled the rice trade to Britain and the reexport business to northern Europe, while colonial merchants in Charleston organized the trades to southern Europe and the Caribbean. By the 1760s, Charleston merchants were also involved in the growing export of grain from the backcountry.[62] Like Bridgetown and Kingston, port towns of similar size, Charleston had sufficient ancillary trades to support a small, indigenous merchant class.[63] Although Charleston was not a "general entrepôt," neither was it a "mere shipping place."

Apart from several long-distance trades, colonial merchants controlled regional and local commerce. Much of the regional trade was by sea, and comprised the distribution of imported manufactured goods from Britain and plantation produce from the West Indies, as well as the collection of staples for export overseas.[64] The general entrepôts—Boston, New York, and Philadelphia—were particularly involved in such distribution and collection, but most port towns played similar roles. Coastwise trade usually was handled by large wholesale merchants—"traders by sea"—who engaged in trans-Atlantic commerce and had access to shipping.[65] Colonial ports also served as central places, retailing goods and services to farm populations and providing markets for foodstuffs. At first, these trades were handled by general, all-purpose wholesale merchants, but as trade expanded in volume and complexity, merchants increasingly specialized in either importing dry goods or exporting provisions.[66] Dry goods merchants imported manufactured items from Britain, and then either sold them through a retail store in the port town or forwarded them to retailers inland. Provisions merchants handled the export of backcountry produce. In the early development of this trade, provisions merchants dealt directly with producers—planters, farmers, fishermen, and lumberers in the hinterland—but as trade became more complicated, specialist dealers began collecting backcountry produce for forwarding to provisions merchants. In Philadelphia, the separation of functions took place after the 1750s, with the rise of specialist flour and timber merchants, well connected to hinterland areas of supply.[67] A similar process marked the development of retailing. Throughout the colonial period, wholesaling and retailing were intertwined, with wholesale merchants retailing goods from their wharves. Nevertheless, retailing gradually emerged as a separate, specialized function, particularly in the larger towns. Retailers began appearing in Boston in the 1640s, New Amsterdam in the 1650s, Newport in the 1660s, and Philadelphia in the 1690s.[68] By the early eighteenth century, retailers were selling all manner of imported and locally produced goods.

Although enormous quantities of manufactured goods were imported from

Britain, the commercial life of the port towns fostered local manufacturing, particularly in the transportation, processing, and consumer goods industries.[69] In areas where water transportation was especially important, such as New England and the Chesapeake, considerable investment was made in shipbuilding and marine trades. Between 1769 and 1771, New England built 68 percent of the entire colonial tonnage, while the Chesapeake, the second-most important shipbuilding area, built 13 percent.[70] Shipbuilding gave employment to shipwrights, blockmakers, sailmakers, ropemakers, and ship chandlers. Significantly, colonial shipbuilders, unwilling to pay for expensive British hardware for outfitting, used as little iron as possible in their vessels, making do with wooden pins or "trunnels" to hold the timbers together.[71] As a result, little leakage from colonial shipbuilding flowed back to British manufacturing; economic benefits were retained in the colonies. In contrast, processing industries were much less developed. The drying of cod in the fishery required only extensive wooden flakes, tobacco was cured on plantations, and rice was exported unprocessed. Only the grain trade through Pennsylvania and Maryland generated ancillary flour milling; most flour mills were concentrated in the main shipping ports of Philadelphia and Baltimore. However, the import of sugar and molasses from the West Indies led to the growth of sugar refineries and rum distilleries.[72] By 1770, twenty-six refineries, concentrated in Boston, Newport, New York, Philadelphia, and Charleston, were supplying about three-quarters of colonial demand. At the same time, some 140 distilleries were located in most urban centers, with the greatest concentrations again in Boston, Newport, New York, and Philadelphia.

Income earned through commerce frequently was used to purchase consumer goods, although demand was distributed unevenly through the American colonies.[73] In the plantation areas of the Chesapeake and South Carolina low country, the concentration of wealth in the hands of a small planter class encouraged the import of high-value luxury goods from Britain, and discouraged local manufacturing of consumer goods.[74] Moreover, strong trade ties between these two areas and Britain ensured a plentiful supply of competitively priced dry goods, further checking colonial manufacturing. In the predominantly family farming areas in New England and the Mid-Atlantic colonies, however, wealth was more evenly distributed among farmers, artisans, and small merchants, and this encouraged the growth of local manufacturing of consumer items. The greater urban population in the New England and Mid-Atlantic colonies also provided a substantial market for locally produced goods. Furthermore, the influx of artisans from continental Europe and the British Isles into the leading port towns, particularly New York and Philadelphia, created a reservoir of skilled labor.[75]

Colonial artisans plied a wide range of trades, including woodworking, leather working, metal working, milling and baking, and construction.[76] Of particular importance in the port towns were the many and varied craftsmen involved in the building and furnishing trades. The considerable growth of population in New York, Philadelphia, the New England outports, and the southern port towns during the middle and late eighteenth century generated demand for new housing, which gave employment to local builders, bricklayers, masons, carpenters, joiners, plumbers, painters, and glaziers. In Annapolis, Maryland, at least fourteen large townhouses and four combined houses/businesses were built between 1764 and 1774, at a total cost of £57,374 currency, a major fillip to the local economy.[77] Towns also erected or rebuilt public buildings, such as colony houses, town houses, law courts, and churches, and these too increased the demand for local craftsmen. Houses and public buildings had to be furnished, and it was frequently much cheaper to purchase locally made furniture than to import it from Britain. Cabinet making became particularly important in Boston, Newport, and Philadelphia.[78] Apart from the building and furnishing trades, demand for consumer goods such as clothing, shoes, wigs, jewelry, silver, and books, supported a wide variety of craftsmen.[79]

The third economic function that developed in the leading American port towns was finance. The concentration of locally controlled trade led to a demand for financial services, particularly merchant exchanges for raising capital, trading bills of exchange, and insuring goods and shipping.[80] Although many colonial merchants gained credit and insurance in Britain, there was a growing demand for these services in the major port towns. By providing basic financial services, leading towns staked their claim to be colonial metropolises, able to exercise significant financial control over their respective hinterlands.[81] As the first English port town in the continental colonies, Boston had an early start in developing what Bowden has called a "financial-exchange-communication" function.[82] During the late seventeenth century, the town became the general entrepôt of New England and an important economic influence over the Chesapeake.[83] As trade developed, local merchants created a mercantile exchange or walk on the first floor of the town house in 1657 (a combination of commercial and government functions common in English town halls), and opened the first coffee shop in 1676. Following the custom in London, coffee shops served as a forum for transacting business, selling goods at auction (public vendues), and acquiring commercial intelligence. By 1690, Boston had two coffee shops, one of which, the London Coffee Shop, was run by the publisher of the first colonial newspaper.[84] As their economies expanded during the late seventeenth and early eighteenth centuries, New York, Philadelphia, and Charleston provided similar financial services. The first coffee houses opened in New York in 1697, in Philadelphia in 1702, and in Charleston about 1724.[85] Situated at the corner of Front and Market streets, overlooking the waterfront and adjacent to the twice-weekly market, the London Coffee House in Philadelphia served as the principal meeting place for the city's merchants for much of the eighteenth century.[86] Eventually, specialist exchanges were built, notably the exchange at the foot of Broad Street in New York in 1752, and the combined customs house and exchange at the intersection of East Bay and Broad Street in Charleston in 1771.[87] A colonial insurance market also emerged, with specialist insurers setting up business in Boston in 1724, in Charleston in 1739, and in New

FIGURE 5.9. *View of Charles Town*, by Thomas Leitch, 1774. The newly built Exchange and Customs House dominates the waterfront. Collection of the Museum of Early Southern Decorative Arts, Winston-Salem.

York in 1740. At about the same time, private merchant banks began appearing in Boston, New York, and Philadelphia.[88] The development of the financial sector marked the increasing commercial maturity and growing economic independence of the leading American port towns from London and the British outports.

The growth and development of American port towns was reflected in the size and complexity of their mercantile landscapes. The general entrepôts of Boston, New York, and Philadelphia, as well as the mini-entrepôts of Newport, Baltimore, and Charleston, all saw considerable landward expansion and functional specialization during the colonial period. Of these various towns, Charleston is probably the best studied.[89] Located on a peninsula at the confluence of the Ashley and Cooper rivers, Charleston was the leading port in the southern colonies, having engrossed the trans-Atlantic trades in rice and indigo, much of the coastal trade from Cape Fear to the Savannah River, and a substantial part of the backcountry trade of the Carolinas. For people arriving by sea, the town made "a very handsome appearance, for it spreads a great deal of ground and there are in it several large capital good looking buildings . . . which being lofty present themselves to your view above the houses many miles off" (figure 5.9).[90] The Cooper River was the center of shipping; it "appear[ed] sometimes a kind of floating market . . . numbers of canoes boats and pettyaguas . . . ply incessantly, bringing down the country produce to town, and returning with such necessary as are wanted by the planters."[91] Ocean-going shipping tied up along nearly a mile of wharves projecting from Bay Street into the river. In the late eighteenth century, these wharves formed the base of a well-developed mercantile triangle (see figure 5.5).[92] Along the wharves were merchant counting houses and warehouses for storing incoming manufactured goods and outgoing staples such as rice, indigo, grain, and naval stores. In the 1760s, almost half of all Charleston merchants had premises on the wharves or "on the bay" (meaning the waterfront street of East Bay).[93] This street was "the principal part of the town for carrying on most kinds of business," containing offices of shippers, brokers, and commission agents, as well as financial and commercial intelligence services, represented by the magnificent Customs House and Exchange.[94] Acting as the central axis of the triangle, Broad Street served as the principal retail street, with a further concentration of merchant outlets. At the intersection of Broad and Meeting Streets lay the apex of the triangle, and perhaps the most well-defined institutional cluster in any British American town. Of the buildings on the four street corners, three comprised public institutions: St. Michael's Anglican Church, the Town Watch House and Public Treasury, and the State House, which also included the principal law courts in the colony. The whole complex formed a Civic Square. In 1767, the town erected a statue of William Pitt at the center of the square in honor of his role in the French and Indian War and in the repeal of the Stamp Act.[95] South of the square lay White Point and the town's elite residential neighborhood, best symbolized by the great Georgian mansion of leading slaver and merchant Miles Brewton.[96] Many of the town's artisans occupied premises scattered along King and Meeting streets, cheaper areas peripheral to the mercantile core. Noxious trades, such as soap making, distilleries, and tallow chandleries, were found on the outskirts of the town.[97]

American port towns also were distinguished by their buildings. Compared to the relatively meager townscapes of the British Atlantic, where fortresses were frequently the most prominent features, American seaboard towns had a wealth of fine housing and public building, much of it modeled on metropolitan prototypes. The English urban renaissance had begun in the late seventeenth century, spurred, in part, by the rebuilding of London after the Great Fire, and spread in the early eighteenth century to many provincial towns, including ports engaged in colonial trade such as Poole, Bristol, Liverpool, and Glasgow.[98] In these and other towns, buildings increasingly were constructed from brick or stone; displayed orderly, harmonious facades; employed sash windows; and showed off classical features and ornament, particularly over front doors, around fireplaces, and on plaster ceilings. By the early eighteenth century, these features began to appear in American port towns.[99] From Portsmouth to Charleston, local merchants, frequently employing immigrant English builders and craftsmen, created strikingly convincing versions of provincial Georgian building. Whether built out of wood, brick, or stone, colonial Georgian houses tended to have symmetrical facades, two stories, single- or double-pile plans, and a classical pediment over the central door.[100] Many of these houses were detached, reflecting the greater amount of land available for building in colonial port towns. Through their houses and furnishings, American merchants were establishing their right to be considered equals of the English commercial class.

Although house styles were derivative of Georgian building in English provincial towns, there were some colonial characteristics, particularly in Charleston and Annapolis. In the early eighteenth century, planters in Charleston developed the "single house," a house type that sat detached on its own lot and comprised a narrow street frontage, one room wide and three rooms deep, two or more stories, and a porch or piazza along one of the long sides (figure 5.10). The piazza shaded the house and opened up the main living rooms to the air. A carriageway led to the back of the house where a courtyard was surrounded by a stable/carriage house,

FIGURE 5.10. Colonel Robert Brewton House, Church Street, Charleston, 1721–1741. One of the oldest surviving single houses in the city. Photograph by the author, 1994.

kitchen, laundry, office, and slave quarter. In essence, low-country planters had created an urban equivalent of the plantation great house and its dependencies.[101] In plan, elevation, and detail, the single house owed much to the terrace houses being built in the squares of the elite residential district of the West End of London, but the single house also reflected the social structure and climate of the low country. In Annapolis, most houses were simple, "Ill Built" structures without "any Degree of Elegance or Taste" until the 1760s, when a post-war boom in the tobacco trade led several planters to build residences in town.[102] A few of these houses had five-part plans (main house, two dependencies, and connecting wings), which replicated the arrangement of great house and dependencies found on tidewater plantations. This plan, in turn, was derived from the grand Palladian mansions of England.

Public buildings, too, owed much to English prototypes. The early town houses were modeled on the town halls of England, which combined commerce with town government.[103] The first town house in Boston, erected in 1657, had two stories: the first floor was an open arcade, serving as a merchant exchange or walk, which quickly became the commercial hub of the town, while the second floor comprised a town meeting room, court room, and library.[104] During the eighteenth century, colonial public buildings consciously drew on metropolitan architecture. Several of the most prominent buildings in Charleston were based on Palladian models, notably the State House and the Customs House and Exchange, perhaps influenced by the Exchange in Bristol. The principal churches in the leading port towns were based on the work of Christopher Wren, particularly his City of London churches, and of James Gibbs, especially his St. Martin-in-the-Fields, the royal parish church in London.[105] Christ Church and the Old South Meetinghouse in Boston and Trinity Church, Newport, were modeled on Wren churches, while Christ Church, Philadelphia; St. Michael's, Charleston; St. Paul's Chapel, New York; and the First Baptist Meeting House in Providence were inspired by St. Martin-in-the-Fields.

Compared to towns in the British Atlantic, port towns along the American eastern seaboard had relatively few military fortifications. To be sure, the principal ports were defended: Boston had Castle Island, which had gun batteries covering the approaches to the harbor; New York had Fort George at the tip of Manhattan; and Charleston had a battery at the entrance to the harbor, as well as walls and bastions in the early eighteenth century to protect the town against French, Spanish, and Indian attacks. Even so, a visitor to the town in 1774 referred to these as "3 apologies for fortifications."[106] Such defenses scarcely compared to those in the West Indies. Moreover, no standing army was maintained in the American colonies until the mid-1750s, which meant that there was no need for barracks, parade grounds, garrison churches, and all the miscellaneous buildings associated with a cantonment. Apart from the occasional visiting warship, the military authority of the imperial power was almost completely absent. As an eighteenth-century governor of Virginia remarked about his predecessor Francis Nicholas, he was "a COMMANDER IN CHIEF without a Single Centinel to defend [him] in this Dominion."[107]

The social composition of American port towns was also much different than that in the towns of the British Atlantic. The economic diversity and accumulation of wealth in seaboard towns, particularly in the general entrepôts of Boston, New York, and Philadelphia, supported a large number of people in service and manufacturing occupations, creating more differentiated and complex urban societies than could be found in Newfoundland or the West Indies.[108] Moreover, a good part of this labor was free. To be sure, American port towns had indentured servants as well as slaves, especially in Charleston, but in smaller proportions than could be found in British Atlantic towns. Labor also had a greater range of employment opportunities; many immigrant workers no doubt moved from temporary jobs in the port towns to take up land on the frontier. Compared to St. John's or Kingston, the options for labor were much greater in New York or Philadelphia.

Colonial merchants also played a more central role in American port towns. Although tobacco planters controlled the economic and political life of the tidewater capitals of Annapolis and Williamsburg and rice planters were heavily involved in the economic and political life of Charleston, merchants dominated the leading towns of Philadelphia, New York, and Boston, as well as the many smaller port towns in the Mid-Atlantic and New England colonies. Such merchants formed the political and economic elite in the northern colonies, and their commercial interests, following patterns of trade, increasingly lay outside the British Atlantic. The imperial presence, too, was much weaker. Government officials were usually drawn from local mercantile elites rather than from the British military establishment, and no permanent British naval or army presence existed until the outbreak of the French and Indian War in the mid-1750s. The Anglican Church was also weak. Although the Church of England was established in several colonies, the leading port towns, with their ethnically diverse populations, were characterized by a proliferation of religious denominations, including Methodists, Quakers, Congregationalists, Presbyterians, Roman Catholics, Dutch Reformed, Lutherans, and Sephardic Jews. The ordered, hierarchical world of the port towns in the British Atlantic simply could not be replicated in the socially and ethnically diverse towns along the American eastern seaboard.

Almost from the beginning of English settlement in the New World, the patterns of urban growth and development in the British Atlantic diverged from those along the eastern edge of the continent. In the British Atlantic, port towns were dependent economically on Britain. At best, they were mini-entrepôts; more commonly, they were little more than shipping places. The towns lacked well-developed manufacturing and service functions, leaving an urban economic base structured around the export of staples and the import of labor, provisions, and manufactured goods. Such a simple economy was reflected in the urban landscape. Few towns had much in the way of a mercantile triangle or fine residential and civic buildings. Societies were highly stratified. Mercantile and planter elites dominated the towns, and limited the freedom of labor through indenture and slavery. To maintain internal control and defend against external attack, the port towns had a strong military and imperial presence. Overall, the towns of the British Atlantic were significant spaces of imperial authority.

In contrast, the leading American port towns were economically more developed. Indeed, Boston, New York, and Philadelphia were sufficiently independent of Britain by the late eighteenth century that they had become general entrepôts of the Atlantic world. These leading towns depended not only on the import-export trades but also on relatively complex manufacturing and financial sectors. Such sophisticated and mature urban economies produced well-developed mercantile triangles and richly textured urban fabrics. The mansions, churches, and town houses of Boston, New York, and Philadelphia were marks of capital accumulation and cultural confidence. Moreover, wealthy merchants in these towns played leading roles in civic and provincial affairs, while labor forces enjoyed considerably more freedom than those in towns of the British Atlantic. Given their relative economic, political, and cultural independence, the port towns along the American eastern seaboard became the principal centers of conflict as Britain tightened its grip on the continental colonies during the 1760s and early 1770s.

Chapter 6

The Fracturing of British America

THE FIRST CRACKS in the English North American empire began appearing between the 1620s and the 1640s with the retreat of metropolitan capital from direct control over colonization in the Chesapeake and the fishery in the Gulf of Maine, but these fissures would not open fully until the early 1760s and the unrest over the Stamp Act. During the intervening 120 years, the geographic patterns that would shape, dominate, and, ultimately, break up colonial British North America became entrenched. In the Atlantic arena, metropolitan merchants and planters established a maritime space containing nodes of staple production and trade; along the eastern edge of the continent, colonial merchants and planters created territories of staple production and port towns; and in the continental interior, European settlers carved out an extensive agricultural frontier based on family farming. During the seventeenth and early eighteenth centuries, these spaces expanded enormously, pushing against the imperial space of the French in North America. After French power on the continent collapsed, the enlarged space of the British Atlantic and the spaces along the American eastern seaboard increasingly grated against one another, creating the friction that would eventually lead to revolution and the remaking of eastern North America.

Early Eighteenth-Century Expansion of British America

In the late seventeenth and early eighteenth centuries, Britain emerged as a major imperial power. During the two wars fought against the French between 1688 and 1713, the British created a formidable "fiscal-military state": the Bank of England was established and deficit financing introduced, the Board of Trade was created to administer commerce and the colonies, the parliament of Scotland was subsumed within that at Westminster, and a sizeable public administration overseeing state finances and the military was put in place.[1] The British also enjoyed considerable military success. In the Mediterranean, they captured Minorca and Gibralter from the Spanish; in the Atlantic, they took Port Royal, the capital of Acadia, from the

FIGURE 6.1. *A Prospect of Annapolis Royal in Nova Scotia*, by J. H. Bastide, 1751. By permission of the British Library, Shelfmark Maps K.Top 11a f83.

French. At the Treaty of Utrecht (1713), Britain gained title to the two Mediterranean territories, as well as Hudson Bay, peninsular Nova Scotia, and the French part of Newfoundland. With the loss of their main fishing bases, the French moved across Cabot Strait and established a new fishery on Île Royale (Cape Breton Island). In 1717, the French began construction of the fortified port town of Louisbourg on the island's Atlantic coast to serve as a base for the new fishery, as an outer bastion of New France, and as an entrepôt of French trade in the North Atlantic.[2] Meanwhile, the British began incorporating mainland Nova Scotia into their Atlantic realm.

At the time of the British conquest, the French population of Nova Scotia comprised about 1,400 people settled in agricultural communities dispersed around the Bay of Fundy.[3] The principal settlements were along the Annapolis Valley and around the Minas and Chignecto basins. Unlike the American settlements farther south, which had expansive agricultural frontiers, the Acadian settlements were little more than enclaves backing into a rocky interior, with little or no room for expansion, and facing outward to the Bay of Fundy and the Gulf of Maine beyond. Transportation by sea connected the settlements together and with the leading trading center of Boston. Much of the rest of the colony lay in the hands of Mi'kmaq Indians, traditional allies of the French. The British established their capital at Port Royal (renaming it Annapolis Royal after Queen Anne), improved the French fortifications, and employed naval vessels and a small army detachment to maintain control over the region (figure 6.1). As in Newfoundland, a military officer was appointed governor.[4] Although the divide between civil and military affairs was respected, the governor ruled without a legislative assembly, an arrangement that lasted until 1758.

As tension increased between Britain and France in the mid-eighteenth century, Nova Scotia became an imperial battleground. In the first year of the War of Austrian Succession (1744–1748), the French captured the fishing station at Canso (see chapter 3), instigated a privateering war against New England vessels, and

FIGURE 6.2. *A Plan of Chebucto Harbour with the Town of Hallefax*, by Moses Harris, 1749. A fortified grid town laid out amid the forest. By permission of the British Library, Shelfmark Maps K.Top 119 f.73.

besieged Annapolis Royal. In response, New England forces, principally from Massachusetts, relieved Annapolis. The following year, a combined force of New England militia and a British naval squadron laid siege to Louisbourg; after a six-week bombardment, the fortress city capitulated. The capture of Louisbourg in 1745 was a stunning demonstration of combined operations, a portent of what would be achieved in the following decade. Nevertheless, the widespread satisfaction in New England at the elimination of the French threat was short lived. At the treaty of Aix-la-Chapelle, Britain handed Louisbourg back to France in return for the English East India Company's trading factory at Madras, which the French had captured during the war. The larger imperial considerations of Britain had overridden the colonial interests of Massachusetts.

In response to the continued threat posed by Louisbourg, the British moved the capital of Nova Scotia from the Fundy backwater of Annapolis Royal to the great Atlantic harbor of Chebucto, where they laid out the new town of Halifax in 1749 (figure 6.2). Even more than St. John's, Bridgetown, or Kingston, Halifax was designed as a fortified garrison town: the governor's residence, parade ground, Anglican church, and citadel dominated the town's grid plan and symbolized British political, military, and religious authority in the colony (figure 6.3).[5] By 1760, a naval dockyard had also been established.[6] For much of the second half of the eighteenth century, imperial expenditure on the colonial government, the army, and the navy provided the main source of income in Halifax as well as in much of the rest of Nova Scotia.[7]

The British also strengthened their position elsewhere in the colony (figure 6.4). Concerned at the rapid increase of the French Acadian population along the Fundy shore, the imperial government sponsored the migration of some 2,500 "Foreign Protestants" (a fragment of the much larger German migration to British

FIGURE 6.3. *Governor's House and Mather's Meeting House, Halifax*, after Dominic Serres, circa 1762. The institutional apex of governor's residence, Congregational meeting house, and Anglican church in the nascent imperial capital of Nova Scotia. National Archives of Canada, Ottawa/C-002482.

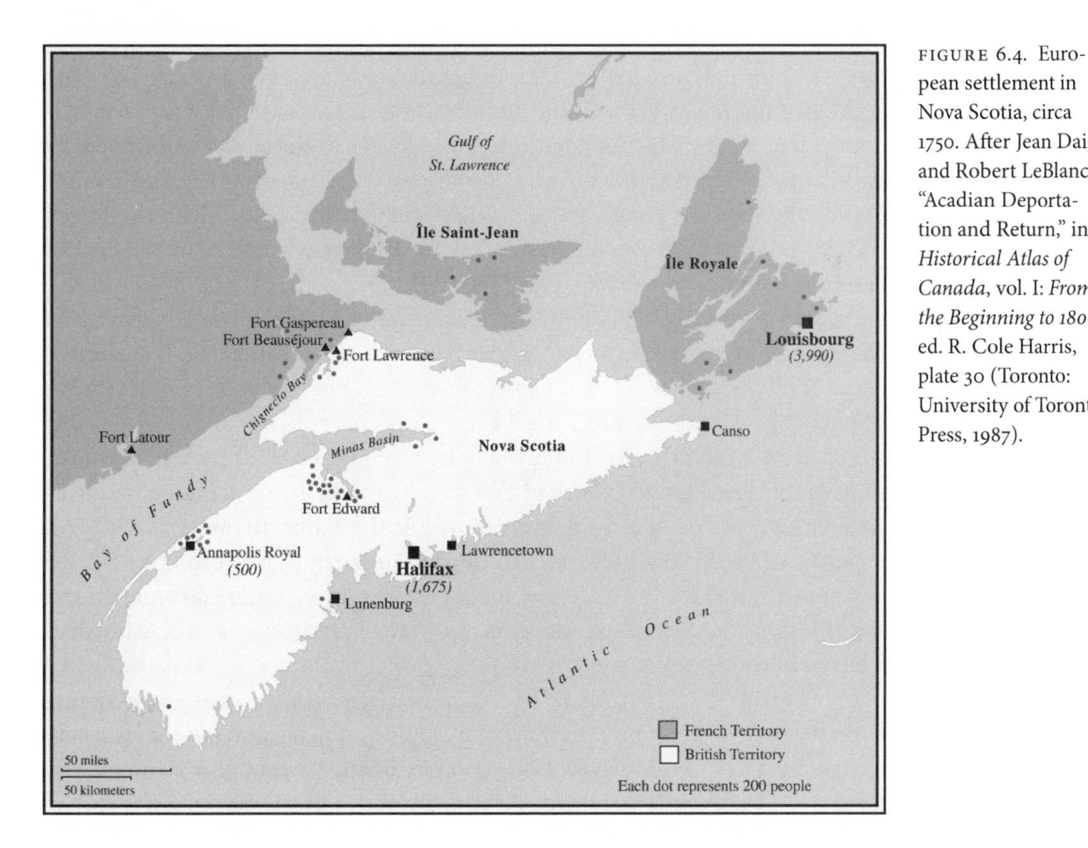

FIGURE 6.4. European settlement in Nova Scotia, circa 1750. After Jean Daigle and Robert LeBlanc, "Acadian Deportation and Return," in *Historical Atlas of Canada*, vol. I: *From the Beginning to 1800*, ed. R. Cole Harris, plate 30 (Toronto: University of Toronto Press, 1987).

America) to Nova Scotia between 1750 and 1752. A group of them was settled at Lunenburg, a new town planted to the west of Halifax, the following year.[8] Situated on the only good patch of agricultural land on the Atlantic shore, Lunenburg quickly became a significant farming settlement and a supplier of produce to Halifax. A less successful plantation was also made to the east of Halifax at Lawrencetown.[9] In addition, forts were placed on the northern periphery of the colony. In 1750, Fort Edward was constructed at the junction of the Avon and St. Croix rivers to control the overland route from the Bay of Fundy to Halifax, while Fort Lawrence was built on the Isthmus of Chignecto to delimit Nova Scotia's northern boundary. The following year, the French countered the British threat by constructing Fort Beauséjour about a mile north of Fort Lawrence. Despite the establishment of forts and agricultural settlements, much of the interior of Nova Scotia was controlled by the Mi'kmaq, who fought an undeclared guerilla war against the British during the early 1750s. Captain John Knox, who was garrisoned at Annapolis Royal for part of the French and Indian War, summarized the situation in 1757: "though we are said to be in possession of Nova Scotia, yet it is in reality of a few fortresses only, the French and Indians disputing the country with us on every occasion, inch by inch."[10] Although the British had managed to establish several nodes of control—Annapolis Royal, Halifax, and Lunenburg—around the periphery of Nova Scotia, which were connected by marine transportation, the British lacked the military power to extend their authority over much of the interior of the province.[11]

Meanwhile, much farther south and west, in the trans-Appalachian space of the upper Ohio Valley, increasing tension developed between the expanding Atlantic system of the French and the growing territorial empire of the American colonists. During the late seventeenth and early eighteenth centuries, the French had flung their waterborne fur trade well west of the Appalachians (figure 6.5).[12] In the 1680s and 1690s, French fur traders, using the waterways of the St. Lawrence, Ottawa, and upper Great Lakes, had pushed a trade route south from Lake Michigan into the headwaters of the Mississippi; in the 1700s and 1710s, a second route was established through the lower Great Lakes, the Wabash, the lower Ohio, and into the Mississippi. At the same time, small French agricultural settlements were set up along the Mississippi in the "Illinois Country," a midway point between French fur posts on the lower Great Lakes and French settlements in Louisiana.[13] The expansion of the fur trade into the heart of North America not only benefitted French merchants but also furthered French imperial designs.[14] From 1700, the French government used the fur trade, the Indian alliances that sustained the trade, and the garrisoned fur posts strategically placed across the interior of the continent as a means of containing the English in their seaboard colonies. Even so, the French never established a forward defensive line in the upper Ohio Valley.

In the early 1720s, American fur traders began venturing into the region, establishing close ties with the local native peoples, the Shawnees and Delawares.[15] By the early 1740s, Virginian land speculators were coveting the area, hoping to acquire

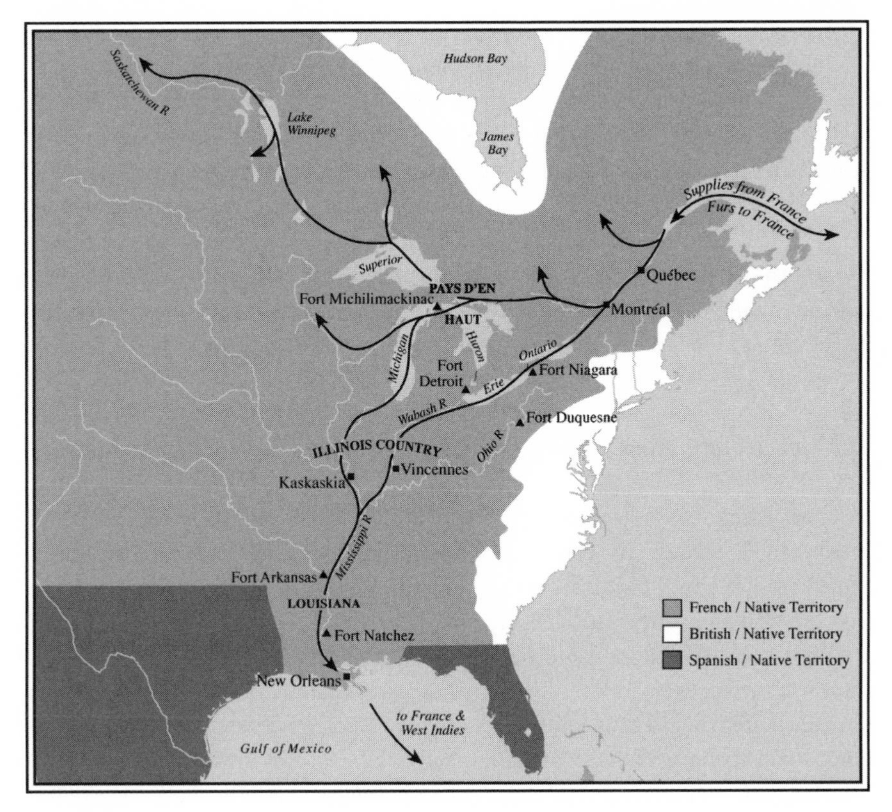

FIGURE 6.5. French fur trade, early 1750s. After Conrad E. Heidenreich and Françoise Noël, "France Secures the Interior, 1740–1755," in *Historical Atlas of Canada*, vol. I: *From the Beginning to 1800*, ed. R. Cole Harris, plate 40 (Toronto: University of Toronto Press, 1987).

the territory from the natives and then sell it to incoming settlers.[16] In 1744, representatives of the Iroquois League and the colonial governments of Pennsylvania, Maryland, and Virginia met in Lancaster, Pennsylvania, to draw up an agreement over the territory. While the Iroquois thought they were giving title to the Shenandoah Valley in return for cash and British recognition of Iroquois overlordship of certain southern tribes, they were, in fact, agreeing to the territorial claims of Virginia and Maryland to the Ohio Country. With the treaty signed, the Virginia House of Assembly granted nearly a third of a million acres in the upper Ohio to a group of tidewater planters and speculators, mostly from the Northern Neck of Virginia, who constituted themselves as the Ohio Company in 1747.[17] Just as the expansion of the French fur trade into the Illinois Country fitted into France's larger imperial ambitions, so the designs of the Ohio Company suited Britain's imperial concerns. Pushing the American frontier beyond the Appalachians was officially welcomed in London; settlement of the Ohio country would provide a buffer between the French in the Illinois Country and the seaboard colonies, as well as disrupt French communication between the Mississippi and the Great Lakes. For the British government, the Ohio Country was strategically important in the global chess match with the French; for American speculators, the lands along the river were a potential economic bonanza.[18]

Recognizing that the Ohio was a shorter route between their settlements along the lower St. Lawrence and those in the Illinois Country and that British encroachment into the valley threatened their continental position, the French moved to control the river. Between 1749 and 1753, the French and their Indian allies attacked American traders and drove them out of the Ohio Valley; in 1753, a French detachment was sent from Quebec to the upper Ohio to claim the area for France and to build a string of forts from Lake Erie to the forks of the Ohio (the confluence of the Monongahela and Allegheny rivers).[19] Four forts were built, with Fort Duquesne controlling the strategically important forks. Such aggressive action, as well as rising tension along the Nova Scotian frontier, soon led to war.

The French and Indian War, 1754 to 1763

Whereas earlier conflicts between Britain and France in North America had been fought largely through the use of naval power (the taking of Port Royal in 1710 and Louisbourg in 1745), the British were now faced not only with commanding the seas but also with launching a continental campaign against the French and their Indian allies.[20] For the first time, the British army was needed on land to defend the western frontier, as well as to launch attacks against the French. The British Atlantic system, which had worked so well in controlling Hudson Bay, Newfoundland, Nova Scotia, and the West Indies, had to be deployed far into the interior of North America, a massive challenge for a preindustrial nation-state.

The problem of projecting military power over land soon became apparent. In reaction to the French seizure of the upper Ohio, the British decided on landward attacks on French North America, launching assaults across the Allegheny Mountains toward Fort Duquesne and up the Champlain corridor toward Montreal (figure 6.6). The attack on Fort Duquesne led by British Commander-in-Chief General Braddock in July 1755 ended in spectacular failure. With insecure supply lines and a failure to appreciate the distinctive nature of wilderness warfare, the inexperienced Braddock and his army were cut to pieces by the French and their Indian allies. Two months later, an assault up the Champlain corridor toward the French Fort Saint-Frédéric (Crown Point) ended in muddled victory and no further gain of territory. Over the course of the war, immense resources had to be devoted to these two campaigns before they met with success. In order to capture the forks of the Ohio, the British had to build a transportation infrastructure across the mountainous wastes of western Pennsylvania, as well as detach the Ohio Indians from the French. The "protected advance" on Fort Duquesne comprised the Forbes road hacked through the forest and a series of forts placed every forty miles to protect the lines of communication.[21] The allegiance of the Delaware and other Ohio Indians was secured through the Treaty of Easton in October 1758, which promised to protect their lands west of the Alleghenies from encroachment. Facing a massive military advance and without Indian allies, the French aban-

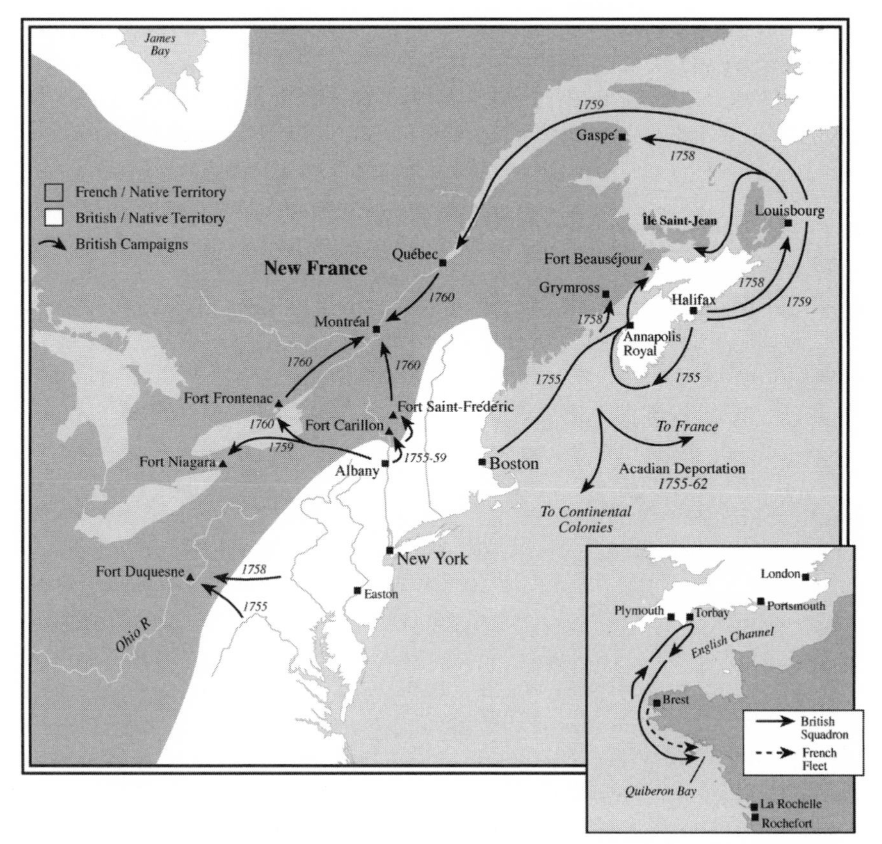

FIGURE 6.6. British campaigns in the French and Indian War, 1754–1763. Inset shows the Battle of Quiberon Bay, 1759. After W. J. Eccles and Susan L. Laskin, "The Seven Years' War," in *Historical Atlas of Canada*, vol. I: *From the Beginning to 1800*, ed. R. Cole Harris, plate 42 (Toronto: University of Toronto Press, 1987).

doned Fort Duquesne in November and pulled out of the Ohio Country.[22] The British consolidated their hold over the forks by building Fort Pitt, an enormous pentagonal fortress commanding the banks of the river.[23] Progress up the Champlain corridor was even slower. British attacks against Fort Carillon (Ticonderoga), which commanded the southern entrance to Lake Champlain, were repulsed with heavy losses in 1758, and it was not until the following summer that the British took the fort and then Fort Saint-Frédéric. Even then, the British hesitated to advance northward toward Montreal, and did not resume their campaign until 1760 when two other British armies were converging on the city from the east and the west. By then, the collapse of New France was a foregone conclusion and the advance up the Champlain corridor scarcely mattered.

More critical to eventual British success was the deployment of maritime power (figure 6.6).[24] Unlike the army's campaigns, which were confined to the territorial spaces of North America, the navy's operations extended over the vast marine surfaces of the Atlantic, Indian, and Pacific oceans. The deployment of a fleet in one theater frequently had strategic implications for naval operations in another. By the mid-eighteenth century, British naval strategists, elaborating on a strategy first used by Drake almost two hundred years earlier, had developed the concept of a Western Squadron stationed in the Western Approaches to the British Isles.[25] With the prevailing westerly wind at its back, the squadron covered the principal French Atlantic port of Brest, located on the western tip of Brittany; protected inbound convoys from North America and the Indies; and was well placed to run up the English Channel to thwart invasion. Moreover, the squadron's vessels could be victualed and refitted by sailing downwind to Plymouth or Torbay. If the French fleet ventured into the Atlantic, the Western Squadron could bring it to battle.

With this maritime watch in place and the French fleet largely restricted to European waters, British naval forces in North America linked up with the army and started laying siege to French bases. The effectiveness of combined operations soon was demonstrated by the British capture of Fort Beauséjour on the Isthmus of Chignecto. Drawing on naval vessels, army regulars, and militia units from New England, the British launched an attack on the French fort in June 1755 and soon captured it. The fort gave the British control over much of Acadia, leaving the French confined to their great fortress at Louisbourg on Île Royale and the Acadian settlers in the region unprotected. Concerned at the large Acadian presence in the hinterland of Halifax and aware that many Acadians had refused to swear loyalty to the British crown, the military governor of the province took the fateful decision to clear the Acadians from their settlements (see figure 6.6). Between October 1755 and the end of 1762, the British army rounded up between 6,000 and 7,000 Acadians, and transported them to several mainland colonies, France, and various islands in the Atlantic.[26] A few hundred also escaped to Île Saint-Jean (Prince Edward Island), the Miramichi and Restigouche valleys in present-day New Brunswick, and the lower St. Lawrence. The effectiveness of the clearance owed much to the accessibility of the Acadian settlements from the sea and the extensive wilderness that lay at the back of these settlements; for a sedentary, farming population, there was no place to hide.[27]

The assault on Fort Beauséjour was but a prelude to further combined naval and army operations against French littoral settlements in the Atlantic region. In the summer of 1758, the British attacked and captured Louisbourg, leaving French settlements in the Gulf of St. Lawrence defenseless. In late summer and fall, attacks were launched against Île Saint-Jean and French fishing stations on the Gaspé peninsula, as well as against Acadian settlements along the north shore of the Bay of Fundy and up the Saint John River (figure 6.7).[28] By the end of the year, French Acadia was in British hands. The following summer, one of the largest British fleets ever assembled, comprising some 200 naval vessels and transports, carried 8,500 troops up the Gulf and into the St. Lawrence River. After laying siege to Quebec for two months, the British drew the French into battle and defeated them in front of the walls of the city.[29] Since Quebec controlled the entrance to the St. Lawrence and the critical riverine transportation system of New France, the British effectively had throttled French power in North America.

Any attempt by the French to strike back depended on their fleet putting to sea from Brest. Although some vessels had managed to give the Western Squad-

FIGURE 6.7. *A View of the Plundering and Burning of the City of Grimross*, by Thomas Davies, 1758. Davies's eyewitness view shows British troops destroying the Acadian settlement at Grymross, present-day Gagetown, fifty miles up the Saint John River. National Gallery of Canada, Ottawa.

FIGURE 6.8. *The Battle of Quiberon Bay*, by Dominic Serres, 1759. In a rising storm, the British Western Squadron chases the French fleet along a rocky lee shore. Such marine paintings helped create a powerful cultural image of a British Atlantic. © National Maritime Museum, London.

ron the slip several times in the early stages of the war, the British blockade had tightened considerably by 1759.[30] After a storm dispersed the Western Squadron in early November, the French fleet left port to rendezvous with military transports waiting to launch an attack against the British Isles. But a rapid regrouping of the British squadron caught the French fleet south of Brittany and, in the dying light of a stormy day, effectively destroyed it at the battle of Quiberon Bay (figure 6.8).[31] Without a fleet, the French had no means of invading Britain or recovering New France. Although a small convoy tried to slip through the British naval blockade in the Gulf of St. Lawrence in July 1760, the French frigate and supply vessels were chased up the Restigouche River and scuttled.[32] By then, however, the French position in North America was hopeless. In addition to the British army advancing up the Champlain corridor toward Montreal, a second army was moving from Quebec up the St. Lawrence while a third army was coming down the river from Lake Ontario. In September, the three armies converged on Montreal and the French capitulated.

After the early disasters of the land-based campaign, naval power had rescued the British from a parlous situation in North America and helped deliver a crushing blow against the French. Indeed, the French defeat in Canada was compounded by their loss of the West Indian sugar islands of Martinique and Guadeloupe in 1759, and the capture of their Spanish allies' principal Caribbean base of Havana and their Asian stronghold of Manilla in 1762. The ability of the British state to project naval power across the world's oceans had been demonstrated stunningly; the British Atlantic system had triumphed.

Integrating Canada into the British Atlantic

Under the terms of the Treaty of Paris of 1763, the French ceded Canada to Britain in return for the sugar islands of Guadeloupe, Martinique, and St. Lucia; the fishing islands of St. Pierre and Miquelon; and fishing rights in parts of Newfoundland. The French also allowed Britain to take over Dominica, St. Vincent, Grenada, and Tobago in the West Indies. The Spanish gave up Florida to Britain in return for Havana and Manila. (The French encouraged the Spanish to participate in the treaty by transferring Louisiana to Spain.) While the French had relinquished their North American colonies, the British had acquired an American empire that stretched unimpeded from Hudson Bay to the Gulf of Mexico and from the Atlantic to the Mississippi. Although the British appeared to have gained a massive continental territory, they had, in fact, acquired a French position in North America that was less territorial than linear. The French position consisted of little more than littoral settlements strung along various seaways and waterways in the northern half of the continent: a handful of fishing stations around the Gulf of St. Lawrence, a patch of rockbound agriculture along the lower St. Lawrence, the two medium-sized towns of Quebec and Montreal, and a network of fur posts scattered along several river systems. All these settlements depended on water transportation, and were easily accessible by metropolitan power.

After the conquest, the British began integrating Canada into the political, military, economic, and ideological infrastructure of their Atlantic empire. The army took political and military control of the conquered territory, instituting the "system of the generals."[33] From 1760 until 1791, military governors administered the province of Quebec, introducing a type of government similar to that in Newfoundland and to the one that formerly operated in Nova Scotia. Regular army

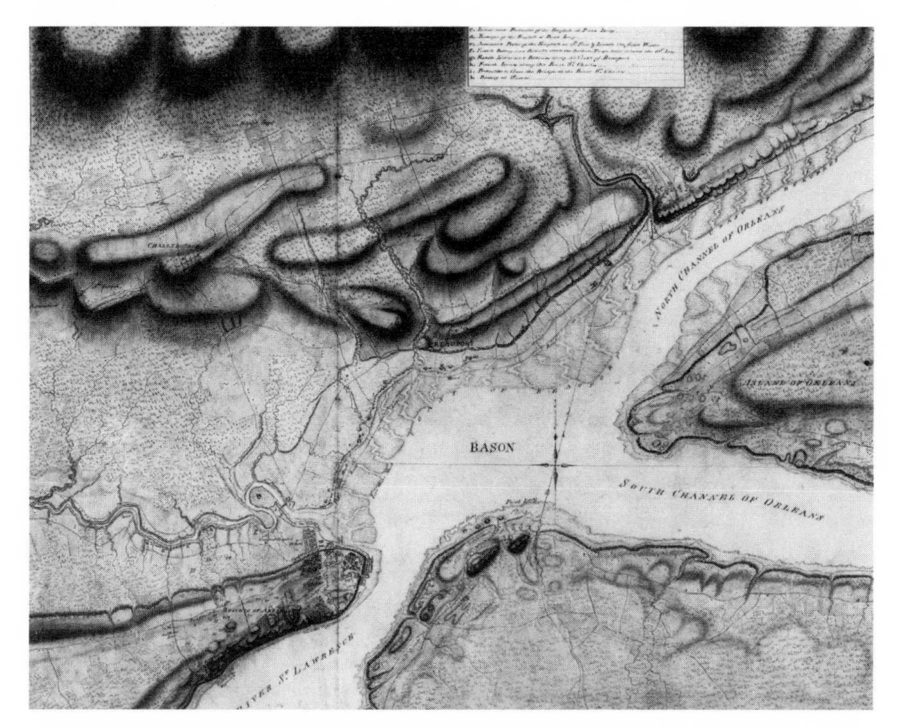

FIGURE 6.9. Murray's Map, Town of Quebec, 1761. National Archives of Canada, Ottawa/NMC 135067.

detachments were stationed across the conquered territories, garrisoning the principal towns of Quebec, Trois-Rivières, and Montreal, as well as the major trading posts of Niagara, Detroit, and Michilimackinac. The Anglican Church was also established in the province, although the Roman Catholic Church was allowed to continue its ministry.

Military surveys of the captured territories were undertaken. The importance of such surveys for controlling territory had become apparent to the British after the Jacobite uprising in the Highlands of Scotland in 1745. Following the defeat of the rebels at the Battle of Culloden in 1746, the British set about pacifying the region: clan warriors were disarmed, garrisons were established at key strategic points, and a comprehensive survey was begun.[34] Between 1747 and 1755, military engineers under the command of cartographer William Roy carried out an extensive survey of Scotland.[35] By the time the French and Indian War broke out in 1754, the British had the cartographic expertise to make detailed, large-scale maps of North America. As maritime power was the key to controlling the northeastern part of the continent, the military surveys concentrated on the coasts and waterways (figure 6.9). In the Atlantic approaches, the Royal Navy and the Royal Engineers undertook several hydrographic surveys, including James Cook's mapping of the St. Lawrence River (1759–1760) and coasts of Newfoundland (1763–1768); Samuel Holland's surveys of the Island of St. John (Prince Edward Island), the Magdalen Islands, and Cape Breton (1764–1766); and J. F. W. DesBarres' survey of coastal Nova Scotia (1764–1774). Many of these maps were collected into a great nautical compendium, *The Atlantic Neptune*, which first began appearing in print in 1777 (see figure 4.5).[36] Meanwhile, several army engineers worked in the province of Quebec on General Murray's survey of the towns and countryside along the lower St. Lawrence. The survey included an accompanying description and census of every parish.[37] These various surveys demonstrated the ability of the British state to survey conquered territory, gather information about subject peoples, and to position the littoral spaces of northeastern North America cartographically within Britain's larger Atlantic empire.

In addition to surveying captured French territory, the British produced a considerable visual record of the towns and countryside. As part of the training at the Royal Military Academy in Woolwich, England, engineers and artillery officers were taught topographical drawing as a means of recording terrain.[38] During their participation in the French and Indian War and subsequent garrisoning of Canada, several of these officers depicted the landscapes of the lower St. Lawrence, providing "a storehouse of information" for the military.[39] Some of these views were engraved and published in the collection *Scenographia Americana* in 1768, the first comprehensive depiction of Britain's North American colonies.[40] Numerous unpublished watercolors also circulated back in England. Among these was the work of artillery officer Thomas Davies, who depicted many of the principal sites of the French and Indian War, including Halifax, Louisbourg, the Hudson River, Fort Ticonderoga, Niagara Falls, the St. Lawrence River, Quebec, and Montreal (figure 6.10).[41] As with mapping, Davies and his fellow officers were surveying captured French territory and providing a visual record that could be useful in further military operations in the province. Indeed, British surveys of the province of Quebec were far better than any that existed for the continental colonies, a shortcoming that soon became apparent in the American War of Independence. Moreover, topographical drawings helped incorporate Quebec into a British cultural milieu. Unfamiliar landscapes were made familiar, key artistic vantage points were established. Such topographic drawings laid the artistic foundations for the picturesque and sublime formulations of Quebec, the lower St. Lawrence, and Niagara Falls that became common in the early nineteenth century.[42]

Britain took a further interest in the natural history of the region. Since the second voyage to Roanoke in 1585 and the production of Hariot's description of Virginia and John White's watercolors, the English had taken a scientific interest in the flora, fauna, and native peoples of the New World, producing, among many works, Mark Catesby's *Natural History of Carolina, Florida and the Bahama Islands* in 1747.[43] With naval survey and fishery protection vessels operating in the western Atlantic during the 1760s, this scientific interest was extended to Newfoundland and Labrador. In 1766, the young Joseph Banks took advantage of a naval cruise along the island's east coast and southern shore of Labrador in order to collect specimens for his developing natural history collection back in London.[44] His voyage coincided with one of Cook's surveying expeditions to Newfoundland, and it

FIGURE 6.10. *A View of Montreal in Canada, Taken from Isle St. Helena in 1762*, by Thomas Davies. Situated at the head of navigation on the St. Lawrence River and at the junction of the St. Lawrence and Ottawa rivers, Montreal served as the principal point of collection and distribution for the Canadian fur trade. The city's role as a place of contact between Europeans and natives is suggested by the well-dressed couple in the foreground and the two Indians paddling a canoe. National Gallery of Canada, Ottawa.

is quite possible that the two men met at a reception hosted by the naval governor in St. John's to celebrate the anniversary of the king's coronation.[45] Although Cook and Banks did not cooperate actively in Newfoundland, their voyages to the island suggest that the imperial state and metropolitan science were beginning to fashion a network of scientific knowledge and cartographic representation that was to have an enormous impact on the future development of the British Empire. The hydrographic surveys, topographic depictions, and scientific research conducted in the littoral realm of northeastern North America in the late 1750s and early 1760s undoubtedly laid the foundations for the triumphs of Cook and Banks in the Pacific in the early 1770s.[46]

Beyond the province of Quebec, British authority extended over the northwest territory (*Pays d'en Haut*). Although several of the colonial governments along the eastern seaboard claimed jurisdiction over the lands beyond the Appalachians, the British government instituted direct rule over the territory and its Indian inhabitants.[47] Unused to the complex system of French-native alliances that made possible a French fur-trading and military presence in the interior, the British quickly aggravated native sensibilities.[48] In an effort to save British government expenditure in North America, General Amherst, the Commander-in-Chief, ordered a reduction in the presents given to interior tribes. The Indians soon suffered privations from the lack of shot, powder, and rum. Meanwhile, American settlers and land speculators, contrary to the terms of the Treaty of Easton, were moving

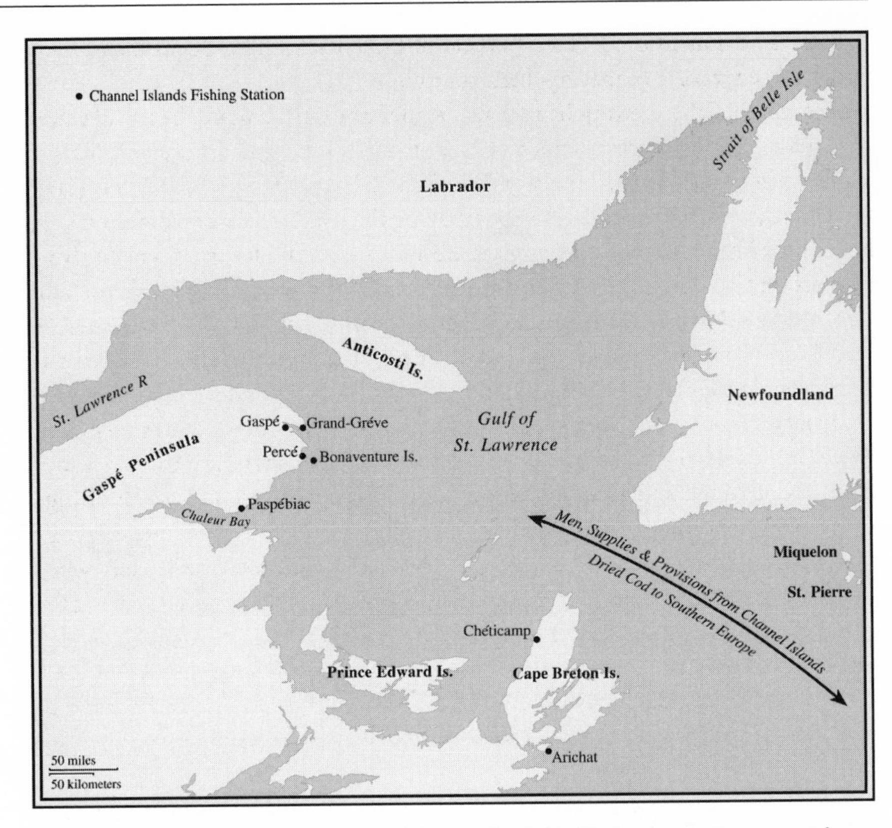

FIGURE 6.11. Channel Islands migratory fishery in the Gulf of St. Lawrence, circa 1770. After Mario Lalancette, "Exploitation of the Gulf of St. Lawrence," in *Historical Atlas of Canada*, vol. I: *From the Beginning to 1800*, ed. R. Cole Harris, plate 54 (Toronto: University of Toronto Press, 1987). Also relevant are David Lee, *The Robins in Gaspé 1766 to 1825* (Markham, Ont.: Fitzhenry & Whiteside, 1984); and Rosemary Ommer, "The Cod Trade in the New World," in *A People of the Sea: The Maritime History of the Channel Islands*, ed. A. G. Jamieson, 245–68 (London: Methuen & Co., 1986).

into the upper Ohio Valley. During the summer of 1763, unrest spread among the Indians of the northwest, and turned into outright rebellion under the leadership of Ottawa war chief Pontiac.[49] Several forts, including the key trading post of Fort Michilimackinac, were captured, while Forts Pitt and Detroit were besieged. Caught by surprise, the British struggled to reimpose their authority. During the fall and over the course of the following summer, the British relieved the sieges, regained the captured forts, and pacified the Indians. As these events were unfolding, the British government issued the Royal Proclamation in November 1763. Consolidating years of evolving British policy toward the trans-Appalachian west, the proclamation set aside all lands west of the Appalachians as a Native Reserve, established garrisons to protect Indian land, and introduced regulations governing the fur trade. By preserving the trans-Appalachian west for native peoples and the fur trade, and with regular army detachments garrisoning the trade posts, the

British had assumed the former position of the French and brought the interior under some measure of metropolitan control.

The hand of the metropolis was also evident in the economic reorganization of Canada. The two great staples of New France—the cod fishery and the fur trade—were soon in the hands of British merchants. By 1765, Channel Island merchants, who had operated previously along the South Coast of Newfoundland, had entered the Gulf of St. Lawrence and taken over the former French fishing stations on Cape Breton Island and around the Gaspé peninsula (figures 6.11 and 6.12).[50] The principal Channel Island firm of Charles Robin & Co. established its New World base at Paspébiac on Chaleur Bay, and opened numerous smaller stations along the Gaspé coast. The company extended the migratory fishery into the Gulf, sending out men, supplies, and provisions to the fishing stations each year, as well as establishing a resident fishery to supply local French and Acadian fishermen. As in Newfoundland, little agricultural potential existed along the Gaspé coast or in Cape Breton, and resident fishermen were largely dependent on fish merchants for provisions and dry goods; many fishermen soon found themselves in debt. Like much of the dried fish from Newfoundland, the cod dried along the coast of the Gaspé and the Gulf shore of Cape Breton was of prime merchantable quality suitable for the lucrative markets in the Iberian Peninsula and the Mediterranean. Although merchants resident in Halifax and Quebec tried to break into the Gulf fishery, they never had the capital, trade connections, or local knowledge

FIGURE 6.12. *A View of the Pierced Island, a Remarkable Rock in the Gulf of St. Lawrence*, engraving after Hervey Smyth, 1760. This view from *Scenographia Americana* shows a British warship cruising past a fishing shallop (right) and the former French fishing stations at Bonaventure Island (left) and Percé (right). These stations were taken over by merchants from the Channel Islands in the mid-1760s. National Archives of Canada, Ottawa/C-000784.

to displace the Jerseymen. The Channel Island monopoly remained intact until the late nineteenth century.

The other great staple—the fur trade—was also tied back to the metropolis. In 1763, the British took over the French fur trade along the St. Lawrence, Ottawa, and Great Lakes (figure 6.13). English, Scottish, and American merchants opened establishments in Montreal, drawing heavily on capital and credit provided by British houses.[51] Commission agents in London forwarded manufactured goods to merchants in Montreal, who in turn supplied traders or wintering partners (*bourgeois*) in the northwest. In return, furs acquired by the traders were sent to Montreal, and then forwarded across the Atlantic to London. As with the fur trade through Hudson Bay, London ultimately controlled the economic system. Unlike the Hudson's Bay Company, however, the Montrealers operated in a highly competitive environment, competing among themselves and with the Company; it was not until the mid-1770s that Montreal merchants, such as McTavish and the Frobisher Brothers, began to pool their resources and form syndicates to rationalize the trade. By reducing competition, the Montrealers could raise prices on their goods and purchase more furs from their native suppliers.[52] It also placed them in a stronger position in dealing with the London capital market. The McTavish and Frobisher syndicate was the beginning of the North West Company, which would dominate the Canadian fur trade after the American Revolution.[53]

While the Hudson's Bay Company "slept by the edge of a frozen sea" waiting for Indians to bring furs to trade each spring, the Montrealers, like the French before them, had to reach far into the continent to trade.[54] Much of the transportation was by birchbark canoe, so cargoes were relatively light: The selection and quantity of trade goods were modest compared to those at the bayside posts, while the furs had to be of high value to cover the transportation costs. Despite these handicaps, the Montrealers maintained the old French trading posts and extended the trade routes deep into the interior. This linear network of rivers, portages, and lakes stretched more than a thousand miles from Montreal to the fur country in the northwest. The first section of the Montreal mainline ran along the Ottawa River, and then across Lake Nipissing, Georgian Bay, and Lake Huron, where it terminated at Michilimackinac situated at the tip of the Michigan peninsula. Michilimackinac served as the great transshipment point "between the upper countries and the lower. Here, the outfits [were] prepared for the countries of Lake Michigan and the Mississippi, Lake Superior and the north-west; . . . here, the returns, in furs [were] collected, and embarked for Montreal."[55] From Michilimackinac, the mainline continued along the north shore of Lake Superior to Grand Portage, and then across Rainy Lake, Lake of the Woods, and Lake Winnipeg to the Saskatchewan River. By the late 1760s, the Montrealers had cut across the headwaters of several rivers draining into Hudson Bay, and were competing directly for bayside trade.[56] As one Company trader ruefully remarked, "the Canada pedlars are got in the very heart of the trading Indians' country."[57]

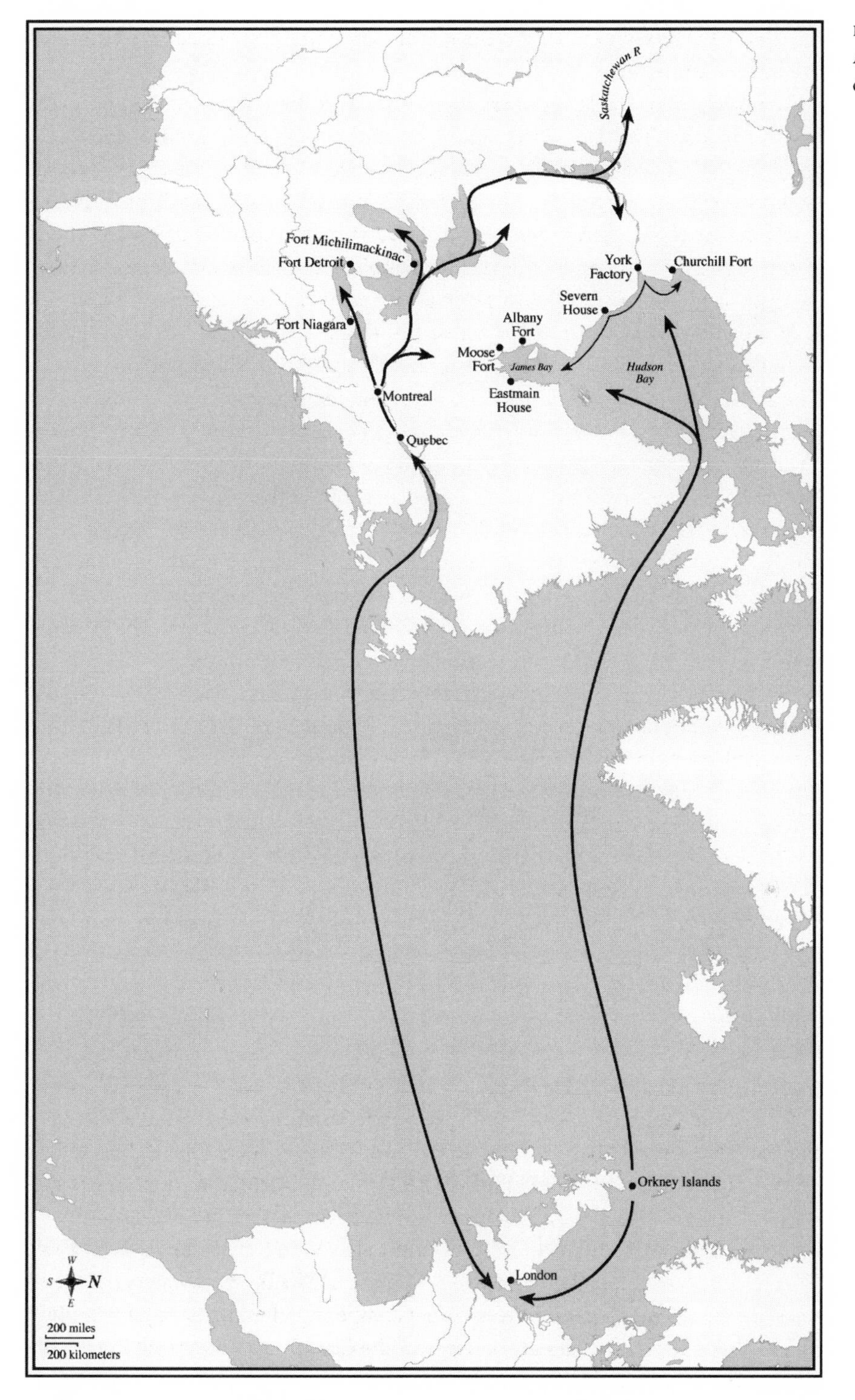

FIGURE 6.13. Trans-Atlantic fur trade, circa 1770.

FIGURE 6.14. *A View of Detroit July 25th 1794*, by E. H. A British garrison and trading post, deep in the continental interior, connected by water transport and portage to Montreal and the St. Lawrence River. Courtesy of the Burton Historical Collection, Detroit Public Library.

The Montrealers inherited the French fur posts and although these were stockaded structures similar to those on the Bay, their internal organization was quite different. Whereas the Hudson's Bay Company maintained strict control over its posts and divided the living areas along hierarchical lines, the old French posts contained numerous buildings belonging to individual traders, the church, and the military. After the British take-over, the posts continued to be populated by traders and Jesuit missionaries, as well as by British army detachments (figure 6.14). The traders formed the largest group; they frequently had native wives and families, and came from diverse backgrounds. Many traders were either Americans from Albany and New York or immigrant Scots, while the canoe men (*engagés*) were a mix of French-Canadians, recruited from the parishes in and around Montreal, and French-Indians.[58] The balanced demographic structure and ethnic mix of these populations was much different from the predominantly male and British world of the Hudson's Bay Company posts, and reflected the long history of interaction between the French and their Indian allies.

Apart from the fur trade and the fishery, British merchants increasingly dominated the economic life of the towns and countryside. Scottish and English merchants established themselves in Quebec and Montreal, controlling the dry goods trade from Britain, the shipping and shipbuilding industries, and the nascent timber trade.[59] Army garrisons in the two towns had to be supplied, and many victualing contracts went to British merchants. By the early 1770s, "Quebec traders" formed a powerful opposition group against the military administration in the province, and lobbied for representative government. But like the merchant communities in the West Indies, criticism of imperial power was tempered by the merchants almost complete economic dependence on Britain.[60] The countryside of Quebec was less important. Unlike the seaboard colonies, which had large rural populations, dynamic frontiers, and extensive export trades, the province had a relatively small rural population of some 60,000 in 1760, a limited frontier, and few agricultural exports. Some produce was marketed in Quebec and Montreal, but very little went beyond the lower St. Lawrence. In these circumstances, the agricultural economy of the region was relatively uncommercial and attracted little British investment. A handful of British officers took up estates or seigneuries, more for social status than for economic gain.

In the fifteen years between the Conquest and the outbreak of the American Revolution, the British integrated the Gulf of St. Lawrence, the province of Quebec, and the northwest into their Atlantic empire. Despite the political and cultural differences between the French and the British, the take-over was relatively smooth, involving the maintenance rather than the disruption of the colony's economic and social structures. The reliance on long-distance staple trades, the weakness of the local agricultural economy, and the accessibility of much of Canada by waterborne transportation ensured a good fit with the British Atlantic world. Essentially, one European metropolitan system had been replaced by another.[61]

The British Atlantic and the American Frontier in the 1760s

The 1760s marked the zenith of British imperial power in North America. For contemporaries in Britain, the triumph of British arms over the French and the Spanish was an affirmation of innate British superiority and a conquest without parallel in history; as William Pitt, the war-time leader, boasted in Parliament, Britain "had over-run more world" in three years than the Romans had "conquered in a century."[62] Indeed, the British increasingly saw themselves as new Romans and London as the new Rome.[63] Yet, in retrospect, the 1760s marked a turning point between the collapse of French power in North America and the rise of American dominance on the continent. It seems an appropriate point, therefore, to take stock of the geographic spaces that had emerged in the different regions of British America in the early modern period.

With the conquest of New France, the British Atlantic had reached its greatest extent (figure 6.15). In North America and the Caribbean, the area that can be delimited as the British Atlantic included the British settlements in Hudson Bay, Newfoundland, Nova Scotia, and the West Indies, as well as the former French settlements that stretched in a great arc from the Gulf of St. Lawrence, through the St. Lawrence–Great Lakes region, to the Ohio and the Mississippi. In other parts of the Atlantic, the British controlled Bermuda, the Bahamas, and several settlements along the coast of West Africa. In the Indian Ocean, the beachheads at Madras, Bombay, and Calcutta had been supplemented, after Clive's victory at Plassey in 1757, with territorial control of Bengal.[64]

All of these various territories comprised islands or littoral spaces. With the exception of the conquered agricultural lands of Quebec and Bengal, these

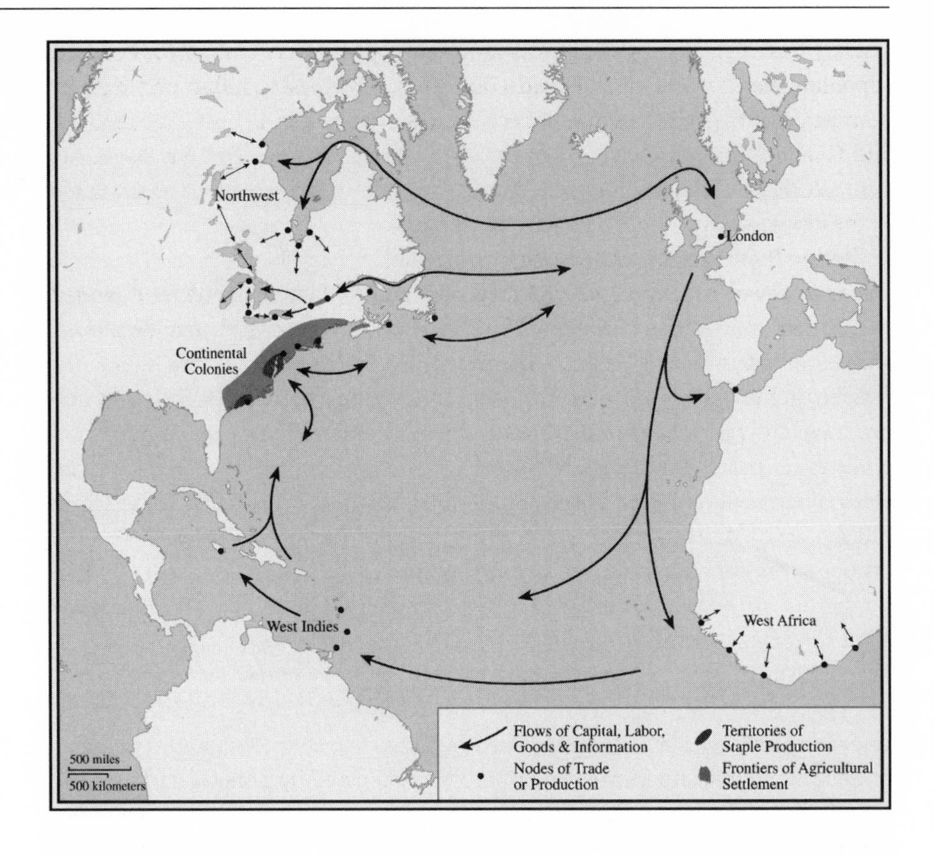

FIGURE 6.15. A geographical model of Britain's Atlantic empire, circa 1770.

maritime enclaves had limited agricultural room. The West Indian islands were completely settled, Nova Scotia had little agricultural potential, and Newfoundland and Hudson Bay were virtually useless for farming. From the British perspective, the British Atlantic was not a world of expanding agricultural frontiers but rather a collection of commercial nodes, consisting of small islands of plantation agriculture in the West Indies, fishing stations in Newfoundland and the Gulf of St. Lawrence, and trading factories along the rivers and lakes of northern North America and along the coasts of Hudson Bay, West Africa, and India.[65] All of these nodes faced outward to the sea and the shipping lanes that tied them, across thousands of miles of ocean, to the British Isles.

The commercial importance of these nodes was striking. From 1772 to 1774, 52 percent of all imports into England came from colonial possessions in North America and India. Of these imports, 71 percent came from areas in the British Atlantic. Almost half the imports from the British Atlantic comprised sugar from the West Indies (49 percent); the other significant import was tobacco from the Chesapeake (11 percent), an area dominated by British firms.[66] In addition, trades in dried fish and slaves contributed further earnings from the British Atlantic. Apart from tobacco, the continental colonies were not a major source of products, although they were taking a major share of English exports by the late eighteenth century.[67] Almost all these various long-distance trades were in the hands of metropolitan merchants and chartered companies. The great London and outport merchants controlled the sugar, tobacco, and slave trades; the West Countrymen and Channel Islanders held the dried fish trade; and the Hudson Bay Company had a monopoly over bayside trade. London and the outports also organized much of the dry goods trade to the American colonies.

Within these various nodes, societies were ordered and hierarchical. The staple trades in sugar, fish, and furs provided opportunity for mercantile and landed capital but little chance for labor.[68] As a result, the societies that developed were dominated by small elites—merchants, planters, factors—who ruled over masses of enslaved and indentured laborers. Keeping these work forces in line produced the slave codes of the West Indies and the rules and regulations of the Hudson's Bay Company. Cultural mores further reinforced this social order. The elites of the British Atlantic were a transient group, as familiar with the world of the metropolis as with the world of the colonies, and well able through their economic wealth and social position to maintain metropolitan customs and culture on the periphery. Moreover, the establishment of the Anglican Church and a general resistance to other religious denominations strengthened the authority of the elites. In contrast, the labor forces in the fishery, fur trade, and sugar plantations were economically so weak that they were in little position to challenge the colonial hierarchy.

Such economic and social structures were reinforced by political and military power. Given the economic importance of the British Atlantic, Britain had little reason to allow much political development; indeed, considerable parts of the British Atlantic—Newfoundland, Quebec, Hudson Bay, parts of India—were under direct administration or corporate government. Centralized political control from the metropolis, rather than local authority in the colonies, marked the British Atlantic empire. Such political control was made possible through the deployment of military power: the navy commanded the sea lanes, strategically placed garrisons controlled important ports and waterways. For American John Adams, glorying in the "Spirit of Liberty" among the population of New England after the Stamp Act crisis of the mid-1760s, the implication of such force was clear. According to Adams, the inhabitants of Quebec were "awed by an Army" while those in Nova Scotia were "kept in fear by a Fleet and an Army."[69]

From the British perspective, the British Atlantic was a commercial, oceanic, ordered, hierarchical, and militarized space. It was a "grand marine empire" that was readily grasped in Britain through the vessels that unloaded their cargoes in London, Bristol, Liverpool, Glasgow, Poole, and other ports; the colonial products of sugar, tobacco, tea, wines, and calicoes consumed at home; the myriad individual experiences of planters, merchants, factors, and servants; the overseas deployment of warships and regiments; and the illustrations and maps representing the Atlantic world that increasingly circulated in printed form.[70] In the 1760s and early 1770s, the words of James Thomson's *Rule Britannia*, composed some twenty years earlier, seemed particularly apposite: Britain had arisen from the azure main and was ruling the waves.[71]

And yet, along the American eastern seaboard, vastly different spaces had emerged (see figure 6.15). From the original core settlements established along the coast in the seventeenth century, several western-moving frontiers had coalesced by the late eighteenth century to create a settlement frontier stretching from Machias, Maine, to St. Augustine, East Florida, and extending west to the Forks of the Ohio in western Pennsylvania.[72] This was an enormous agricultural area, far larger than anything that existed in the British Atlantic, and it appeared to have almost unlimited room for expansion. By the early 1760s, the barrier of the Appalachian Mountains had been breached, and European and American settlers were pouring into the western reaches of New York, Pennsylvania, Virginia, the Carolinas, and Georgia. Ostensibly, the western boundary of British America was the Mississippi River, many hundreds of miles farther west. For European immigrants and landless Americans, the agricultural potential of the trans-Appalachian west appeared limitless.

Within this continental frontier, several territories of staple production and nodes of trade had emerged over the course of the seventeenth and early eighteenth centuries. The principal staples of the New England fishery, the Chesapeake tobacco plantations, and the Carolina rice plantations supported colonial elites of merchants and planters, while the port towns that handled several of these trades sustained an urban mercantile class. Economically and politically, these elites straddled the worlds of the Atlantic and the continental interior; they were as much concerned with trans-Atlantic connections to Britain as with expansion of the frontier. But these links to the continental interior gave colonial elites a much greater geographic context in which to maneuver than existed on the islands of the British Atlantic. Well able to see the immense economic opportunities that lay in the backcountry, merchants and planters along the seaboard enjoyed considerably more economic independence than elites in Newfoundland or the West Indies.[73] Furthermore, these seaboard elites, through their dominance of the colonial assemblies, had acquired a good deal of political power and increasingly saw themselves as equal partners with London in the running of the British American empire.

The composition of American society was also much different. Although the towns and the areas of staple production supported hierarchical societies, only Charleston and the low country came close to approximating the extreme social divisions common in the West Indies and other parts of the British Atlantic. Moreover, American social hierarchies were set within a larger population that was much less stratified and deferential. Unlike the British Atlantic, the agricultural spaces of the seaboard provided European immigrants and American settlers with an opportunity to achieve a measure of economic independence. Widespread ownership of property created a more egalitarian society, and allowed male freeholders to participate in the political process, which led to a more democratic form of government.[74] The stratified, deferential society of the British Atlantic or Britain itself simply could not be replicated in the American colonies.

Social hierarchy and authority were weakened further by the American cultural mix. In a population that was increasingly polyglot, the imposition of British cultural norms was especially difficult. To be sure, Americans were happy to participate in British material culture, consuming English manufactured goods and constructing houses in the Georgian style, but there were also well-developed regional vernacular cultures that drew little on Britain. Throughout the seaboard colonies, dialects, agricultural techniques, foodways, and methods of building differed from those across the Atlantic. Moreover, there was considerable resistance to British cultural institutions. The numerous dissenting churches opposed any strengthening of Anglican religious authority in the colonies.[75] Unlike the Anglican hierarchy of England, with its "degrees of Eminence among the Clergy," the colonial Anglican Church had only "Priests and nothing more." The intimate relationship between the Anglican Church and monarchy, which was so strong in the British Atlantic world, had been "totally dissolved." For British imperial official William Knox, writing in the late 1770s, religious toleration in the colonies had allowed "Every Man . . . to be his own Pope, [thus] he becomes disposed to wish to be his own King, and so great a latitude in the choice of a religious system naturally begets republican and independent ideas in politics."[76]

Of course, the worlds of the British Atlantic and the American eastern seaboard were not self-contained. American merchants participated in trans-Atlantic and interregional commerce, particularly with the West Indies and Newfoundland; American consumers bought British manufactured goods; and American colonial legislatures lobbied Parliament in Britain. Likewise, British merchants controlled much of the dry goods trades to the colonies and dominated the staple economy of the Chesapeake; British consumers purchased American tobacco; and British officials oversaw colonial governments. Nevertheless, these were different spaces occupied by different peoples with different societies and cultures; the norms of the British Atlantic world were not those of the American continental frontier.[77] The British and American elites that dominated this bifurcated world had disparate experiences and perceptions; each group viewed British America through its own, particular lens and beheld a different vision of empire.

Tightening the Cordon

British victory in the French and Indian War dramatically changed the geopolitical context in North America. With the French and Spanish removed, the British had responsibility for managing the former French territories of Acadia, Canada, and the trans-Appalachian west, as well as the former Spanish Florida. Within these territories, imperial authority had to be exercised over alien populations, comprising mostly French Canadians and native peoples. At the same time, the overwhelming British victory almost inevitably meant that the French and the Spanish would go to war again to redress the balance of power.

For the British, the new geopolitical situation carried immense financial cost. Not only had they to deal with the huge expenses generated by the war, but also with the costs associated with garrisoning troops in North America and maintaining naval supremacy over the French and the Spanish. By 1760, Britain had spent more than £160 million fighting the war, twice the country's gross national product.[78] Much of this expenditure had been borrowed, leaving the government with a national debt of approximately £146 million.[79] In 1762, the interest charges on this debt were consuming almost half of all government net income.[80] Meanwhile, the costs of maintaining the army in North America were consuming more than £200,000 a year, and the navy and dockyards also required substantial sums.[81]

By the mid-eighteenth century, British politicians increasingly had become aware of the link between geopolitical ascendancy and commercial power.[82] Any weakness in Britain's commercial economy, it was thought, would inevitably lead to a decline in the nation's political and military status. As much of the government's revenue came from customs and excise duties, the health of the country's commercial economy was of critical importance. Reluctant to impose further taxes on trade, the British government decided in the early 1760s to turn to its North American colonies for extra revenue. As the French and Indian War had been fought to defend the colonies and British army units were keeping the peace on the frontier, this action had widespread political support in Britain. Moreover, the only revenues raised from the colonies were customs duties, which were widely evaded and scarcely paid for their collection. In an attempt to improve its fiscal position, the British government decided to reduce expenditure in the colonies and to enhance the collection of customs duties as well as levy new taxes.

The series of measures that were implemented—the Royal Proclamation of 1763, the Sugar Act of 1764, the Stamp Act of 1765, the Townshend Acts of 1767—revealed the strengths and weaknesses of metropolitan government in North America. The integration of Canada into the British Atlantic gave the imperial government a geographic position on the continent that stretched in an enormous arc from the trans-Appalachian west through the Great Lakes, the St. Lawrence River, the Gulf of St. Lawrence to Nova Scotia and Newfoundland. Metropolitan control over Quebec and Nova Scotia was particularly important. The acquisition of the fortified town of Quebec and the establishment of Halifax provided the British government with army and navy bases from which to project military power. From Quebec, the army could maintain garrisons deep into the interior; from Halifax, the navy could patrol the eastern approaches to the continent as well as the eastern seaboard. Although peripheral to the American colonies, Nova Scotia and Quebec allowed the British government to deploy military power along both seaward and landward sides of the continental colonies.[83] The significance of these bases would soon become apparent as the government began to implement its legislative program.

The two most effective measures—the Royal Proclamation and the Sugar Act—could be enforced because Britain had the military power to hand. The Proclamation aimed to reduce friction between American settlers and Indians, control the fur trade, and save the imperial government from an expensive frontier war. Although the British government could not prevent American settlers trickling into the upper Ohio Valley, it could withhold legal title to land, which effectively stopped speculation by the Virginia land companies.[84] Moreover, American settlers were exposed to Indian attack and pressure from the British garrison at Fort Pitt to leave the area. The imperial government also had considerable success in implementing the Sugar Act. During the eighteenth century, American merchants honored the navigation acts more in the breach than in the observance, and this was particularly true of the sugar trade. Molasses was in great demand for making rum in the northern colonies and was cheaper to purchase from the French West Indies than the English West Indies, so American merchants traded with the French, even when Britain and France were at war. Although French imports were liable to stiff duties, colonial merchants frequently evaded them through smuggling. To put teeth into the colonial customs service and to increase revenue from the sugar trade, the imperial government introduced the Sugar Act and deployed the navy to enforce it.[85] With the naval dockyard and the Court of Vice-Admiralty for All America established in Halifax, the navy had a secure base from which to patrol the eastern seaboard and a sympathetic court in which to try cases. Although the navy did not have sufficient vessels to close down the trade, it had enough to prevent the transshipment of goods at the French islands of St. Pierre and Miquelon and to deter many American smugglers along the coast of New England. Despite bitter complaints from colonial merchants, enforcement soon produced increased revenue. The successful implementation of the Royal Proclamation and the Sugar Act showed that the metropolitan government could wield power reasonably effectively in the continental interior and along the eastern seaboard.

The Stamp Act and Townshend Acts, on the other hand, were failures simply because there was no way to readily enforce them. The purpose of the Stamp Act was to raise revenue by charging a fee for stamped paper in commercial transactions, legal actions, and newspaper publications. Yet the major centers of commerce—Boston, New York, Philadelphia—were not spaces that could be easily controlled by either the British army or navy. In 1765, the great bulk of Britain's troops in North America were garrisoning the Atlantic empire: Newfoundland, Nova Scotia, Quebec, the northwest, and East and West Florida (figure 6.16). These troops, as General Gage, the Commander-in-Chief, ruefully observed were "at a great distance and a good deal Dispersed."[86] Boston and Philadelphia had no British troops, while New York had only one company. Moreover, the military bases in the cities were far from being effective platforms to project power. Fort George in New York was in a ramshackle state, while Castle William in Boston was located in the outer harbor. Of more use were the naval vessels moored in New York harbor, which acted as floating fortresses and refuges for government officials. As

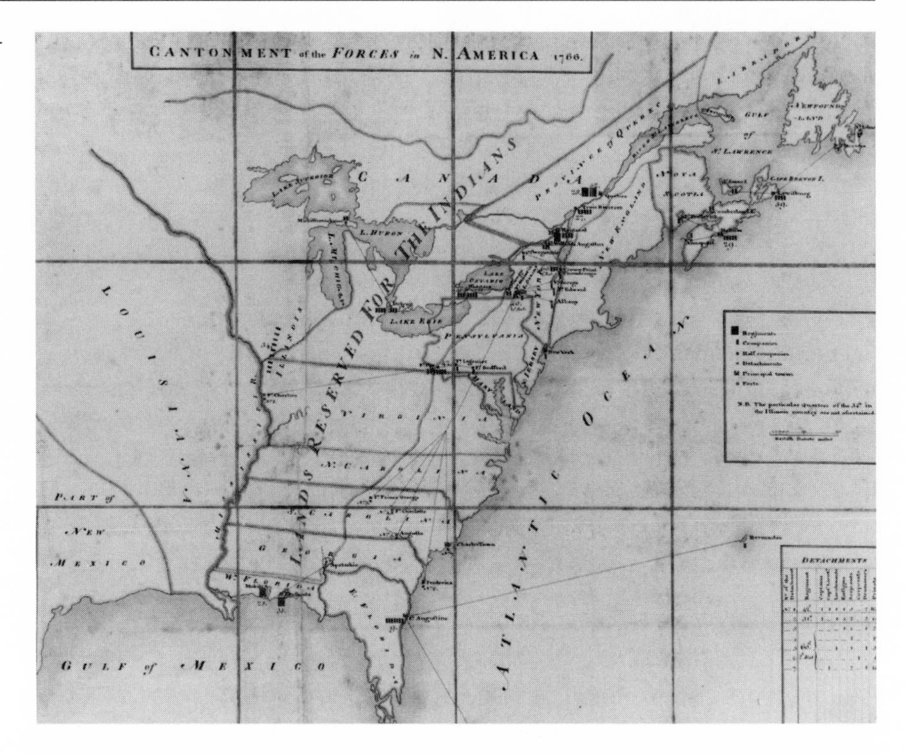

FIGURE 6.16. *Cantonment of the Forces in N. America 1766.* A British military map showing the deployment of British troops in North America, the Proclamation Line running from the Gulf of Mexico to Lake Ontario, and the lands set aside for the Indians. Clements Library, University of Michigan.

opposition to the Stamp Act developed, the inability of the imperial government to control the seaboard towns became obvious. In 1766, the British government, in an attempt to regain control over the continental colonies, repealed the act. Nevertheless, the government, still desperate for revenue, introduced the Townshend Acts the following year. The purpose of the legislation was to tax certain imported goods—paper, paint, lead, glass, and tea—and to establish an American Board of Customs Commissioners under metropolitan supervision. The headquarters of the Board was located in Boston, one of the leading centers of opposition to the Stamp Act. Yet an outbreak of rioting against the commissioners in 1768 led to their evacuation by the navy and the introduction of troops from Florida to restore order. Meanwhile, a nonimportation movement among American merchants stifled the import of taxed goods and stimulated domestic production of the same items. Given these circumstances, the British government was forced to back down again, repealing all the taxes except for the one on tea, a symbolic assertion of the government's right to tax the colonies.

Apart from these imperial pieces of legislation, other flash points flared between Britain and the continental colonies. From the start of the French and Indian War, the British navy needed crews, and regularly swept American ports to impress seamen; even after the war, the practice continued.[87] Of doubtful legality, impresses frequently provoked riots against imperial authority, pitting ordinary seamen fighting for their "right" and "liberty" against naval officers and colonial officials.

Such seamen were only too willing to join with other rioters in the agitation against the Stamp Act in 1765. Colonial elites were also becoming increasingly entangled in metropolitan economic power. In the Chesapeake, the growing indebtedness of tobacco planters to metropolitan merchants during the 1760s and early 1770s (see chapter 3) threatened the long-established social and economic position of the planter class, and helped sway an increasing number of them behind the growing resistance movement to British imperial impositions.[88]

In the early 1770s, pressure from the British Atlantic system on the continental colonies increased. In 1773, the British government, in an attempt to shore up the precarious financial state of the East India Company, granted the company a monopoly of tea exports to British America. For the government, the East India Company was of critical importance. Apart from being the only instrument of British rule in India and a major prop of Britain's commercial economy, the East India Company provided the government with substantial revenue. But while the monopoly helped solve a major problem within Britain's Atlantic empire, it riled colonial merchants dealing in tea. When the first shipments of tea arrived, they were either turned back from American ports or ceremoniously dumped into the harbor, as was the case in Boston. For the British, this was the last straw. In early 1774, the imperial government responded with the Coercive Acts, a series of measures designed to increase metropolitan administrative control in Massachusetts and to strengthen the military presence in Boston.[89] To implement these acts, General Gage, the Commander-in-Chief, was appointed governor of the province. Meanwhile, in an attempt to maintain its hold over Quebec, the British government passed the Quebec Act, which guaranteed French civil law and protection of the Catholic religion, and extended the boundaries of the province into the trans-Appalachian west.[90] The preservation of French law and defense of the Catholic Church reassured the majority French inhabitants of the province, and helped ensure that they remained neutral during the coming conflict with the Americans. Extension of the province's boundaries to its "ancient limits" reaffirmed the old French position in the heart of the continent, and confirmed the sway of the Montreal-based fur trade over the interior. For Americans, the Quebec Act was yet another example of metropolitan interference in the internal affairs of the colonies, and was quickly labeled one of the "intolerable Acts" of 1774.

The long-developing differences between the British Atlantic and the continental colonies, which had been masked to some extent by the conflicts with France during the 1740s and 1750s, had come to the fore with the new geopolitical situation in eastern North America in the 1760s and early 1770s. In attempting to govern the enlarged spaces of the British Atlantic, the British government had tightened imperial authority over the seaboard colonies, which, in turn, had provoked a fierce political backlash. As early as the Stamp Act, the conflict between the British government and American legislatures had crystalized over the issue of sovereignty versus representation.[91] On the British side, the American colonies came under British sovereignty and thus could be taxed; on the American side, colonial legislatures rejected taxation from Westminster because they were not represented in the British Parliament. Within the British Atlantic, raw power soon overrode discussions of political principle. In Quebec, Nova Scotia, and the West Indies, British taxation policies were implemented successfully because the metropolitan government had the power to enforce them; in the American colonies, this power was lacking.[92] The British government controlled the oceanic and trans-Appalachian spaces that surrounded the eastern seaboard, but not the colonies themselves. The cry of "no taxation without representation" could be made in the continental colonies simply because imperial power was too weak to silence it; in the British Atlantic, that cry could be stifled. The strength of British power over its Atlantic empire reinforced the imperial government's desire to rein in the continental colonies, while at the same time spurring the colonists to throw off the imperial harness.

The War of American Independence

The revolt of the American colonies posed immense military and logistical problems for Britain. Although the principal theater of operations was in North America, the American Revolution was a fundamentally different conflict than the earlier French and Indian War. In the struggle against France, Britain was essentially trying to capture nodes—Louisbourg, Quebec, Montreal, the sugar islands—that controlled oceanic or riverine spaces. British naval power could be brought to bear on these nodes, frequently with decisive results. Moreover, the British had a continental base—the American colonies—to draw upon for manpower, supplies, and provisions. In the American conflict, however, the geographical context was reversed. Unlike the situation in New France, where Quebec controlled the St. Lawrence and much of the interior, the American colonies had no central node that dominated the continental hinterland. Even though the British held Boston, Newport, New York, Philadelphia, and Charleston at various times during the course of the war, no one town gave control over the interior. To dominate the seaboard colonies, the British needed more than command of the seas; they required an army large enough to control enormous continental spaces and dispersed, hostile populations. Furthermore, British bases in Nova Scotia and the province of Quebec could not supply much in the way of manpower or materiel; virtually all army units, supplies, and provisions had to be shipped across the Atlantic from the British Isles. Through herculean efforts, the British solved the logistical problem of waging war three thousand miles from home, but they never fully grasped the challenge of capturing and holding the eastern edge of the continent.

The military conundrum facing the British soon became apparent.[93] In April 1775, General Gage in Boston decided on a military sweep of the interior to gather up and destroy weapons that could be used by rebel militia. Although Gage sent troops to Concord, only twenty miles from Boston, they were forced to beat a bloody retreat in the face of attacks from militia and patriots.[94] The difficulties of projecting military power over land into hostile territory had become dramatically obvious. With the surrounding countryside in rebel hands, Gage and his troops were confined to the Boston peninsula from where they had to be evacuated by the navy to Halifax in March 1776. With no bases left in the continental colonies, the British focused their attention on capturing New York, a city that had an ice-free port and was strategically located at the mouth of the Hudson River–Champlain corridor. In August 1776, the navy ferried General Howe's army from Halifax to New York. Although Howe forced the continental army under George Washington to withdraw from Brooklyn and New York, the British failed to position warships on the East River or troops on Harlem Heights to cut off the rebel retreat. As a result, much of the continental army escaped. Having missed an excellent opportunity to destroy the rebel army, the British quickly compounded their error. Control of New York should have allowed them to create a defensive pale around the city and across southern New Jersey, which would have provided a base from which to launch combined naval-army operations against coastal New England and the Chesapeake. Such a strategy would have made full use of the flexibility afforded by the navy, and caused havoc up and down the seaboard.[95] Moreover, control of New Jersey would have provided an agricultural hinterland from which to feed the army. Instead, the British weakened their position in New York by sending reinforcements to Canada in order to launch an offensive headed by General Burgoyne down the Champlain corridor in 1777. The strategic aim was to cut off the rebel colonies in New England from those in the Mid-Atlantic region. But the lessons of Amherst's laborious advance up the Champlain Valley as well as Braddock's disastrous foray into the Appalachian Mountains during the French and Indian War had not been learned. Burgoyne's advance got bogged down in rugged terrain north of Albany, while his ever-lengthening lines of communication and supply back to Quebec became ever more vulnerable to attacks from New England militia. The slowly unfolding disaster of the British advance reached its denouement at the Battle of Saratoga and the capitulation of Burgoyne's army.[96] At the same time, Howe took the army in New York on a lengthy sea voyage down the coast and into the Chesapeake Bay in order to launch an overland attack on Philadelphia. Although Howe took the city, the campaign served little strategic purpose; he also failed to support Burgoyne's advance by driving north from New York up the Hudson Valley. Both campaigns revealed a failure to coordinate strategy and to solve the logistical difficulties of overland warfare.

The American victory at Saratoga provided enough encouragement for the French, keen to revenge their earlier defeat, to join the rebels and declare war on Britain. The involvement of the French changed the strategic situation completely. Now the British Isles, as well as valuable colonies in the West Indies and India, were open to attack. As the theater of conflict spread, the war in the continental colonies was quickly relegated to a sideshow. For the British, the strategic priorities were to defend the country against French invasion and to protect the West Indies.

Naval vessels were redeployed to home waters, while troops and vessels were sent to the Caribbean. These redeployments left the units of the army still operating in the continental colonies with much less naval support.[97] Nevertheless, the British pressed ahead with a new campaign in the southern colonies. In 1779, they attacked Savannah and quickly overran Georgia; the following year, they moved on to take Charleston and gained control over South Carolina. By spring 1781, the British, having fought a frustrating campaign in North Carolina, decided to move north into Virginia. Despite considerable military success in the Chesapeake during the summer, the British army was gradually being boxed in. By late summer, Washington's continental army, reinforced by French troops who had marched overland from Rhode Island, had pinned the British into their camp on the Yorktown peninsula. Meanwhile, a French fleet positioned itself at the mouth of the Chesapeake, cutting off the British army's lines of supply and retreat. Although a British squadron sailed from the West Indies and joined forces with naval vessels from New York in order to engage the French, the British were beaten off at the Battle of the Virginia Capes. With this temporary loss of maritime supremacy, the army at Yorktown had no means of resupply or evacuation—a situation faced by the French in Quebec twenty-one years earlier—and had little alternative but surrender.[98] After defeat in the Chesapeake, the British had lost their principal army in North America and thus effective control over much of the eastern seaboard. Nevertheless, the navy managed to salvage the British position in the West Indies by destroying the French fleet the following year at the Battle of the Saintes.[99] At the Treaty of Paris in 1783, Britain recognized the independence of the United States of America, but held onto its islands in the Caribbean as well as Newfoundland, Nova Scotia, Quebec, and parts of the northwest. Although the loss of the thirteen colonies seemed like a great disaster for Britain at the time, American independence had, in fact, shorn the British empire of its immense geographic anomaly.

Diverging Empires

The Treaty of Paris confirmed a geographic division of North America that had been developing over the previous 150 years. The new British North America comprised all land within the colonial boundaries of Nova Scotia and Quebec north of the Great Lakes, while the United States comprised all land south of the Great Lakes, east of the Mississippi, and north of Florida and New Orleans. By agreeing to a boundary through the Great Lakes, the British had given up the old French territories in the Ohio and Mississippi that formerly belonged to Quebec, although Britain held onto several fur posts south of the lakes until 1796. Such a division of the continent reflected different visions of empire. The British position in North America and the Caribbean represented a metropolitan, commercial, and maritime view. Command of the sea allowed the British to control islands in the West

Indies, as well as littoral spaces in the northeastern part of North America: the rim of Hudson Bay, Newfoundland and Labrador, Nova Scotia, the Gulf of St. Lawrence, and Quebec. British control of the upper St. Lawrence and the Great Lakes extended their imperial reach deep into the heart of the continent, protecting the northwest from American encroachment. In contrast, the American position on the continent rested on a territorial, agricultural vision of empire. From their base along the eastern seaboard, Americans looked westward to the lands that stretched from the Appalachian Mountains to the Mississippi. This enormous new space represented an immense opportunity to expand an agricultural economy, provide support for a growing population, and extend a democratic society.

Although British and American visions of empire had been protected by the Treaty of Paris, the aftermath of the American war produced significant geographical change. In British North America and parts of the Caribbean, considerable growth of population and expansion of settlement occurred. The influx of some 40,000 to 50,000 Loyalist refugees from the United States to British North America created new settlements in Nova Scotia and Quebec. The migration of almost 40,000 Loyalists to Nova Scotia led to the division of the province in the early 1780s and the creation of new Loyalist colonies in New Brunswick and Cape Breton Island.[100] Although several thousand refugees soon left the region, about 13,500 migrants settled in New Brunswick, approximately 19,000 stayed in peninsular Nova Scotia, and a few hundred moved to Cape Breton. Almost 6,000 Loyalists moved to Quebec, most of them settling along the upper St. Lawrence, with other pockets in the Eastern Townships, Montreal, Sorel, and along the bleak shore of the Gaspé peninsula.[101] The settlement of a large English-speaking, Protestant group in the western portion of Quebec led to the division of the province in 1791 into Upper and Lower Canada. Farther south, a few thousand Loyalists, a mix of white planters as well as free and enslaved blacks, moved to the Bahamas.[102]

In British North America, the influx of Loyalists led to significant economic growth and development. Agricultural settlement pushed up the Saint John River valley in New Brunswick, and along the upper St. Lawrence and Niagara rivers in Upper Canada. New towns were established at St. Andrews, Saint John, and Fredericton in New Brunswick; while a string of settlements, many bearing Hanoverian names, stretched along the upper St. Lawrence and Niagara rivers. In the Bahamas, Nassau became the center for island trade and government. Although the major staple exports of the British Atlantic economy remained in place, the West Indies and Newfoundland were no longer able to import produce legally from New England and the Mid-Atlantic, and needed new sources of supply. Despite importing agricultural produce from Nova Scotia, New Brunswick, and Quebec, West Indian merchants could not meet the demand and illicit trade with the United States soon developed. A similar situation existed in Newfoundland. As the resident population increased during the last quarter of the eighteenth century, there was a growing demand for provisions, which was satisfied by imports from Nova Scotia, Prince Edward Island, and southeastern Ireland.

The influx of Loyalists introduced a new social and cultural mix to the remaining British colonies. The migration to Nova Scotia reinforced the American presence among the New Englanders settled in the colony since the early 1760s, while the migration to the province of Quebec introduced a significant English-speaking minority to Lower Canada. Although these immigrants were loyal to the Crown, they had generations of American living behind them, and were less socially deferential than British officials in Quebec, Halifax, and Fredericton expected. In Nova Scotia and New Brunswick, Loyalists strengthened the houses of assembly and resisted centralized, imperial control.[103] Moreover, the Loyalists introduced several Protestant denominations, which weakened British designs to establish the Anglican Church as the principal church in each of the Maritime colonies.[104] The Loyalists also brought with them an American material culture, best exemplified by the delicate Anglican churches, modeled on New England prototypes, that were built in Nova Scotia in the 1790s.[105]

Yet for all the demographic, economic, social, and cultural changes engendered by the Loyalist influx, the political and military superstructure of the British Atlantic remained in place. The British imperial historian P. J. Marshall has observed that the imperial response to the loss of the American colonies "was less than strictly logical . . . Much survived unchanged . . . In the most important of the colonies that remained . . . the West Indian islands, the old system which had failed so disastrously from Britain's point of view remained in operation."[106] Yet this imperial framework, represented by the Crown, the military, and the Anglican Church, was indigenous to the British Atlantic and had served the empire well; the northern and southern flanks of North America had been protected from rebellion and the critically important sugar islands and fishery retained for Britain. Indeed, this imperial framework was maintained and enhanced as the empire expanded and developed in the late eighteenth and early nineteenth centuries.[107]

The new United States also changed considerably. The imperial restriction on westward expansion had been swept away, allowing settlers to move into western New York, the Ohio Valley, and the Kentucky country. What had been Crown land became the public domain controlled by the federal government. In 1785, Congress passed the Northwest Land Ordinance, allowing the survey and sale of this huge area. The survey began in the upper Ohio Valley. Through Jefferson's influence, land was sold cheaply to encourage immigration and settlement and to extend the American empire of yeoman farmers; the guarantors, thought Jefferson, of American democracy. The navigation acts also had been repudiated, allowing American merchants to trade around the Atlantic. In 1778, trade was opened with France, and this was soon extended to the German states, the Netherlands, and Scandinavia.[108] Americans also became more involved in the African slave trade, and developed a new trade to China. Other remnants of the old imperial system were dismantled. The Anglican Church was disestablished, and a division between church and state enshrined in the new constitution. State governments created new capitals, more central to their expanding backcountry populations: in South Carolina, the capital

FIGURE 6.17. *View of the City of Richmond from the Bank of James River*, by Benjamin Henry Latrobe, 1798. An American vision of a continental, republican empire represented by Jefferson's Capitol building dominating the fledgling town of Richmond, the new capital of Virginia, situated on the fall line between the tidewater and the Piedmont. The Maryland Historical Society, Baltimore, Maryland.

was moved from Charleston to Columbia; in Virginia, it was shifted from Williamsburg to Richmond. Jefferson's state capitol at Richmond represented the new nation's republican vision. Modeled on the Roman temple at Nimes, the capitol dominated the Piedmont town, emphasizing the importance of a democratically elected legislature (figure 6.17). The building was a direct repudiation of Nicholson's layout of Williamsburg, which had emphasized imperial authority, and served as a bold assertion of American political and cultural independence.[109]

More generally, the years after the American Revolution saw the British and American empires move farther and farther apart (figure 6.18). For Americans, the

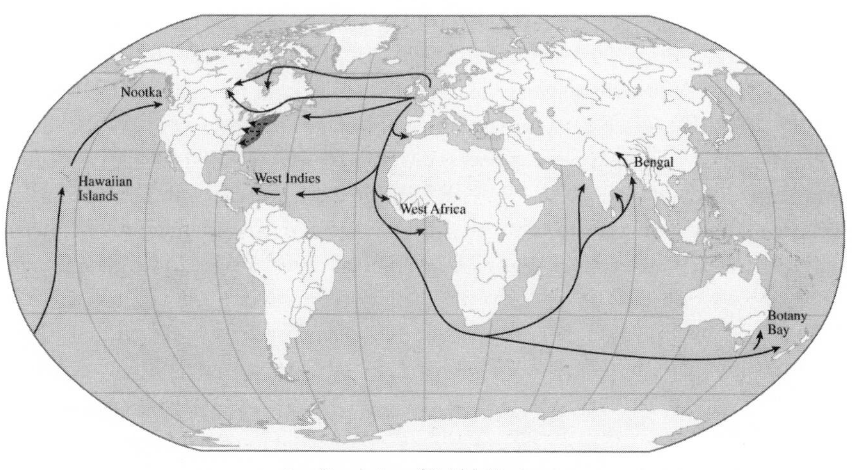

FIGURE 6.18. British and American empires in the 1780s.

FIGURE 6.19. *A View in Ship Cove*, by John Webber, 1778. A British vision of a maritime, commercial empire represented by James Cook's *Resolution* trading with the Nootka on the Northwest Coast. A comparison with John White's view (figure 1.4) of Frobisher's men encountering the Eskimo on the other side of the continent almost exactly two hundred years earlier suggests the continuity of the English maritime enterprise. By permission of the British Library, Shelfmark Add. 15514 f.10.

course of empire lay westward into the interior of the continent. Another century would pass and half of North America would be resettled before the American frontier officially closed in 1890. In the process, Americans would turn away from the Atlantic and become a transcontinental power. For the British, the course of empire lay eastward into the Indian and Pacific oceans. During the 1760s and 1770s, the British had expanded up the Ganges—South Asia's equivalent of the St. Lawrence—turning Calcutta into a version of Quebec and tightening their imperial grip over the princely states of north India. Meanwhile, James Cook had been exploring the Pacific, allowing the British to lay claim to Australia, New Zealand, and the Northwest Coast (figure 6.19). American expansion was internal, landward, and isolationist; British expansion was external, seaward, and engaged. The legacies of these continental and oceanic empires continue to influence our world today.

Notes

Introduction (pages 1–7)

1 Marc Bloch, "Toward a Comparative History of European Societies," in *Enterprise and Secular Change: Readings in Economic History*, ed. Frederic C. Lane and Jelle C. Riemersma, 494–521, quotation on 515 (Homewood, Ill.: Richard D. Irwin, Inc., 1953).

2 For transnational approaches to the past, see the AHR forum on "American Exceptionalism in an Age of International History," *American Historical Review* 96, no. 4 (1991): 1031–72; Michael Adas, "From Settler Colony to Global Hegemon: Integrating the Exceptionalist Narrative of the American Experience into World History," *American Historical Review* 106, no. 5 (2001): 1692–1720; and Thomas Bender, ed., *Rethinking American History in a Global Age* (Berkeley: University of California Press, 2002).

3 Donald W. Meinig, *The Shaping of America: A Geographical Perspective on 500 Years of History*, vol. I, *Atlantic America, 1492–1800* (New Haven: Yale University Press, 1986). This epic work has generated many reviews and discussions; a good place to start is Cole Harris, "Taking on a Continent," *Journal of Historical Geography* 14, no. 4 (1988): 416–19.

4 Bernard Bailyn, "The Idea of Atlantic History," *Itinerario* 20, no. 1 (1996): 19–44, quotation from 32. For a commentary on Bailyn's contribution to Atlantic history, see Peter A. Coclanis, "*Drang Nach Osten*: Bernard Bailyn, the World-Island, and the Idea of Atlantic History," *Journal of World History* 13, no. 1 (2002): 169–82. For an alternative view of the development of Atlantic history, see Nicholas Canny, "Atlantic History: What and Why?" *European Review* 9, no. 4 (2001): 399–411. Canny stresses the influence of Fernand Braudel's *The Mediterranean and the Mediterranean World in the Age of Philip II* (London: Fontana/Collins, 1975) in the development of an Atlantic perspective. Canny has also written on the British Atlantic in "Writing Atlantic History; or, Reconfiguring the History of Colonial British America," *Journal of American History* 86, no. 3 (1999): 1093–1114. Apart from Bailyn and Canny, see the "Round Table Conference: The Nature of Atlantic History," which includes articles on the Dutch, French, Iberian, British, and Black Atlantics, in *Itinerario* 23, no. 2 (1999); David Armitage, "Three Concepts of Atlantic History," in *The British Atlantic World, 1500–1800*, ed. Armitage and Michael Braddick, 11–27 (London: Palgrave, 2002); and the forum on "The New British History in Atlantic Perspective," *American Historical Review* 104, no. 2 (1999): 426–500. For an intriguing collection of essays focusing on maritime basins as a way of conceptualizing the world, see the "Oceans Connect" issue of the *Geographical Review* 89, no. 2 (1999). Of particular relevance are Martin W. Lewis and Kären Wigen, "A Maritime Response to the Crisis in Area Studies"; Jerry H. Bentley, "Sea and Ocean Basins as Frameworks

of Historical Analysis"; and Elizabeth Mancke, "Early Modern Expansion and the Politicization of Oceanic Space." For substantive examples focusing on the Atlantic, see Paul Butel, *The Atlantic* (London: Routledge, 1999), and Barry Cunliffe, *Facing the Ocean: The Atlantic and Its Peoples 8000 BC–AD 1500* (Oxford: Oxford University Press, 2001).

5 Daniel T. Rodgers, "Exceptionalism," in *Imagined Histories: American Historians Interpret the Past*, ed. Anthony Molho and Gordon S. Wood, 21–40 (Princeton, N.J.: Princeton University Press, 1998).

6 Frederick J. Turner, "The Significance of the Frontier in American History." *Annual Report of the American Historical Association for the Year 1893* (Washington, D.C.: American Historical Association, 1894). The literature on Turner is immense. For recent introductions, see William Cronon, "Turner's First Stand: The Significance of Significance in American History" in *Writing About Western History: Essays on Major Western Historians*, ed. Richard W. Etulain, 73–101 (Albuquerque: University of New Mexico Press, 1991); and Cronon, "Revisiting the Vanishing Frontier: The Legacy of Frederick Jackson Turner," *Western Historical Quarterly* 18 (1987): 157–76. Harold Innis's work includes *The Fur Trade in Canada* (New Haven: Yale University Press, 1930; rev. ed., Toronto: University of Toronto Press, 1956); *The Cod Fisheries: The History of an International Economy* (New Haven: Yale University Press, 1940; rev. ed.: Toronto: University of Toronto Press, 1954); *Problems of Staple Production in Canada* (Toronto: Ryerson Press, 1933); and "Significant Factors in Canadian Economic Development." *Canadian Historical Review* 18, no. 4 (1937): 374–84. Many of Innis's essays are collected in Daniel Drache, ed., *Staples, Markets, and Cultural Change* (Montreal and Kingston: McGill-Queen's University Press, 1995). For an introduction to the ideas of Innis, see Carl Berger, *The Writing of Canadian History: Aspects of English-Canadian Historical Writing since 1900* (Toronto: University of Toronto Pres, 1986), 85–111; and R. Douglas Francis, "The Anatomy of Power: A Theme in the Writings of Harold Innis" in *Nation, Ideas, Identities: Essays in Honour of Ramsay Cook*, ed. Michael D. Behiels and Marcel Martel, 26–40 (Oxford: Oxford University Press, 2000). Innis's relationship to geography is explored in "Focus: A Geographical Appreciation of Harold A. Innis," *Canadian Geographer* 37, no. 4 (1993): 352–64. Further essays on Innis from a variety of disciplinary perspectives are in William H. Melody, Liora Salter, and Paul Heyer. eds., *Culture, Communication, and Dependency: The Tradition of H. A. Innis* (Norwood, N.J.: Ablex Publishing Corporation, 1981); and Charles R. Acland and William J. Buxton, eds., *Harold Innis in the New Century: Reflections and Refractions* (Montreal and Kingston: McGill-Queen's University Press, 1999).

7 John J. McCusker and Russell R. Menard, *The Economy of British America 1607–1789* (Chapel Hill: University of North Carolina Press, 1985), especially 5–34.

8 M. H. Watkins, "A Staple Theory of Economic Growth," in *Approaches to Canadian Economic History*, edited by W. T. Easterbrook and M. H. Watkins, 49–73 (Toronto: McClelland and Stewart, 1967); Albert O. Hirschman, "A Generalized Linkage Approach to Development, with Special Reference to Staples," in *Essays on Economic Development and Cultural Change in Honor of Bert F. Hoselitz*, ed. M. Nash, 67–98 (Chicago: University of Chicago Press, 1977).

9 James E. Vance, Jr., *The Merchant's World: The Geography of Wholesaling* (Englewood Cliffs, N.J.: Prentice-Hall, Inc., 1970).

10 C. Earle and R. Hoffman, "Staple Crops and Urban Development in the Eighteenth-Century South," *Perspectives in American History* 10 (1976): 7–78.

11 Cole Harris, "European Beginnings in the Northwest Atlantic: A Comparative View," in *Seventeenth-Century New England*, ed. David D. Hall and David Grayson Allen, 119–52 (Boston: Colonial Society of Massachusetts, 1984); and Harris, "Making an Immigrant Society," in *The Resettlement of British Columbia: Essays on Colonialism and Geographical Change* (Vancouver: University of British Columbia Press, 1997), 250–75. For much of the Canadian evidence for this model, see R. Cole Harris, ed., *The Historical Atlas of Canada*, vol. I: *From the Beginning to 1800* (Toronto: University of Toronto Press, 1987). Harris's model, combining both the demand and supply side, seems both more comprehensive and nuanced than the interpretation of "free land" put forward in Barbara L. Solow, "Slavery and Colonization," in *Slavery and the Rise of the Atlantic System*, 21–42 (Cambridge: Cambridge University Press, 1991).

12 R. Cole Harris, "The Simplification of Europe Overseas," *Annals of the Association of American Geographers* 67, no. 4 (1977): 469–83; and Harris and Leonard Guelke, "Land and Society in Early Canada and South Africa," *Journal of Historical Geography* 3, no. 2 (1977): 135–53.

13 J. M. S. Careless, "Frontierism, Metropolitanism, and Canadian History," *Canadian Historical Review* 35, no. 1 (1954): 1–21; see also Careless, *Frontier and Metropolis: Regions, Cities, and Identities in Canada before 1914* (Toronto: University of Toronto Press, 1989).

14 Careless, "Frontierism, Metropolitanism, and Canadian History," 20–21.

15 William Cronon, *Nature's Metropolis: Chicago and the Great West* (New York: W.W. Norton & Company, 1991), 46–54.

16 Harold A. Innis, *Empire and Communications* (Oxford: Clarendon Press, 1950); see also Graeme Patterson, *History and Communications: Harold Innis, Marshall McLuhan, the Interpretation of History* (Toronto: University of Toronto Press, 1990), 3–12. Of particular importance is the critical work on Innis by James W. Carey in *Communication as Culture: Essays on Media and Society* (Boston: Unwin Hyman, 1989), 142–72; and "Culture, Geography, and Communications: The Work of Harold Innis in an American Context," in *Culture, Communication, and Dependency*, ed. Melody et al., 73–91. See also Jody Berland, "Space at the Margins: Colonial Spatiality and Critical Theory after Innis," *Topia* 1 (1997): 60–78, reprinted in Acland and Buxton, *Harold Innis in the New Century*, 281–308.

17 Innis discusses "Transportation as a Factor in Canadian Economic History," in *Problems of Staple Production*, 1–17.

18 This point is developed in a suggestive article by Donald W. Meinig, "A Macrogeography of Western Imperialism: Some Morphologies of Moving Frontiers of Political Control," in *Settlement and Encounter: Geographical Studies Presented to Sir Grenfell Price*, ed. Fay Gale and Graham H. Lawton, 213–40 (Melbourne: Oxford University Press, 1969).

19 Innis, *Fur Trade*, xvi, 386.

20 The relationship between power and space has been addressed by several modern social theorists, including Foucault, who memorably observed that "A whole history remains to be written of *spaces*—which would at the same time be the history of *powers* . . . from the great strategies of geo-politics to the little tactics of the habitat . . ." in *Power/Knowledge: Selected Interviews and Other Writings 1972–1977*,

ed. Colin Gordon, 149 (New York: Pantheon Books, 1980). Foucault further noted that "it took Marc Bloch and Fernand Braudel to develop a history of rural and maritime spaces." It seems to me that Turner and Innis were developing similar types of history for North America.

21 Michael Mann, *The Sources of Social Power*, vol. I (Cambridge: Cambridge University Press, 1986), 1.

22 In addition to Mann's ideas about different categories of power, I have drawn on the suggestive discussion of networks in Bruno Latour, *We Have Never Been Modern* (Cambridge, Mass.: Harvard University Press, 1993), 117–22.

1. Creating an English Atlantic, 1480–1630 (pages 8–25)

1 Janet L. Abu-Lughod, *Before European Hegemony: The World System A.D. 1250–1350* (Oxford: Oxford University Press, 1989); Fernand Braudel, *Civilization and Capitalism 15th–18th Century*, vol. III: *The Perspective of the World* (Berkeley: University of California Press, 1992), 116–38.

2 Braudel, *Perspective of the World*, 143–57.

3 E. Misselden, *Free Trade* (1622), 40, cited in B. E. Supple, *Commercial Crisis and Change in England 1600–1642* (Cambridge: Cambridge University Press, 1964), 6. See also Astrid Friis, *Alderman Cockayne's Project and the Cloth Trade* (London: Oxford University Press, 1927), especially 1–130; and G. D. Ramsay, *English Overseas Trade during the Centuries of Emergence* (London: Macmillan & Co., 1957), 1–33. For general context, see Lawrence Stone, "Elizabethan Overseas Trade," *Economic History Review* 11, no. 1 (1949–1950): 30–58.

4 Michael Postan, "The trade of Medieval Europe: The North," in the *Cambridge Economic History of Europe*, vol. II: *Trade and Industry in the Middle Ages*, 2nd ed., ed. M. M. Postan and Edward Miller, 282–305 (Cambridge: Cambridge University Press, 1987). See also Brian Dietz, "Antwerp and London: The Structure and Balance of Trade in the 1560s," in *Wealth and Power in Tudor England: Essays Presented to S. T. Bindoff*, ed. E. W. Ives, R. J. Knecht, and J. J. Scarisbrick, 186–203 (London: Athlone Press, 1978).

5 Maryanne Kowaleski, "The Expansion of the South-Western Fisheries in Late Medieval England," *Economic History Review* 53, no. 3 (2000): 429–54.

6 Patrick McGrath, "Bristol and America 1480–1631," in *The Westward Enterprise: English Activities in Ireland, the Atlantic, and America 1480–1650*, ed. K. R. Andrews et al., 81–102 (Detroit: Wayne State University Press, 1979).

7 Harold A. Innis, *The Cod Fisheries: The History of an International Economy* (New Haven: Yale University Press, 1940), 11–51. Although Innis and other scholars of the Newfoundland cod fishery have assumed that the Portuguese were involved in the sixteenth-century fishery, Darlene Abreu-Ferreira argues that the Portuguese were not a significant presence. Abreu-Ferreira, "Notes and Comments: Terra Nova through the Iberian Looking Glass: The Portuguese-Newfoundland Cod Fishery in the Sixteenth Century," *Canadian Historical Review* 79, no. 1 (1998): 100–115. For the French and Basque fisheries in Newfoundland, see John Mannion and Selma Barkham, "The 16th Century Fishery," in *The Historical Atlas of Canada*, vol. I: *From*

the Beginning to 1800, ed. R. Cole Harris, plate 22 (Toronto: University of Toronto Press, 1987); John Mannion and Gordon Handcock, "The 17th Century Fishery," ibid., plate 23.

8 Kenneth R. Andrews, *Trade, Plunder and Settlement: Maritime Enterprise and the Genesis of the British Empire, 1480–1630* (Cambridge: Cambridge University Press, 1984), 59–63.

9 Robert Brenner, *Merchants and Revolution: Commercial Change, Political Conflict, and London Overseas Traders, 1550–1653* (Princeton: Princeton University Press, 1993), 3–50; David Harris Sacks, *The Widening Gate: Bristol and the Atlantic Economy, 1450–1700* (Berkeley: University of California Press, 1991), 36–48. See also Brian Dietz, "Overseas Trade and Metropolitan Growth," in *London 1500–1700: The Making of the Metropolis*, ed. A. L. Beir and Roger Finlay, 115–40 (London: Longmans, 1986).

10 Ralph Davis, "England and the Mediterranean, 1570–1670" in *Essays in the Economic and Social History of Tudor and Stuart England in Honour of R. H. Tawney*, ed. F. J. Fisher, 117–37 (Cambridge: Cambridge University Press, 1961); Davis, *The Rise of the English Shipping Industry* (London: Macmillan & Co., 1962), 1–21, 228–55; and Ramsay, *English Overseas Trade*, 34–62.

11 Fernand Braudel, *The Mediterranean and the Mediterranean World in the Age of Philip II*, vol. I (London: Fontana/Collins, 1975), 621–29. For the decline of Venice, see Domenico Sella, "Crisis and Transformation in Venetian Trade," in *Crisis and Change in the Venetian Economy in the Sixteenth and Seventeenth Centuries*, ed. Brian Pullen, 88–105 (London: Methuen & Co., 1968).

12 Pauline Croft, *The Spanish Company* (London: London Record Society, 1973); Croft, "English Commerce with Spain and the Armada War, 1558–1603," in *England, Spain and the Gran Armada 1585–1604*, ed. M. J. Rodriguez-Salgado and Simon Adams, 236–63 (Edinburgh: John Donald Publishers, 1991).

13 Brenner, *Merchants and Revolution*, 16–19.

14 In 1931, Innis observed that "It is probably not too much to say that in the sixteenth and seventeenth centuries the cornerstone of the British Empire had been well and truly laid in Spanish trade." Innis, "The Rise and Fall of the Spanish Fishery in Newfoundland," *Proceedings and Transactions of the Royal Society of Canada*, Section 2 (1931): 51–70, quotation from 67.

15 Despite the work by Innis and many other scholars on the cod fishery, it is still common to find European writers ignoring the importance of the fishery to the colonial powers. For example, Niels Steensgaard, in his overview of English and Dutch trade, discusses the New World trades in tobacco and sugar but makes no mention of codfish; see Steensgaard, "The Growth and Composition of the Long-Distance Trade of England and the Dutch Republic before 1750," in *The Rise of Merchant Empires: Long-Distance Trade in the Early Modern World, 1350–1750*, ed. James D. Tracy, 102–152 (Cambridge: Cambridge University Press, 1990).

16 Anthony Parkhurst to Richard Hakluyt the elder, 13 November 1578, in David B. Quinn, ed., *New American World: A Documentary History of North America to 1612*, vol. IV (New York: Arno Press, 1979), 7.

17 Gillian T. Cell, *English Enterprise in Newfoundland 1577–1660* (Toronto: University of Toronto Press, 1969), 31.

18 State Papers Domestic, Charles II, cited in W. B. Stephens, "The West-Country Ports and the Struggle for the Newfoundland Fisheries in the Seventeenth Century," *Report and Transactions: Devonshire Association for the Advancement of Science, Literature and Art* 88 (1956): 90.

19 Davis, *Rise of English Shipping Industry*, 2–6.

20 Ibid., 6–7; Andrews, *Trade, Plunder and Settlement*, 24.

21 By focusing only on London in *Merchants and Revolution*, Brenner misses the rise of the small, independent merchants in the West Country in the 1570s and their pioneering role in colonial trade and commitment to free enterprise. Innis recognized their significance; see *Cod Fisheries*, 64–68.

22 The term "sack" probably refers to *vino de sacco* or wine set aside for export. See Peter E. Pope, "The South Avalon Planters, 1630 to 1700: Residence, Labour, Demand and Exchange in Seventeenth-Century Newfoundland" (Ph.D. diss., Memorial University, 1992), 93–143.

23 After the end of the war with Spain in 1604, London merchants attempted to reestablish the old chartered Spanish Company with its monopoly of trade to the Iberian peninsula, but they were defeated by outport merchants dealing in the Newfoundland fishery who pressed for free trade. See Croft, *Spanish Company*, xlv–xlvii, 68, 124; Croft, "Free Trade and the House of Commons, 1605–1606," *Economic History Review* 28, no. 1 (1975): 17–27; Robert Ashton, "The Parliamentary Agitation for Free Trade in the Opening Years of the Reign of James I," *Past and Present* 38 (1967): 40–55; and Friis, *Alderman Cockayne's Project*, 149–51.

24 Andrews, *Trade, Plunder and Settlement*, 101–26.

25 Thomas H. B. Symons, ed., *Meta Incognita: A Discourse of Discovery: Martin Frobisher's Arctic Expeditions, 1576–1578*, 2 vols. (Hull, Quebec: Canadian Museum of Civilization, 1999).

26 Andrews, *Trade, Plunder and Settlement*, 167–82.

27 Braudel, *Mediterranean World*, vol. II, 1176.

28 Geoffrey Parker, "David or Goliath? Philip II and His World in the 1580s," in *Spain, Europe and the Atlantic World: Essays in Honour of John H. Elliott*, ed. Richard L. Kagan and Geoffrey Parker, 245–66 (Cambridge: Cambridge University Press, 1995); and Parker, *The Grand Strategy of Philip II* (New Haven: Yale University Press, 1998).

29 Braudel, *Mediterranean World*, vol. II, 1176; Andrews, *Trade, Plunder and Settlement*, 230. See also Geoffrey W. Symcox, "The Battle of the Atlantic, 1500–1700," in *First Images of America: The Impact of the New World on the Old*, vol. I, ed. Fredi Chiappelli, 265–77, especially 271–72 (Berkeley: University of California Press, 1976).

30 Colin Martin and Geoffrey Parker, *The Spanish Armada* (New York: W.W. Norton & Company, 1988).

31 Cell, *English Enterprise*, 47–48; David B. Quinn, "Sir Bernard Drake," in *Dictionary of Canadian Biography*, vol. I: *1000 to 1700* (Toronto: University of Toronto Press, 1979), 278–80.

32 Simon Adams, "The Outbreak of the Elizabethan Naval War against the Spanish Empire: The Embargo of May 1585 and Sir Francis Drake's West Indies Voyage," in *England, Spain and the Gran Armada*, ed. Rodriguez-Salgado and Adams, 45–69.

33 Andrews, *Trade, Plunder and Settlement*, 235–45.

34 Ibid., 250.

35 Kenneth R. Andrews, *Elizabethan Privateering: English Privateering during the Spanish War 1585–1603* (Cambridge: Cambridge University Press, 1964), 32. Of the 206 vessels operating as privateers from known ports, 101 came from ports between Poole and Bristol. Andrews makes the point that many privateering vessels were simply merchantmen armed with guns and more men. See also Davis, *Rise of English Shipping*, 4.

36 Andrews, *Trade, Plunder and Settlement*, 248–49.

37 Andrews, *Elizabethan Privateering*, 128–29.

38 Ibid., 147–48, 229–30.

39 For the loss of vessels on the Indies run, see Parker, "Philip II and His World," 262. For a detailed discussion of the decline of the Spanish Basque presence in Newfoundland and Labrador, see Michael M. Barkham, "Shipowning, Shipbuilding and Trans-Atlantic Fishing in Spanish Basque Ports, 1560–1630, A Case Study of Motrico and Zumaya" (Ph.D. diss., Cambridge University, 1990). A more accessible account, outlining the general argument, is available in Barkham, "French Basque 'New Found Land' Entrepreneurs and the Import of Codfish and Whale Oil to Northern Spain, c.1580 to c.1620: The Case of Adam de Chibau, Burgess of Saint-Jean-de-Luz and 'Sieur de St. Julien,'" *Newfoundland Studies* 10, no. 1 (1994): 1–43. The importance of the Basques or "Biscayans" to the Spanish maritime empire is stressed in José Luis Casado Soto, "Atlantic Shipping in Sixteenth-Century Spain and the 1588 Armada," in *England, Spain and the Gran Armada*, ed. Rodriguez-Salgado and Adams, 95–133.

40 Innis, *Cod Fisheries*, 52; also Innis, "Spanish Fishery in Newfoundland."

41 Andrews, *Trade, Plunder and Settlement*, 183–222.

42 Even during the war years, significant contraband trade passed between England and Spain. This trade included dried fish from Newfoundland, shipped through French ports, to Spanish markets. See Pauline Croft, "Trading with the Enemy 1585–1604," *Historical Journal* 32, no. 2 (1989): 281–302, especially 292.

43 Howard Mumford Jones, *O Strange New World* (New York: Viking, 1964).

44 G. D. Ramsay, "Clothworkers, Merchant Adventurers and Richard Hakluyt," *English Historical Review* 92 (1977): 504–21.

45 Hariot's *True Report* is reprinted in Peter C. Mancall, ed., *Envisioning America: English Plans for the Colonization of North America, 1580–1640* (Boston: Bedford Books, 1995). White's watercolors are reproduced in Paul Hulton, *America 1585: The Complete Drawings of John White* (London: British Museum, 1984).

46 Thomas Richards, *The Imperial Archive: Knowledge and the Fantasy of Empire* (London: Verso, 1993).

47 Thomas Dunbabin, "George Waymouth," in *Dictionary of Canadian Biography*, vol. I, 667–68.

48 Andrews, *Trade, Plunder and Settlement*, 341–55.

49 David B. Quinn and Alison M. Quinn, eds., *The English New England Voyages 1602–1608* (London: Hakluyt Society, 1983).

50 Edmund S. Morgan, *American Slavery American Freedom: The Ordeal of Colonial Virginia* (New York: W.W. Norton, 1975), 71–91.

51 Faith Harrington, "'Wee Tooke Great Store of Cod-fish': Fishing Ships and First

Settlements on the Coast of New England, 1600–1630," in *American Beginnings: Exploration, Culture, and Cartography in the Land of Norumbega*, ed. Emerson W. Baker et al., 191–216 (Lincoln: University of Nebraska, 1994); Todd Gray, "Devon's Fisheries and Early-Stuart Northern New England," in *The New Maritime History of Devon*, ed. Michael Duffy et al., 139–44 (London: Conway Maritime Press, 1992).

52 Charles M. Andrews, *The Colonial Period of American History*, vol. I (New Haven: Yale University Press, 1934), 261–66.

53 Ibid., 344–74; Brenner, *Merchants and Revolution*, 148–53.

54 Andrews, *Colonial Period*, vol. I, 214–35; Brenner, *Merchants and Revolution*, 153–56.

55 Cell, *English Enterprise*, 53–96, quotation from 95.

56 Andrews, *Trade, Plunder and Settlement*, 280–300; Joyce Lorimer, "The English Contraband Trade in Trinidad and Guiana 1590–1617," in *Westward Enterprise*, ed. Kenneth R. Andrews et al., 124–50.

57 Andrews, *Trade, Plunder and Settlement*, 300–303.

58 Brenner, *Merchants and Revolution*, 92–112.

2. **Atlantic Staple Regions** (pages 26–72)

1 The average annual values of exports from areas under metropolitan control are approximate but they give a sense of magnitude. The figures are as follows: sugar, rum, molasses from the West Indies for the years 1768–1772 = £3,910,600; fish from Newfoundland for the years 1764–1774 = £453,000; furs from Hudson Bay for 1770 = £9,213 and from Quebec for 1770 = £28,433. The sources for these figures are John J. McCusker and Russell R. Menard, *The Economy of British America 1607–1789* (Chapel Hill: University of North Carolina Press, 1985), 160, Table 7.3; James F. Shepherd, "Staples and Eighteenth-Century Canadian Development: The Case of Newfoundland," in *Explorations in the New Economic History: Essays in Honor of Douglass C. North*, ed. Roger L. Ransom et al., 97–124 (New York: Academic Press, 1982), 111, Table 5; 117, Table 5.8.

For total exports from British America, I began with the average total for the years 1768 to 1772 in James F. Shepherd, "Commodity Exports from the British North American Colonies to Overseas Areas, 1768–1772: Magnitudes and Patterns of Trade," *Explorations in Economic History* 8 (1970–1971), 56, Table 7. I then deducted the Newfoundland figures contained in the article (which refer to exports from St. John's only) and added the more comprehensive estimates in Shepherd, "Staples and Eighteenth-Century Canadian Development," as well as the estimate for furs from Hudson Bay. The total was £7,268,694.

2 According to Innis, "The Avalon Peninsula [of Newfoundland] became in some sense a cornerstone of the British Empire from the standpoint of territory, trade, shipping, seamen, industry, agriculture, and finances." Harold A. Innis, "An Introduction to the Economic History of the Maritimes, Including Newfoundland and New England," *Canadian Historical Association Report* (1931): 85–95, reprinted in *Essays in Canadian Economic History* (Toronto: University of Toronto Press, 1956), 27–42.

3 Drawing on McCusker and Menard, *Economy of British America*, 160, Table 7.3;

Shepherd, "Staples and Eighteenth-Century Canadian Development," 11, Table 5.5; and Shepherd, "Commodity Exports," 65, the ranking is as follows:

1.	Sugar, rum, and molasses	£3,910,600
2.	Tobacco	£766,000
3.	Newfoundland fish	£453,000
4.	Bread and flour	£412,000*
5.	Rice	£312,000
6.	New England fish	£152,000
7.	Indigo	£117,000

*This figure includes all the continental colonies. For the Mid-Atlantic colonies, the principal colonial breadbasket, the figure was £171,044.

4 Drawing again on Shepherd's work, the average annual value of dried fish was as follows:

Newfoundland	£453,000 (1764–1774)
New England	£152,000 (1768–1772)
Nova Scotia	£9,314 (1768–1772)

5 John Mannion and Gordon Handcock, "The 17th Century Fishery," in *The Historical Atlas of Canada*, vol. I: *From the Beginning to 1800*, ed. R. Cole Harris, plate 23 (Toronto: University of Toronto Press, 1987).

6 John Mannion, Gordon Handcock and Alan Macpherson, "The Newfoundland Fishery, 18th Century," in *Historical Atlas of Canada*, vol. I, plate 25.

7 John J. Mannion, "Settlers and Traders in Western Newfoundland," in *The Peopling of Newfoundland: Essays in Historical Geography*, 234–75 (St. John's: Memorial University, 1977); Patricia Thornton, "The Transition from the Migratory to the Resident Fishery in the Strait of Belle Isle," in *Merchant Credit and Labour Strategies in Historical Perspective*, ed. Rosemary E. Ommer, 138–66 (Fredericton, N.B.: Acadiensis Press, 1990).

8 T. S. Willan, *Studies in Elizabethan Foreign Trade* (New York: Augustus M. Kelley, 1968), 65–91. Even when London merchants participated in the Newfoundland trade, they often relied on West Country connections. For the example of the London merchant Sir David Kirke and his West Country alliances, see Peter E. Pope, "The South Avalon Planters, 1630 to 1700: Residence, Labour, Demand and Exchange in Seventeenth-Century Newfoundland" (Ph.D. diss., Memorial University, 1992), 146.

9 John J. Mannion, "Irish Merchants Abroad: the Newfoundland Experience, 1750–1850," *Newfoundland Studies* 2, no. 2 (1986): 127–90; Rosemary Ommer, *From Outpost to Outport: A Structural Analysis of the Jersey-Gaspé Cod Fishery, 1767–1886* (Montreal and Kingston: McGill-Queen's University Press, 1991).

10 W. Gordon Handcock, *Soe Longe As There Comes Noe Women: Origins of English Settlement in Newfoundland* (St. John's: Breakwater, 1989), 219–32; D. F. Beamish, "Benjamin Lester," in *Dictionary of Canadian Biography*, vol. V: *1801 to 1820* (Toronto: University of Toronto Press, 1983), 490–92; Gordon Handcock and Alan Macpherson, "Trinity, 18th Century," in *Historical Atlas of Canada*, vol. I, plate 26.

11 C. Grant Head, Christopher Moore and Michael Barkham, "The Fishery in Atlantic Commerce," in ibid., plate 28.

12 Ralph Davis, "English Foreign Trade, 1700–1774," *Economic History Review* 15 (1962–1963): 285–303, especially 302–303. English woollen exports in 1772 to 1774 to

Southern Europe were worth £1,667,000; Newfoundland dried fish exports in 1764 to 1774 were valued at £453,000, although a small part of these exports was shipped to the West Indies. For the Mediterranean market, see Sari R. Hornstein, *The Restoration Navy and English Foreign Trade, 1674–1688: A Study in the Peacetime Use of Sea Power* (Aldershot: Scolar Press, 1991), 33–52.

13 H. E. S. Fisher, *The Portugal Trade: A Study of Anglo-Portuguese Commerce 1700–1770* (London: Methuen & Co., 1971), 92–106.

14 Pope, "South Avalon Planters," 95.

15 Ralph Pastore, "The Collapse of the Beothuk World," *Acadiensis* 19, no. 1 (Fall 1989): 52–71; and "The Sixteenth Century: Aboriginal Peoples and European Contact," in *The Atlantic Region to Confederation: A History*, ed. Phillip A. Buckner and John G. Reid, 22–39 (Toronto: University of Toronto Press, 1994).

16 Carl O. Sauer, *The Early Spanish Main* (Berkeley: University of California Press, 1966).

17 Handcock, *Soe Longe As There Comes Noe Women*, 145–216.

18 Ibid., 243–263.

19 Ibid., 249.

20 D. W. Prowse, *A History of Newfoundland* (London: Eyre and Spottiswoode, 1896), 298.

21 Daniel Vickers, "Merchant Credit and Labour Strategies in the Cod Fishery of Colonial Massachusetts," in *Merchant Credit and Labour Strategies*, 36–48.

22 Harold A. Innis, *Cod Fisheries: The History of an International Economy* (New Haven: Yale University Press, 1940), 95–111; C. Grant Head, *Eighteenth-Century Newfoundland* (Ottawa: Carleton University Press, 1976), 30–41.

23 Head, *Eighteenth-Century Newfoundland*, 24; Mannion and Handcock, "17th Century Fishery"; Mannion, Handcock, and Macpherson, "Newfoundland Fishery, 18th Century."

24 Benjamin Lester diary, 31 May 1762. Originals in the Dorset Records Office, Dorchester, England; microfilm copies in the National Archives of Canada, Ottawa. I am grateful to Gordon Handcock for making available a transcribed copy of the diaries.

25 Head, *Eighteenth-Century Newfoundland*, 3; John Mannion and C. Grant Head, "The Migratory Fisheries," in *Historical Atlas of Canada*, vol. I, plate 21.

26 Gerald L. Pocius, "The House that Poor-Jack Built: Architectural Stages in the Newfoundland Fishery," in *The Sea and Culture of Atlantic Canada*, ed. Larry McCann with Carrie MacMillan, 62–105 (Sackville, N.B.: Mount Allison University, 1992).

27 Harold Kalman, *A History of Canadian Architecture*, vol. I (Toronto: Oxford University Press, 1994), 88–90; Shane O'Dea, "The Tilt: Vertical-Log Construction in Newfoundland," in *Perspectives in Vernacular Architecture*, vol. I, ed. Camille Wells, 55–64 (Columbia: University of Missouri Press, 1987).

28 Head, *Eighteenth-Century Newfoundland*, 19.

29 E. A. Churchill, "A Most Ordinary Lot of Men: The Fishermen at Richmond Island, Maine in the Early Seventeenth Century," *New England Quarterly* 57, no. 2 (June 1984): 185. See also the discussion of socioeconomic differentiation in Pope, "South Avalon Planters," 265–67.

30 See map of Lester and Company Trading System, 1760–1770, in Handcock and Macpherson, "Trinity, 18th Century."

31 Pope, "South Avalon Planters," 201–56.

32 Head, *Eighteenth-Century Newfoundland*, 82; Mannion and Handcock, "17th Century Fishery"; Mannion, Handcock, and Macpherson, "Newfoundland Fishery, 18th Century." See also Pope, "South Avalon Planters," 212, 217.

33 John Mannion and W. Gordon Handcock, "Origins of the Newfoundland Population, 1836," in *Historical Atlas of Canada*, vol. II: *The Land Transformed*, ed. R. Louis Gentilcore, plate 8 (Toronto: University of Toronto Press, 1993). See also Mannion, "Old World Antecedents, New World Adaptations: Inistioge (Co. Kilkenny) Immigrants in Newfoundland," *Newfoundland Studies* 5, no. 2 (1989): 103–73; and Mannion, "Irish Migration and Settlement in Newfoundland: The Formative Phase, 1697–1732," *Newfoundland Studies* 17, no. 2 (2001): 257–93.

34 Innis, *Cod Fisheries*, 159.

35 Keith Matthews, *Lectures on the History of Newfoundland 1500–1830* (St. John's: Breakwater Books, 1988), 110; and Matthews, "History of the West of England—Newfoundland Fisheries (D. Phil. thesis, Oxford, 1968), which served as the basis for the *Lectures*.

36 Cited in *Historical Atlas of Canada*, vol. I, 50.

37 Head, *Eighteenth-Century Newfoundland*, 100–37; John Mannion, "Victualling a Fishery: Newfoundland Diet and the Origins of the Irish Provisions Trade, 1675–1700," *International Journal of Maritime History* 12, no. 1 (2000): 1–60; and Mannion, "The Waterford Merchants and the Irish-Newfoundland Provisions Trade, 1770–1820," in *Canadian Papers in Rural History*, vol. 3, ed. D. H. Akenson, 178–203 (Gananoque, Ont.: Langdale Press, 1983).

38 Benjamin Lester diary, 3 September 1767.

39 Head, *Eighteenth-Century Newfoundland*, 142; Pope, "South Avalon Planters," 257–317.

40 Matthews, *Lectures on the History of Newfoundland*, 85.

41 Pope, "South Avalon Planters," 267; Head, *Eighteenth-Century Newfoundland*, 143.

42 Pope, "South Avalon Planters," 270.

43 Ibid., 272.

44 Matthews, *Lectures on the History of Newfoundland*, 88.

45 Pope, "South Avalon Planters," 71–74.

46 Handcock, *Soe Longe As There Comes No Women*, 145–216.

47 Cited in ibid., 32.

48 Handcock and Macpherson, "Trinity, 18th Century."

49 Ibid.; Gordon Handcock, "The Poole Mercantile Community and the Growth of Trinity 1700–1839," *Newfoundland Quarterly* 15, no. 3 (1984): 19–30.

50 John Thomas, cited in Peter Pope, "'A True and Faithful Account': Newfoundland in 1680," *Newfoundland Studies* 12, no. 1 (1996): 32–49.

51 Derek Beamish et al., *Mansions and Merchants of Poole and Dorset* (Poole: Poole Historical Trust, 1976); Mark Girouard, *Town and Country* (New Haven: Yale University Press, 1992), 149–68.

52 John J. Mannion, *Point Lance in Transition: The Transformation of a Newfoundland Outport* (Toronto: McClelland and Stewart, 1976). Although Mannion focuses on an outport established by Irish settlers in the early nineteenth century, his findings are generally applicable to the rise of the resident population in the late eighteenth century.

53 G. M. Story et al., eds., *Dictionary of Newfoundland English* (Toronto: University of Toronto Press, 1982).

54 E. R. Seary, *Place Names of the Avalon Peninsula of the Island of Newfoundland* (Toronto: University of Toronto Press, 1971), 81–98, 121–36.

55 I would like to thank Elizabeth Mancke for clarifying the role of the military in civil government. See also Jerry Bannister, "The Naval State in Newfoundland, 1749–1791," *Journal of the Canadian Historical Association* n.s. 11 (2000): 17–50; and Bannister, "Convict Transportation and the Colonial State in Newfoundland, 1789," *Acadiensis* 27, no. 2 (1998): 95–123.

56 Matthews, *Lectures on the History of Newfoundland*, 89–102; see also Christopher English, "The Development of the Newfoundland Legal System to 1815," *Acadiensis* 20, no. 1 (Autumn 1990): 89–119.

57 Matthews, *Lectures on the History of Newfoundland*, 122; English, "Newfoundland Legal System," 103.

58 Benjamin Lester served as the MP for Poole between 1790 and 1796.

59 Gerald S. Graham, "Fisheries and Sea Power," in *Historical Essays on the Atlantic Provinces*, ed. G. A. Rawlyk, 7–16 (Toronto: McClelland and Stewart, 1971); Hornstein, *Restoration Navy and English Foreign Trade*, 55–59, 81, 260.

60 Handcock and Macpherson, "Trinity, 18th Century"; John Mannion, "St John's," in *Historical Atlas of Canada*, vol. I, plate 27.

61 Prowse, *History of Newfoundland*, 579–615.

62 McCusker and Menard, *Economy of British America*, 114; Donald W. Meinig, *The Shaping of America: A Geographical Perspective on 500 Years of History*, vol. I, *Atlantic America, 1492–1800* (New Haven: Yale University Press, 1986), 100–109.

63 McCusker and Menard, *Economy of British America*, 160, Table 7.3; B. W. Higman, "The Sugar Revolution," *Economic History Review* 53, no. 2 (2000): 213–36.

64 David Watts, *The West Indies: Patterns of Development, Culture and Environmental Change since 1492* (Cambridge: Cambridge University Press, 1987), 177–84.

65 Richard S. Dunn, *Sugar and Slaves: The Rise of the Planter Class in the English West Indies, 1624–1713* (New York: W.W. Norton, 1973), 83.

66 T. G. Burnard, "'Prodigious Riches': The Wealth of Jamaica Before the American Revolution," *Economic History Review* 54, no. 3 (2001): 506–24.

67 Watts, *West Indies*, 182–84, 223–28, 285–96.

68 Richard B. Sheridan, *Sugar and Slaves: An Economic History of the British West Indies 1623–1775* (Baltimore: The Johns Hopkins University Press, 1973), Appendix 1.

69 James Walvin, *Fruits of Empire: Exotic Produce and British Taste, 1660–1800* (New York: New York University Press, 1997), 117–31.

70 Davis, "English Foreign Trade, 1700–1774," 300–301.

71 Sheridan, *Sugar and Slaves: An Economic History*, 309.

72 Robert Brenner, *Merchants and Revolution: Commercial Change, Political Conflict, and London Overseas Traders, 1550–1653* (Princeton: Princeton University Press, 1993), 109–12.

73 Richard Pares, *Merchants and Planters* (Cambridge: Economic History Review, 1960), 5; Brenner, *Merchants and Revolution*, 161–63; Ronnie Hughes, "The Origin of Barbadian Sugar Plantations and the Role of the White Population in Sugar Plantation Society," in *Emancipation I*, ed. Alvin O. Thompson, 26–32 (Barbados: University of West Indies, 1984); Dunn, *Sugar and Slaves: The Rise of the Planter Class*, 58.

74 Pares, *Merchants and Planters*, 29–37; Sheridan, *Sugar and Slaves: An Economic History*, 262–305.

75 Pares, *Merchants and Planters*, 38–50; see also Pares, *A West-India Fortune* (London: Longmans, Green and Co., 1950); Sheridan, *Sugar and Slaves: An Economic History*, 295–98; and Sheridan, "The Wealth of Jamaica in the Eighteenth Century," *Economic History Review* 18 (1965): 292–311. For an alternative view, see S. D. Smith, "Merchants and Planters *Revisited*," *Economic History Review* 55, no. 3 (2002): 434–65.

76 Hughes, "Barbadian Sugar Plantations."

77 For the rise in the price of land, see Dunn, *Sugar and Slaves: The Rise of the Planter Class*, 66.

78 Ibid., 78.

79 Hilary Beckles, *A History of Barbados* (Cambridge: Cambridge University Press, 1990), 24.

80 Dunn, *Sugar and Slaves: the Rise of the Planter Class*, 58, 78.

81 Ibid., 131, 140–143; R. B. Sheridan, "The Rise of a Colonial Gentry: A Case Study of Antigua, 1730–1775," *Economic History Review* 13 (1961): 342–57.

82 Cited in Dunn, *Sugar and Slaves: The Rise of the Planter Class*, 148.

83 Ibid., 155.

84 Ibid., 165.

85 Douglas Hall, "Absentee-Proprietorship in the British West Indies, to about 1850," *Jamaican Historical Review* 4 (1964): 15–35; Frank W. Pitman, "The West Indian Absentee Planter as a British Colonial Type," *American Historical Association Pacific Coast Branch* 2 (1927): 113–27; Lowell Joseph Ragatz, "Absentee Landlordism in the British Caribbean, 1750–1833," *Agricultural History* 5 (1931): 7–24.

86 Pares, *West-India Fortune*.

87 Beckles, *History of Barbados*, 6; Kenneth F. Kiple and Kriemhild C. Ornelas, "After the Encounter: Disease and Demographics in the Lesser Antilles," in *The Lesser Antilles in the Age of European Expansion*, ed. Robert L. Paquette and Stanley L. Engerman, 50–67 (Gainesville: University Press of Florida, 1996).

88 Richard S. Dunn, "Servants and Slaves: The Recruitment and Employment of Labor," in *Colonial British America: Essays in the New History of the Early Modern Era*, ed. Jack P. Greene and J. R. Pole, 157–94 (Baltimore: Johns Hopkins University Press, 1984).

89 Hilary McD. Beckles, *White Servitude and Black Slavery in Barbados, 1627–1715* (Knoxville: University of Tennessee Press, 1989), 39.

90 David W. Galenson, *White Servitude in Colonial America: An Economic Analysis* (Cambridge: Cambridge University Press, 1981), 220–22.

91 Ibid., 48.

92 David Souden, "'Rogues, Whores and Vagabonds'? Indentured Servant Emigrants in North America, and the Case of Mid-Seventeenth-Century Bristol," *Social History* 3, no. 1 (1978): 23–41; John Wareing, "Migration to London and Trans-Atlantic Emigration of Indentured Servants, 1683–1775," *Journal of Historical Geography* 7, no. 4 (1981): 356–78. Dunn notes that two-thirds of the 2,678 migrants from Bristol to Barbados between 1654 and 1686 "came from the south-western corner of England, from the counties adjacent to Bristol"; see *Sugar and Slaves: The Rise of the Planter Class*, 71.

93 L. M. Cullin, "The Irish Diaspora of the Seventeenth and Eighteenth Centuries,"

in *Europeans on the Move: Studies on European Migration, 1500–1800*, ed. Nicholas Canny, 113–49, especially 126–27 (Oxford: Oxford University Press, 1994).

94 Riva Berleant-Schiller, "Free Labor and the Economy in Seventeenth-Century Montserrat," *William and Mary Quarterly* 46, no. 3 (1989): 539–64, especially 544.

95 Alan L. Karras, *Sojourners in the Sun: Scottish Migrants in Jamaica and the Chesapeake, 1740–1800* (Ithaca: Cornell University Press, 1992), 46–80.

96 Edward Long, *The History of Jamaica*, vol. 1 (London: 1774; new ed., Frank Cass & Co., 1970), 286.

97 Watts, *West Indies*, 200.

98 Beckles, *White Servitude and Black Slavery*, 46–58.

99 Watts, *West Indies*, 542.

100 Dunn, *Sugar and Slaves: The Rise of the Planter Class*, 55, 87, 127, 155; Watts, *West Indies*, 311.

101 Galenson, *White Servitude in Colonial America*, 174.

102 Watts, *West Indies*, 216–17.

103 Ibid., 311, 313.

104 Galenson, *White Servitude in Colonial America*; Beckles, *White Servitude and Black Slavery*, 115–39.

105 Dunn, "Servants and Slaves," 163–64.

106 James A. Rawley, *The Trans-Atlantic Slave Trade: A History* (New York: W.W. Norton, 1981), 167.

107 Long, *History of Jamaica*, vol. 1, 491.

108 K. G. Davies, *The Royal African Company* (London: Longmans, 1957); Rawley, *Trans-Atlantic Slave Trade*, 149–246; Nigel Tattersfield, *The Forgotten Trade* (London: Jonathan Cape, 1991).

109 Davies, *Royal African Company*, 240–64; David Hancock, *Citizens of the World: London Merchants and the Integration of the British Atlantic Community, 1735–1785* (Cambridge: Cambridge University Press, 1995), 172–220.

110 David Eltis, Stephen D. Behrendt, David Richardson, and Herbert S. Klein, eds., *The Trans-Atlantic Slave Trade: A Database on CD-ROM* (Cambridge: Cambridge University Press, 1999). For recent accounts drawing on the slave trade database, see David Eltis, *The Rise of African Slavery in the Americas* (Cambridge: Cambridge University Press, 2000), and David Richardson, "The British Empire and the Atlantic Slave Trade, 1660–1807," in the *Oxford History of the British Empire*, vol. II: *The Eighteenth Century*, ed. P. J. Marshall, 440–64 (Oxford: Oxford University Press, 1998).

111 Philip D. Curtin, *The Atlantic Slave Trade: A Census* (Madison: University of Wisconsin Press, 1969), 57–58; Karl Watson, "The Trans-Atlantic Slave Trade (with special reference to Barbados)," in *Emancipation I*, 20–21; Tattersfield, *Forgotten Trade*, 350–56.

112 Aaron S. Fogleman, "From Slaves, Convicts, and Servants to Free Passengers: The Transformation of Immigration in the Era of the American Revolution," *Journal of American History* (1998): 43–76. Fogleman's estimate for the period 1607 to 1699 is significantly higher than Galenson's figure of 90,000 (derived from the residual method) in *White Servitude in Colonial America*, 216–17. See also Jim Potter, "Demographic Development and Family Structure," in *Colonial British America*, 135–36.

113 Philip D. Morgan, "The Cultural Implications of the Atlantic Slave Trade: African

Regional Origins, American Destinations and New World Developments," in *Routes to Slavery: Direction, Ethnicity, and Mortality in the Atlantic Slave Trade*, ed. David Eltis and David Richardson, 122–45 (London: Frank Cass, 1997).

114 Watson, "Trans-Atlantic Slave Trade," 19. See also John Thornton, *Africa and Africans in the Making of the Atlantic World, 1400–1800* (Cambridge: Cambridge University Press, 1992).

115 Curtin, *Atlantic Slave Trade*, 150.

116 Codrington cited by Michael Craton, "Reluctant Creoles: The Planters' World in the British West Indies," in *Strangers within the Realm: Cultural Margins of the First British Empire*, ed. Bernard Bailyn and Philip D. Morgan, 330 (Chapel Hill: University of North Carolina Press, 1991).

117 Curtin, *Atlantic Slave Trade*, 154–61; Watson, "Trans-Atlantic Slave Trade," 19–20; Jerome S. Handler and Frederick W. Lange, *Plantation Slavery in Barbados: An Archaeological and Historical Investigation* (Cambridge: Harvard University Press, 1978), 20–28.

118 Cited in Dunn, *Sugar and Slaves: The Rise of the Planter Class*, 236.

119 David W. Galenson, *Traders, Planters, and Slaves: Market Behavior in Early English America* (Cambridge: Cambridge University Press, 1985), 94–95.

120 B. W. Higman, "Economic and Social Development of the British West Indies, from Settlement to ca. 1850," in *The Cambridge Economic History of the United States*, vol. I: *The Colonial Era*, ed. Stanley L. Engerman and Robert E. Gallman, 308–309 (Cambridge: Cambridge University Press, 1996).

121 Watts, *West Indies*, 311, 313.

122 Dunn, *Sugar and Slaves: The Rise of the Planter Class*, 237.

123 Watts, *West Indies*, 311, 313.

124 Thomas M. Truxes, *Irish-American Trade, 1660–1783* (Cambridge: Cambridge University Press, 1988), 76–77, 89–105; and Joseph Leydon, "The Irish Provisions Trade to the Caribbean, c.1650–1780: An Historical Geography" (Ph.D. diss., University of Toronto, 1995).

125 Shepherd, "Commodity Exports," Table 4, 37–48.

126 Watts, *West Indies*, 219–28, 393–97, 434–43.

127 Patrick Leigh-Fermor, *The Traveller's Tree* (New York: Harper & Brothers, 1950), 136.

128 B. W. Higman, *Jamaica Surveyed: Plantation Maps and Plans of the Eighteenth and Nineteenth Centuries* (Kingston: Institute of Jamaica Publications, 1988).

129 B. W. Higman, "The Spatial Economy of Jamaican Sugar Plantations: Cartographic Evidence from the Eighteenth and Nineteenth Centuries," *Journal of Historical Geography* 13, no. 1 (1987): 17–39.

130 Pares, *West-India Fortune*, 103.

131 David Buisseret, *Historic Architecture of the Caribbean* (London: Heinemann, 1980); Edward E. Crain, *Historic Architecture in the Caribbean Islands* (Gainesville: University Press of Florida, 1994); and A. W. Ackworth, *Treasure in the Caribbean: A First Study of Georgian Buildings in the British West Indies* (London: Pleides Books, 1949). See also Pamela Gosner, *Caribbean Georgian: The Great and Small Houses of the Caribbean* (Washington, D.C.: Three Continents Press, 1982); and Henry Fraser and Ronnie Hughes, *Historic Houses of Barbados* (Barbados: Barbados National Trust,

1986). The similarities between Jacobean and Georgian architecture in the Caribbean and the architecture of planter houses in Ireland are sometimes very close. For example, St. Nicholas Abbey, Barbados, a Jacobean house with Dutch gables dating from around 1650 to 1660 (perhaps the oldest English house in the New World), is very similar to Richhill Castle, County Armagh, illustrated in Brian de Breffny and Rosemary Ffolliott, *The Houses of Ireland: Domestic Architecture from the Medieval Castle to the Edwardian Villa* (New York: Viking, 1975), 65.

132 Evangeline Walker Andrews, ed., *Journal of a Lady of Quality* (New Haven: Yale University Press, 1934), 124.

133 Ibid., 83.

134 Pares, *West-India Fortune.*

135 Watts, *West Indies*, 195–96, 399–405.

136 Pares, *Merchants and Planters*, 23; Crain, *Historic Architecture*, 51–57; Higman, *Jamaica Surveyed*, 80–83.

137 Handler and Lange, *Plantation Slavery in Barbados*, 72–83.

138 Richard S. Dunn, "A Tale of Two Plantations: Slave Life at Mesopotamia in Jamaica and Mount Airy in Virginia, 1799 to 1828," *William and Mary Quarterly* 34, no. 1 (1977): 52.

139 Handler and Lange, *Plantation Slavery in Barbados*, 81.

140 William Dickson, *Letters on Slavery* (London: 1789) cited in Handler and Lange, *Plantation Slavery in Barbados*, 77.

141 Andrews, ed., *Journal of a Lady*, 127.

142 David Bary Gaspar, *Bondmen and Rebels: A Study of Master-Slave Relations in Antigua* (Baltimore: Johns Hopkins University Press, 1985), 255–58.

143 Jill Sheppard, *The "Redlegs" of Barbados: Their Origins and History* (New York: KTO Press, 1977).

144 Edward Brathwaite, *The Development of Creole Society in Jamaica 1770–1820* (Oxford: Oxford University Press, 1971), 105–75.

145 Franklin W. Knight refers to the plantocracy as "psychologically transients" in *The Caribbean: The Genesis of a Fragmented Nationalism* (New York: Oxford University Press, 1978), 65. See also Andrew Jackson O'Shaughnessy, *An Empire Divided: The American Revolution and the British Caribbean* (Philadelphia: University of Pennsylvania Press, 2000).

146 Newton D. Mereness, ed., *Travels in the American Colonies* (New York: Macmillan Company, 1916), 380.

147 Alexander Boyd, *England's Wealthiest Son: A Study of William Beckford* (London: Centaur Press, 1962); for another example, see Alistair Hennessy, "Penryhn Castle," *History Today* (January 1995): 40–45.

148 Robert McCrum et al., *The Story of English* (London: BBC Books, 1986), 195–208. More detailed studies include Thornton, *Africa and Africans*, 211–18; Maureen Warner Lewis, "The African Impact on Language and Literature in the English-speaking Caribbean," in *Africa and the Caribbean: The Legacies of a Link*, ed. Margaret E. Crahan and Franklin W. Knight, 101–23 (Baltimore: The Johns Hopkins University Press, 1979).

149 James Walvin, *Black Ivory: A History of British Slavery* (Washington, D.C.: Howard University Press, 1992), 157–75; Brathwaite, *Creole Society*, 212–39; and Thornton, *Africa and Africans*, 221–34.

150 Lydia M. Pulsipher, "Galways Plantation, Montserrat," in *Seeds of Change*, ed. Herman J. Viola and Carolyn Margolis, 138–159 (Washington, D.C.: Smithsonian Institution Press, 1991); and Robert L. Hall, "Savoring Africa in the New World," in ibid., 160–69.

151 Brathwaite, *Creole Society*, 23–25.

152 Keith D. Hunte, "Church and Society in Barbados in the Eighteenth Century," in *Social Groups and Institutions in the History of the Caribbean*, Annual Conference of the Association of Caribbean Historians, 15 (Puerto Rico: 1974).

153 Ibid., 17.

154 Barbara Hill, *Historic Churches of Barbados* (Barbados: Art Heritage Publications, 1984). It is worth remembering that very few of the churches existing in the West Indies today date to the eighteenth century, mainly because so many of the earlier buildings were destroyed by hurricanes.

155 Buisseret, *Architecture of the Caribbean*, 46–71.

156 David H. Makinson, *Barbados: A Study of North-American-West-Indian Relations 1739–1789* (London: Mouton & Co., 1964), 26, 74; and George Metcalf, *Royal Government and Political Conflict in Jamaica 1729–1783* (London: Longmans, 1965), 2, 228.

157 The fur trade through Hudson Bay receives one brief mention in McCusker and Menard, *Economy of British America*, and no mention at all in Greene and Pole, *Colonial British America*, or Jack P. Greene, *Pursuits of Happiness: The Social Development of Early Modern British Colonies and the Formation of American Culture* (Chapel Hill: University of North Carolina Press, 1988).

158 Shepherd, "Staples and Eighteenth-Century Canadian Development," Table 5.8, 117.

159 The centralization of bayside trade is discussed in Harold A. Innis, *The Fur Trade in Canada* (New Haven: Yale University Press, 1930), 142–43.

160 Ibid., 127–28.

161 Ibid., 127.

162 D. Wayne Moodie, Victor P. Lytwyn, and Barry Kaye, "Trading Posts, 1774–1821," in *Historical Atlas of Canada*, vol. I, plate 62.

163 Glyndwr Williams, ed., *Andrew Graham's Observations on Hudson's Bay 1767–91* (London: The Hudson's Bay Record Society, 1969), 315.

164 Innis, *Fur Trade*, 139.

165 Eric W. Morse, *Fur Trade Canoe Routes of Canada* (Toronto: University of Toronto Press, 1969), 42; Alexander Henry, *Travels and Adventures in Canada and the Indian Territories Between the Years 1760 and 1776* (Edmonton: Hurtig, 1969), 261–62.

166 D. Wayne Moodie, and Victor P. Lytwyn, "Fur-trade Settlements," in *Historical Atlas of Canada*, vol. I, plate 64.

167 D. Wayne Moodie, Barry Kaye, Victor P. Lytwyn, and Arthur J. Ray (Indians), "Peoples of the Boreal Forest and Parkland," in ibid., plate 65.

168 Innis, *Fur Trade*, 141–42.

169 Ibid., 125.

170 Williams, *Andrew Graham's Observations*, 244–45.

171 Moodie et al., "Peoples of the Boreal Forest and Parkland"; John Nicks, "Orkneymen in the HBC 1780–1821," in *Old Trails and New Directions: Papers of the Third North American Fur Trade Conference*, ed. Carol M. Judd and Arthur J. Ray, 102–26 (Toronto: University of Toronto Press, 1980); Edith I. Burley, *Servants of the*

Honourable Company: Work, Discipline, and Conflict in the Hudson's Bay Company, 1770–1879 (Toronto: Oxford University Press, 1997), 64–108.

172 Innis, *Fur Trade*, 125–26; Nicks, "Orkneymen."

173 Jennifer S. Brown, *Strangers in Blood: Fur Trade Company Families in Indian Country* (Vancouver: University of British Columbia, 1980), 29–35.

174 Joseph Robson, *An Account of Six Years Residence in Hudson's-Bay from 1733 to 1736 and 1744 to 1747* (London: 1752; reprinted S. R. Publishers, 1965), 38.

175 Brown, *Strangers in Blood*, 13.

176 Robson, *Six Years Residence*, 17.

177 Sylvia Van Kirk, *"Many Tender Ties" Women in Fur-Trade Society in Western Canada, 1670–1870* (Winnipeg: Watson & Dwyer, 1980), 41.

178 Burley, *Servants of the Honourable Company*, 175–76.

179 Ibid., 131–39.

180 Harold A. Innis, "An Introduction to the Economic History of the Maritimes, Including Newfoundland and New England," in *Essays in Canadian Economic History* (Toronto: University of Toronto Press, 1956), 39.

181 Arthur J. Ray, *Indians in the Fur Trade* (Toronto: University of Toronto Press, 1974), 105–107.

182 Arthur J. Ray, D. Wayne Moodie, and Conrad E. Heidenreich, "Rupert's Land," in *Historical Atlas of Canada*, vol. I, plate 57.

183 Ray, *Indians in the Fur Trade*, 12–23.

184 Innis, *Fur Trade*, 134.

185 Letter from Thomas McCleish, Albany Fort, 16 July 1716, in *Letters from Hudson Bay 1703–40*, ed. K. G. Davies, 43 (London: Hudson's Bay Record Society, 1965).

186 Williams, *Andrew Graham's Observations*, 322.

187 Daniel A. Baugh, "Maritime Strength and Atlantic Commerce: The Uses of 'a Grand Marine Empire'" in *An Imperial State at War: Britain from 1689 to 1815*, ed. Lawrence Stone, 185–223 (London: Routledge, 1994); Baugh, "Trade and Colonies: Financial and Maritime Strength, ca. 1714–1790," in *Maritime History*, vol. 2: *The Eighteenth Century and the Classic Age of Sail*, ed. John D. Hattendorf, 145–62 (Malabar, Fla.: Krieger Publishing Company, 1997).

188 This is contrary to the argument put forward by Greene that there was "a common developmental process that took all of the new societies established by the English overseas after 1600 in the direction of an ever more complex, differentiated, and Old World-style society"; see Greene, *Pursuits of Happiness*, 167. For an alternative interpretation, see R. Cole Harris, "The Simplification of Europe Overseas," *Annals of the Association of American Geographers* 67, no. 4 (1977): 469–83.

189 John J. McCusker, "British Mercantilist Policies and the American Colonies," in *The Cambridge Economic History of the United States*, vol. I: *The Colonial Era*, ed. Stanley L. Engerman and Robert E. Gallman, 337–62 (Cambridge: Cambridge University Press, 1996).

190 N. A. M. Rodger, "Sea-Power and Empire, 1688–1793," in *The Oxford History of the British Empire*, vol. II, 169–83. See also Daniel A. Baugh, "Elements of Naval Power in the Eighteenth Century," in *Maritime History*, vol. 2, 119–35; and N. A. M. Rodger, "The Exercise of Seapower and its Challenges," in ibid., 175–84.

191 For an account of some parts of this oceanic empire, see Hancock, *Citizens of the World*.

192 For the underside of this ordered world, see Peter Linebaugh and Marcus Rediker, *The Many-Headed Hydra: Sailors, Slaves, Commoners, and the Hidden History of the Revolutionary Atlantic* (Boston: Beacon Press, 2000), especially 143–247.

193 For the influence on artistic representation in Britain, see Edward W. Said, *Culture and Imperialism* (New York: Vintage Books, 1994), particularly his discussion of Jane Austen's *Mansfield Park*.

194 This qualifies the argument put forward by Greene in "Negotiated Authorities: The Problem of Governance in the Extended Polities of the Early Modern Atlantic World," in *Negotiated Authorities: Essays in Colonial Political and Constitutional History* (Charlottesville: University Press of Virginia, 1994), 1–24. For a treatment that highlights the different measures of control that the metropolitan government exercised over oceanic and territorial spaces, see Elizabeth Mancke, "Negotiating an Empire: Britain and Its Overseas Peripheries, c.1550–1780," in *Negotiated Empires: Centers and Peripheries in the Americas, 1500–1820*, ed. Christine Daniels and Michael V. Kennedy, 235–65 (New York: Routledge, 2002).

3. **Continental Staple Regions** (pages 73–125)

1 Apart from these major staple economies, lesser staples consisted of the whale fishery based in Nantucket and Providence, the fur trade through Albany and New York, and the trade in deer skins in the southern colonies. A comparative treatment of the fur trades through Hudson Bay and the Hudson Valley might be enlightening.

2 According to figures in Shepherd, the average annual value of dried fish exports between 1768 and 1772 comprised 32.3 percent of all exports from New England, and 54.5 percent of exports from Massachusetts. James F. Shepherd, "Commodity Exports from the British North American Colonies to Overseas Areas, 1768–1772: Magnitudes and Patterns of Trade," *Explorations in Economic History* 8 (1970–1971): 5–76.

3 Massachusetts General Court, House of Representatives, Committee on the History of the Emblem of the Codfish, *A History of the Emblem of the Codfish in the Hall of the House of Representatives* (Boston: Wright & Potter, 1895).

4 Todd Gray, "Devon's Fisheries and Early-Stuart Northern New England," in *The New Maritime History of Devon*, vol. I: *From Early Times to the Late Eighteenth Century*, ed. Michael Duffy et al., 139–144 (London: Conway Maritime Press, 1992); Faith Harrington, "'Wee Tooke Great Store of Cod-fish': Fishing Ships and First Settlements on the Coast of New England, 1600–1630," in *American Beginnings: Exploration, Culture, and Cartography in the Land of Norumbega*, ed. Emerson W. Baker et al., 191–216 (Lincoln: University of Nebraska, 1994). See also Gray, "Richmond Island: An Early-Stuart Devon Fishing Station in the Gulf of Maine," in *New Maritime History of Devon*, 145–46; and Alaric Faulkner, "Archaeology of the Cod Fishery: Damariscove Island," *Historical Archaeology* 19, no. 2 (1985): 57–86.

5 James Phinney Baxter, ed., *The Trelawny Papers* (Portland, Maine: Hoyt, Fogg, and Donham, 1884), 164, 33, 108, 137, 146.

6 Bernard Bailyn, *The New England Merchants in the Seventeenth Century* (Cambridge: Harvard University Press, 1955), 77.

7 Daniel Vickers, *Farmers and Fishermen: Two Centuries of Work in Essex County, Massachusetts, 1630–1850* (Chapel Hill: University of North Carolina Press, 1994); see also William I. Davisson and Dennis J. Dugan, "Commerce in Seventeenth-Century Essex County, Massachusetts," *Essex Institute Historical Collections* 107, no. 2 (1971): 113–42.

8 Martyn J. Bowden, "Culture and Place: English Sub-Cultural Regions in New England in the Seventeenth Century," *Connecticut History* 35, no. 1 (1994): 68–146, especially 113–15, 124–25.

9 Vickers, *Farmers and Fishermen*, 103–107.

10 Bailyn, *New England Merchants*, 78–82.

11 Vickers, *Farmers and Fishermen*, 130.

12 W. Gordon Handcock, *Soe Longe As There Comes Noe Women: Origins of English Settlement in Newfoundland* (St. John's: Breakwater, 1989), 29–30.

13 Christine Leigh Heyrman, *Commerce and Culture: The Maritime Communities of Colonial Massachusetts 1690–1750* (New York: W.W. Norton & Co., 1984), 214–15.

14 Harold A. Innis, *The Cod Fisheries: The History of an International Economy* (New Haven: Yale University Press, 1940; rev. ed., Toronto: University of Toronto Press, 1954), 117, 161.

15 Shepherd, "Commodity Exports." For the trade to southern Europe, see James G. Lydon, "Fish and Flour for Gold: Southern Europe and the Colonial American Balance of Payments," *Business History Review* 39, no. 2 (1965): 171–83; and "Fish for Gold: The Massachusetts Fish Trade with Iberia, 1700–1773," *New England Quarterly* 54, no. 4 (1981): 539–82.

16 Byron Fairchild, *Messrs. William Pepperrell: Merchants at Piscataqua* (Ithaca, N.Y.: Cornell University Press, 1954).

17 Lydon, "Fish for Gold."

18 *Historical Statistics of the United States: Colonial Times to 1970* (Washington, D.C.: U.S. Bureau of the Census, 1975), 1195, Series Z 534–538, "State of the Cod Fishery of Massachusetts: 1765 to 1775." The data are displayed cartographically in Richard W. Wilkie and Jack Tager, eds., *Historical Atlas of Massachusetts* (Amherst: University of Massachusetts Press, 1991), 25. Although the map of "Cod Fisheries 1765–1775" is for the late eighteenth century, the pattern of urban development was in place earlier, according to Vickers, *Farmers and Fishermen*, 145–47.

19 According to figures in Shepherd, "Commodity Exports," 86 percent of the average annual value of cod exports between 1768 and 1772 was controlled by Massachusetts (including Maine), 6.5 percent by New Hampshire, and 5.7 percent by Rhode Island.

20 Vickers, *Farmers and Fishermen*, 148–49; David B. Flemming, *The Canso Islands: An 18th Century Fishing Station* (Halifax, N.S.: Parks Canada Manuscript Report 308, 1977); Judith Tulloch, "The New England Fishery and Trade at Canso, 1720–1744," in *How Deep is the Ocean: Historical Essays on Canada's Atlantic Fishery*, ed. James E. Candow and Carol Corbin, 65–73 (Sydney, N.S.: University College of Cape Breton Press, 1997).

21 Innis, *Cod Fisheries*, 186; Graeme Wynn and Debra McNabb, "Pre-Loyalist Nova Scotia," in *The Historical Atlas of Canada*, vol. I: *From the Beginning to 1800*, ed. R. Cole Harris, plate 31 (Toronto: University of Toronto Press, 1987).

22 As Innis observed, "The inherently divisive character of the fishing industry showed itself in the emergence of a divided control of the fishing region." *Cod Fisheries*, 194. For the Channel Islands migratory fishery, see Rosemary Ommer, *From Outpost to Outport: A Structural Analysis of the Jersey-Gaspé Cod Fishery, 1767–1886* (Montreal and Kingston: McGill-Queen's University Press, 1991).

23 William A. Baker, "Fishing Under Sail in the North Atlantic," in *The Atlantic World of Robert G. Albion*, ed. Benjamin W. Labaree, 51 (Middletown, Conn.: Wesleyan University Press, 1975).

24 Innis, *Cod Fisheries*, 200.

25 Samuel Roads, Jr., *The History and Traditions of Marblehead* (Boston: Houghton, Mifflin and Company, 1881), 60.

26 For Salem yards, see James Duncan Phillips, *Salem in the Eighteenth Century* (Boston: Houghton Mifflin Company, 1937), 28, 30, 80, 121, 275–76, 465.

27 Samuel Eliot Morison, *The Maritime History of Massachusetts 1783–1860* (Boston: Houghton Mifflin Company, 1921), 15; Douglas R. McManis, *Colonial New England: A Historical Geography* (New York: Oxford University Press, 1975), 117.

28 Innis, *Cod Fisheries*, 134.

29 James G. Lydon, "North Shore Trade in the Early Eighteenth Century," *American Neptune* 28, no. 4 (1968): 261–74.

30 James F. Shepherd and Samuel H. Williamson, "The Coastal Trade of the British North American Colonies, 1768–1772," *Journal of Economic History* 32, no. 4 (1972): 783–810, especially 801; David C. Klingaman, "The Coastwise Trade of Colonial Massachusetts," *Essex Institute Historical Collections* 108, no. 3 (1972): 217–34. For the development of the Caribbean trade, see Richard Pares, *Yankees and Creoles: The Trade between North America and the West Indies before the American Revolution* (London: Longmans, Green and Co., 1956), especially 24–34.

31 Phyllis Whitman Hunter, *Purchasing Identity in the Atlantic World: Massachusetts Merchants, 1670–1780* (Ithaca, N.Y.: Cornell University Press, 2001), 46–55.

32 W. T. Baxter, *The House of Hancock: Business in Boston, 1724–1775* (New York: Russell & Russell, 1965), 209; James S. Leamon, "Maine in the American Revolution, 1763–1787," in *Maine The Pine Tree State from Prehistory to the Present*, ed. Richard W. Judd, Edwin A. Churchill, and Joel W. Eastman, 145 (Orono: University of Maine Press, 1995).

33 Vickers, *Farmers and Fishermen*, 156–57.

34 Daniel Vickers and Vince Walsh, "Young Men and the Sea: The Sociology of Seafaring in Eighteenth-Century Salem, Massachusetts," *Social History* 24, no. 1 (1999): 17–38.

35 Heyrman, *Commerce and Culture*, 226–27.

36 Ibid., 330–32; Lydon, "Fish for Gold."

37 Heyrman, *Commerce and Culture*, 343.

38 Ibid., 415–16.

39 James Birket, *Some Cursory Remarks Made by James Birket in His Voyage to North America 1750–1751* (New Haven: Yale University Press, 1936), 16.

40 William Hammond Bowden, "The Commerce of Marblehead, 1665–1775," *Essex Institute Historical Collections* 68 (1932): 138.

41 Vickers, *Farmers and Fishermen*, 168–69.

42 Heyrman, *Commerce and Culture*, 415–16.

43 Ibid., 354.

44 Ibid., 77, 336–37.

45 Vickers and Walsh, "Young Men and the Sea," especially 23.

46 Roads, *Marblehead*, 7, 43–45.

47 Richard L. Bushman, *The Refinement of America: Persons, Houses, Cities* (New York: Vintage Books, 1993), 100–27; Hunter, *Purchasing Identity*, 114–16.

48 A particularly stimulating and original discussion about the transfer of architectural style and its reworking in a different context is in Deborah Howard, *Venice and the East: The Impact of the Islamic World on Venetian Architecture 1100–1500* (New Haven: Yale University Press, 2000), especially 218.

49 W. Hammond Bowden, *The Jeremiah Lee Mansion* (Marblehead, Mass.: Marblehead Historical Society, n.d.); National Park Service, *Salem: Maritime Salem in the Age of Sail* (Washington, D.C.: U.S. Department of the Interior, 1987), 26–27.

50 Child's *A New Discourse of Trade* (c.1669) cited in Innis, *Cod Fisheries*, 111–12. For background on Child, see William Letwin, *Sir Josiah Child: Merchant Economist* (Boston: Harvard Graduate School of Business Administration, 1959).

51 Calculated from John J. McCusker and Russell R. Menard, *The Economy of British America 1607–1789* (Chapel Hill: University of North Carolina Pres, 1985), Table 5.2, 108; Table 6.1, 130; Table 8.2, 174; Table 9.3, 199.

52 J. Frederick Fausz, "An 'Abundance of Blood Shed on Both Sides': England's First Indian War, 1609–1614," *Virginia Magazine of History and Biography* 98, no. 1 (1990): 3–56.

53 Carville V. Earle, "Environment, Disease, and Mortality in Early Virginia," in *The Chesapeake in the Seventeenth Century: Essays in Anglo-American Society*, ed. Thad W. Tate and David L. Ammerman, 96–125 (Chapel Hill: University of North Carolina Press, 1979).

54 Ivor Noël Hume, *Martin's Hundred* (Charlottesville: University Press of Virginia, 1991), 65–66.

55 Allan M. Fraser, "Sir George Calvert," in *Dictionary of Canadian Biography*, vol. I (Toronto: University of Toronto Press, 1966), 162–63; Aubrey C. Land, *Colonial Maryland: A History* (New York: KTO Press, 1981), 4–8.

56 Russell R. Menard and Lois Green Carr, "The Lords Baltimore and the Colonization of Maryland," in *Early Maryland in a Wider World*, ed. David B. Quinn, 167–215 (Detroit: Wayne State University Press, 1982); Gary Wheeler Stone, "Manorial Maryland," *Maryland Historical Magazine* 82, no. 1 (1987): 3–36.

57 Land, *Colonial Maryland*, 33–56.

58 Robert Brenner, *Merchants and Revolution: Commercial Change, Political Conflict, and London Overseas Traders, 1550–1653* (Princeton: Princeton University Press, 1993), 92–112.

59 Jacob M. Price and Paul G. E. Clemens, "A Revolution in Scale in Overseas Trade: British Firms in the Chesapeake Trade, 1675–1775," *Journal of Economic History* 47, no. 1 (1987): 1–43, especially 2–8; Jacob M. Price, "Merchants and Planters: The Market Structure of the Colonial Chesapeake Reconsidered," in *Lois Green Carr: The Chesapeake and Beyond—A Celebration*, 96–97 (Crownsville: Maryland Historical & Cultural Publications, 1992); and Price, *Perry of London: A Family and Firm on the Seaborne Frontier, 1615–1753* (Cambridge: Harvard University Press, 1992), 29–30.

60 Cited in McCusker and Menard, *Economy of British America*, 119.

61 Lorimer, "English Contraband Trade," in *The Westward Enterprise: English Activities in Ireland, the Atlantic, and America 1480–1650*, ed. K. R. Andrews et al., 124–50 (Detroit: Wayne State University Press).

62 Russell R. Menard, "The Tobacco Industry in the Chesapeake Colonies, 1617–1730: An Interpretation," *Research in Economic History* 5 (1980): 109–77.

63 There is no consensus over the scale of white migration to the Chesapeake in the seventeenth century. Russell R. Menard estimates a "best" figure of 90,000 for the period 1625 to 1680 in Table 1 of "British Migration to the Chesapeake Colonies in the Seventeenth Century," in *Colonial Chesapeake Society*, ed. Lois Green Carr et al., 99–132, especially 102 (Chapel Hill: University of North Carolina Press, 1988). However, David W. Galenson reckons about 75,000 in Table 4.5 of "The Settlement and Growth of the Colonies: Population, Labor, and Economic Development," in *The Cambridge Economic History of the United States*, vol. I: *The Colonial Era*, ed. Stanley L. Engerman and Robert E. Gallman, 178 (Cambridge: Cambridge University Press, 1996). Alternatively, Nicholas Canny estimates 116,000 migrants for the entire seventeenth century in Table 4.1 of "English Migration into and across the Atlantic during the Seventeenth and Eighteenth Centuries," in *Europeans on the Move: Studies on European Migration, 1500–1800*, 39–75 (Oxford: Oxford University Press, 1994).

64 Durand of Dauphiné, *A Huguenot Exile in Virginia* (New York: The Press of the Pioneers, Inc., 1934), 110.

65 James Horn, *Adapting to a New World: English Society in the Seventeenth-Century Chesapeake* (Chapel Hill: University of North Carolina Press, 1994), 19–77; but see also Menard, "British Migration"; David Souden, "'Rogues, Whores and Vagabonds?': Indentured Servant Emigrants in North America, and the Case of Mid-Seventeenth-Century Bristol," *Social History* 3, no. 1 (1978): 23–41; and John Wareing, "Migration to London and Trans-Atlantic Emigration of Indentured Servants, 1685–1775," *Journal of Historical Geography* 7, no. 4 (1981): 356–78.

66 David Hackett Fischer makes much of the southern English contribution to the culture of the Chesapeake in his *Albion's Seed: Four British Folkways in America* (Oxford: Oxford University Press, 1989), 236–46, but fails to acknowledge that the south of England was not an homogeneous region. There were important differences between subregions, such as Gloucestershire and Kent, as Horn explains in *Adapting to a New World*, 78–120.

67 Galenson, "Settlement and Growth of Colonies," 171.

68 Menard, "Tobacco Industry," 118–19.

69 Quotation from James Horn, "Tobacco Colonies: The Shaping of English Society in the Seventeenth-Century Chesapeake," in *The Oxford History of the British Empire*, vol. I: *The Origins of Empire*, ed. Nicholas Canny, 171 (Oxford: Oxford University Press, 1998).

70 Peter C. Mancall, "Native Americans and Europeans in English America, 1500–1700," in ibid., 328–50.

71 Richard S. Dunn, "Masters, Servants, and Slaves in the Colonial Chesapeake and the Caribbean," in *Early Maryland*, 250.

72 Lois Green Carr and Russell R. Menard, "Immigration and Opportunity: The Freedman in Early Colonial Maryland," in *Chesapeake in the Seventeenth-Century*, 206–42. More generally, see Harris, "Simplification of Europe Overseas."

73 Lois Green Carr, "Emigration and the Standard of Living: The Seventeenth Century Chesapeake," *Journal of Economic History* 52, no. 2 (1992): 271–91, especially 283.

74 Russell R. Menard, P. M. G. Harris, and Lois Green Carr, "Opportunity and Inequality: The Distribution of Wealth on the Lower Western Shore of Maryland, 1638–1705," *Maryland Historical Magazine* 69 (1974): 169–84, especially 182.

75 Carville V. Earle, *The Evolution of a Tidewater Settlement System: All Hallow's Parish, Maryland, 1650–1783* (Chicago: University of Chicago Press, 1975), 19; David W. Jordan, "'In dispers'd Country Plantations': Settlement Patterns in Seventeenth-Century Surry County, Virginia," in *Chesapeake in the Seventeenth Century*, 183–205.

76 Quotation from Arthur Pierce Middleton, *Tobacco Coast: A Maritime History of Chesapeake Bay in the Colonial Era* (Newport News, Va.: The Mariners' Museum, 1953), 42.

77 Earle, *Tidewater Settlement System*, 60.

78 Such "earthfast" construction was once thought to be peculiar to the Chesapeake, but now appears to have been much more widespread. For seventeenth-century building practice in the Chesapeake, see Cary Carson et al., "Impermanent Architecture in the Southern American Colonies," *Winterthur Portfolio* 16, nos. 2/3 (1981): 135–78. See also Dell Upton, "New Views of the Virginia Landscape," *Virginia Magazine of History and Biography* 96, no. 4 (1988): 403–70.

79 Dauphiné, *Huguenot in Exile*, 119.

80 Cary Carson, "The 'Virginia House' in Maryland," *Maryland Historical Magazine* 69 (1974): 185–96. Dauphiné observed: "Whatever their rank . . . they build only two rooms with some closets on the ground floor, & two rooms in the attic above." Dauphiné, *Huguenot in Exile*, 119.

81 Lois Green Carr, Russell R. Menard, and Lorena S. Walsh, *Robert Cole's World: Agriculture and Society in Early Maryland* (Chapel Hill: University of North Carolina Press, 1991), 90–117.

82 Earle, *Tidewater Settlement System*, 18.

83 Ibid., 29.

84 Dauphiné, *Huguenot in Exile*, 116; Carr et al., *Robert Cole's World*, 55–75.

85 A French traveler cited in Earle, *Tidewater Settlement System*, 24.

86 Sally V. McGrath and Patricia J. McGuire, eds., *The Money Crop: Tobacco Culture in Calvert County, Maryland* (Crownsville: Maryland Historical and Cultural Publications, 1992); Carr et al., *Robert Cole's World*, 55–65.

87 Dauphiné, *Huguenot in Exile*, 115–16.

88 Henry M. Miller, "An Archaeological Perspective on the Evolution of Diet in the Colonial Chesapeake, 1620–1745," in *Colonial Chesapeake Society*, 176–99.

89 Carr, "Emigration and the Standard of Living," 279.

90 Carr et al., *Robert Cole's World*, 21–25.

91 Franklin W. Knight, *The Caribbean, the Genesis of a Fragmented Nationalism* (Oxford: Oxford University Press, 1978), 65. A similar argument is put forward by Dunn in an excellent essay comparing the Chesapeake and the Caribbean, "Masters, Servants, and Slaves," in *Early Maryland*, 242–66. Dunn observes that "By the time of the American Revolution, the two regions had diverged very strikingly. In the Chesapeake were rebels against the crown . . . Americans who thought of the New World as home. In the Caribbean were loyalists to the crown, many of whom were absentee proprietors, and who thought of England rather than the islands as home."

92 Carr et al., *Robert Cole's World*, 21–25.
93 Carr and Menard, "Immigration and Opportunity."
94 Fischer, *Albion's Seed*, 256–64.
95 The list is from Fischer, *Albion's Seed*, 240.
96 Maryland was divided into parishes in 1692.
97 Horn, *Adapting to a New World*, 187–98; Warren M. Billings, "The Growth of Political Institutions in Virginia, 1634 to 1676," *William and Mary Quarterly* 31, no. 2 (1974): 225–35.
98 Menard, "Chesapeake Tobacco Industry," 138–40. For a contrary view, see P. M. G. Harris, "Economic Growth and Demographic Perspective: The Example of the Chesapeake, 1607–1775," in *Lois Green Carr*, 55–92.
99 McCusker and Menard, *Economy of British America*, Figure 6.1, 121.
100 Jacob M. Price, "The Economic Growth of the Chesapeake and the European Market, 1697–1775," *Journal of Economic History* 24 (1964): 496–511; and Price, *France and the Chesapeake*, vol. 1 (Ann Arbor: University of Michigan Press, 1973), especially 509–648.
101 Price and Clemens, "Revolution of Scale," 24.
102 Price, "Economic Growth of the Chesapeake," 508–509; Price, "The Rise of Glasgow in the Chesapeake Tobacco Trade, 1707–1775," *William and Mary Quarterly* 11 (1954): 179–99. See also W. Iain Stevenson, "Some Aspects of the Geography of the Clyde Tobacco Trade in the Eighteenth Century," *Scottish Geographical Magazine* 89 (1973): 19–35; and T. M. Devine, *The Tobacco Lords: A Study of the Tobacco Merchants of Glasgow and their Trading Activities c.1740–90* (Edinburgh: John Donald Publishers, 1975). For Liverpool, see Paul G. E. Clemens, "The Rise of Liverpool, 1665–1750," *Economic History Review* 29, no. 2 (1976): 211–25.
103 Price, *Perry of London*, 28–40.
104 John W. Tyler, "Foster Cunliffe and Sons: Liverpool Merchants in the Maryland Tobacco Trade, 1738–1765," *Maryland Historical Magazine* 73, no. 3 (1978): 246–79; Price, "Glasgow in the Chesapeake Tobacco Trade"; J. H. Soltow, "Scottish Traders in Virginia, 1750–1775," *Economic History Review* 12 (1959): 83–98; and Felix Morley, "Pre-Revolutionary Letters to Great Britain from the Eastern Shore," *Maryland Historical Magazine* 76, no. 1 (1981): 45–61.
105 Soltow, "Scottish Traders," 85.
106 Michael L. Nicholls, "Competition, Credit and Crisis: Merchant-Planter Relations in Southside Virginia," in *Merchant Credit and Labour Strategies in Historical Perspective*, ed. Rosemary E. Ommer, 273–289 (Fredericton, N.B.: Acadiensis Press, 1990).
107 Price, "Rise of Glasgow," 197.
108 Jacob M. Price, *Capital and Credit in British Overseas Trade: The View from the Chesapeake, 1700–1776* (Cambridge: Harvard University Press, 1980), 10; see also Price's discussion of the role of credit on the settlement frontier in "Conclusion," in *Merchant Credit*, 360–73; T. H. Breen, *Tobacco Culture: The Mentality of the Great Tidewater Planters on the Eve of the Revolution* (Princeton: Princeton University Press, 1985), 128.
109 Cited in Breen, *Tobacco Culture*, 141.
110 For general context, see Marc Egnal and Joseph A. Ernst, "An Economic Interpretation of the American Revolution," *William and Mary Quarterly* 29, no. 1 (1972): 3–32, especially 24–28. For more detailed discussion, see Breen, *Tobacco Culture*, especially

124–203; and Tommy R. Thompson, "Personal Indebtedness and the American Revolution in Maryland," *Maryland Historical Magazine* 73, no. 1 (1978): 13–29.

111 Aubrey C. Land, "Economic Base and Social Structure: The Northern Chesapeake in the Eighteenth Century," *Journal of Economic History* 25, no. 4 (1965): 639–54.

112 Robert D. Mitchell, *Commercialism and Frontier: Perspectives on the Early Shenandoah Valley* (Charlottesville: University Press of Virginia, 1977), 61–63. The expansive views of tidewater planters, particularly those from the Northern Neck of Virginia, are well treated in Marc Egnal, *A Mighty Empire: The Origins of the American Revolution* (Ithaca, N.Y.: Cornell University Press, 1988), 87–101.

113 Louis Morton, *Robert Carter of Nomini Hall: A Virginia Tobacco Planter of the Eighteenth Century* (Williamsburg, Va.: Colonial Williamsburg, Inc., 1941), 62–71.

114 Russell Menard, "From Servants to Slaves: The Transformation of the Chesapeake Labor System," *Southern Studies* 16, no. 4 (1977): 355–90.

115 Dunn, "Servants and Slaves," in *Early Maryland*, 251–52.

116 These figures are calculated from David Eltis, Stephen D. Behrendt, David Richardson, and Herbert S. Klein, eds., *The Trans-Atlantic Slave Trade: A Database on CD-ROM* (Cambridge: Cambridge University Press, 1999). See also Philip D. Morgan, *Slave Counterpoint: Black Culture in the Eighteenth-Century Chesapeake and Low Country* (Chapel Hill: University of North Carolina Press, 1998), 58–59; and Darold D. Wax, "Black Immigrants: The Slave Trade in Colonial Maryland," *Maryland Historical Magazine* 73, no. 1 (1978): 30–45.

117 Curtin makes this point in *The Atlantic Slave Trade: A Census* (Madison: University of Winsconsin Press, 1969), 86–93, 215.

118 Morgan, *Slave Counterpoint*, 62–66; Wax, "Black Immigrants," 35–36. For the general context, see Curtin, *Atlantic Slave Trade*, 127–62, especially 144 and 161.

119 Daniel C. Littlefield, *Rice and Slaves: Ethnicity and the Slave Trade in Colonial South Carolina* (Urbana: University of Illinois Press, 1991), 11.

120 Morgan, *Slave Counterpoint*, 70–71; David W. Galenson, *Traders, Planters, and Slaves: Market Behavior in Early English America* (Cambridge: Cambridge University Press, 1985), 94–95.

121 Morgan, *Slave Counterpoint*, 62, 81–101.

122 Ibid., 41.

123 Peter Kolchin, *American Slavery 1619–1877* (New York: Hill and Wang, 1993), 33.

124 Galenson, "Settlement and Growth in the Colonies," 171–73.

125 Allan Kulikoff, *Tobacco and Slaves: The Development of Southern Cultures in the Chesapeake, 1680–1800* (Chapel Hill: University of North Caroline Press, 1986), 95, Map 2; Mitchell, *Commercialism and Frontier*, 26–31; Harry Roy Merrens, *Colonial North Carolina in the Eighteenth Century: A Study in Historical Geography* (Chapel Hill: University of North Carolina Press, 1964), 68, 73, 124.

126 Camille Wells, "The Planter's Prospect: Houses, Outbuildings, and Rural Landscapes in Eighteenth-Century Virginia," *Winterthur Portfolio* 28, no. 1 (1993): 1–31; Ashli White, "The Character of a Landscape: Domestic Architecture and Community in Late Eighteenth-Century Berkeley Parish, Virginia," *Winterthur Portfolio* 34, nos. 2/3 (1999): 109–38. More generally, see John Michael Vlach, *Back of the Big House: The Architecture of Plantation Slavery* (Chapel Hill: University of North Carolina Press, 1993).

127 Dauphiné, *Huguenot in Exile*, 120.

128 Daniel D. Reiff, *Small Georgian Houses in England and Virginia: Origins and Development through the 1750s* (London: Associated University Presses, 1986).

129 Houses that typify these three categories include Pemberton Hall, Maryland (1741); Beverly, Maryland (1774); and Mount Airy, Virginia (1748–1758). For the Maryland houses, see Michael Bourne et al., *Architecture and Change in the Chesapeake: A Field Tour of the Eastern and Western Shores* (Crownsville, Md.: Vernacular Architecture Forum and the Maryland Historical Trust Press, 1998), 124–26, 131–34. For discussions of the gentry landscape, see Carter L. Hudgins, "Robert 'King' Carter and the Landscape of Tidewater Virginia in the Eighteenth Century," in *Earth Patterns: Essays in Landscape Archaeology*, ed. William M. Kelso and Rachel Most, 59–70 (Charlottesville: University Press of Virginia, 1990); and Dell Upton, "Imagining the Early Virginia Landscape," in ibid., 71–86.

130 For example, Stratford and Mount Vernon in Virginia.

131 Paul G. E. Clemens, *The Atlantic Economy and Colonial Maryland's Eastern Shore: From Tobacco to Grain* (Ithaca, N.Y.: Cornell University Press, 1980), 168–205.

132 Kulikoff, *Tobacco and Slaves*, 275.

133 Richard L. Bushman, "American High-Style and Vernacular Cultures," in *Colonial British America: Essays in the New History of the Early Modern Era*, ed. Jack P. Greene and J. R. Pole, 345–83 (Baltimore: Johns Hopkins University Press, 1984), and his more detailed examination in *Refinement of America.*

134 Kulikoff, *Tobacco and Slaves*, 205–60.

135 Wilbur Zelinsky, "Some Problems in the Distribution of Generic Terms in the Place-Names of the Northeastern United States," *Annals of the Association of American Geographers* 45, no. 4 (1955): 319–49; Craig M. Carver, *American Regional Dialects: A Word Geography* (Ann Arbor: University of Michigan Press, 1987), 93–123.

136 Morgan, *Slave Counterpoint*, 560–80.

137 Ibid., 580–94.

138 Rhys Isaac, *The Transformation of Virginia 1740–1790* (Chapel Hill: University of North Carolina Press, 1982) especially 9–94; Dell Upton, "White and Black Landscapes in Eighteenth-Century Virginia," in *Material Life in America 1600–1860*, ed. Robert Blair St. George, 357–69 (Boston: Northeastern University Press, 1988); Upton, "Imagining the Early Virginia Landscape," in *Earth Patterns*, 71–86; and Upton, *Holy Things and Profane: Anglican Parish Churches in Colonial Virginia* (Cambridge: MIT Press, 1986).

139 Alan Gowans, "Paradigmatic Social Function in Anglican Church Architecture of the Fifteen Colonies," in *Retaining the Original: Multiple Originals, Copies, and Reproductions*, Center for Advanced Study in the Visual Arts Symposium Papers VII, 75–95 (Washington, D.C.: National Gallery of Art, 1989). For the Anglican Church in eighteenth-century Virginia, see John K. Nelson, *A Blessed Company: Parishes, Parsons, and Parishioners in Anglican Virginia, 1690–1776* (Chapel Hill: University of North Carolina Press, 2001).

140 Jack P. Greene, *Pursuits of Happiness: The Social Development of Early Modern British Colonies and the Formation of American Culture* (Chapel Hill: University of North Carolina Pres, 1988), 99–100.

141 Fischer writes movingly about the similarities between the Chesapeake and southern

England, "the cradle of Virginia," in *Albion's Seed*, 246. Nevertheless, it is worth remembering Carl Bridenbaugh's observation that slaves "as the largest single element in the population started and held the Chesapeake civilization on its unique course—a course that forever prevented the possibility of its becoming the exact replica of the English rural society its votaries so eagerly sought"; see Bridenbaugh, *Myths and Realities: Societies of the Colonial South* (Baton Rouge: Louisiana State University Press, 1952), 6.

142 For general accounts of colonial South Carolina, see Greene, *Pursuits of Happiness*, 141–51; Donald W. Meinig, *The Shaping of America: A Geographical Perspective on 500 Years of History*, vol. I: *Atlantic America, 1492–1800* (New Haven: Yale University PRess, 1986), 172–90; and Robert M. Weir, "'Shaftesbury's Darling': British Settlement in The Carolinas at the Close of the Seventeenth Century," in *Origins of Empire*, 375–97.

143 Robert M. Weir, *Colonial South Carolina: A History* (Millwood, N.Y.: KTO Press, 1983), 47–73; Wesley Frank Craven, *The Southern Colonies in the Seventeenth Century 1607–1689* (Louisiana State University, 1949), 310–59.

144 Converse D. Clowse, *Economic Beginnings in Colonial South Carolina 1670–1730* (Columbia: University of South Carolina Press, 1971), 46, 78.

145 Ibid., 48–50, 80.

146 Warren Alleyne and Henry Fraser, *The Barbados-Carolina Connection* (London: Macmillan Caribbean, 1988), 20–23.

147 Cited in Peter H. Wood, *Black Majority: Negroes in Colonial South Carolina from 1670 through the Stono Rebellion* (New York: W.W. Norton, 1974), 24.

148 Weir, *Colonial South Carolina*, 75–103.

149 David J. Weber, *The Spanish Frontier in North America* (New Haven: Yale University Press, 1992), 141–45.

150 R. C. Nash, "South Carolina and the Atlantic Economy in the Late Seventeenth and Eighteenth Centuries," *Economic History Review* 45, no. 4 (1992): 677–702; McCusker and Menard, *Economy of British America*, 171–75, 179–80, Table 8.2.

151 James Glen cited in Wood, *Black Majority*, 35.

152 Clowse, *Economic Beginnings in Colonial South Carolina*, 244.

153 Stephen G. Hardy, "Colonial South Carolina Rice Industry and the Atlantic Economy," in *Money, Trade, and Power: The Evolution of Colonial South Carolina's Plantation Society*, ed. Jack P. Greene, Rosemary Brana-Shute, and Randy J. Sparks, 108–40, Table 5.2 (Columbia: University of South Carolina Press, 2001).

154 McCusker and Menard, *Economy of British America*, 186–87.

155 Richard Waterhouse, *A New World Gentry: The Making of a Merchant and Planter Class in South Carolina, 1670–1770* (New York: Garland Publishing Inc., 1989), 55–56.

156 Peter A. Coclanis, *The Shadow of a Dream: Economic Life and Death in the South Carolina Low Country 1670–1920* (New York: Oxford University Press, 1989), 70, 98.

157 Governor James Glen, cited in Bridenbaugh, *Myths and Realities*, 57.

158 Morgan, *Slave Counterpoint*, 35, 40.

159 Alice Hanson Jones, *Wealth of a Nation to Be: The American Colonies on the Eve of the Revolution* (New York: Columbia University Press, 1980); Coclanis, *Shadow of a Dream*, 88–91.

160 Leila Sellers, *Charleston Business on the Eve of the American Revolution* (Chapel Hill: University of North Carolina Press, 1934), 49–78; R. C. Nash, "Urbanization in the Colonial South: Charleston, South Carolina as a Case Study," *Journal of Urban History* 19, no. 1 (1992): 3–29, especially 14, 20; and Russell R. Menard, "Financing the Lowcountry Export Boom: Capital and Growth in Early South Carolina," *William and Mary Quarterly* 51, no. 4 (1994): 659–76.

161 R. C. Nash, "Trade and Business in Eighteenth-Century South Carolina: The Career of John Guerard, Merchant and Planter," *South Carolina Historical Magazine* 96, no. 1 (1995): 6–29, especially 28; Egnal, *Mighty Empire*, 102–25.

162 Weir, *Colonial South Carolina*, 11–32.

163 "Governor William Bull's Representation of the Colony, 1770," in *The Colonial South Carolina Scene: Contemporary Views, 1697–1774*, ed. H. Roy Merrens, 253–70, quotation from 268. (Columbia: University of South Carolina Press, 1977).

164 Wood, *Black Majority*, 31.

165 These figures are calculated from Eltis et al., eds., *Trans-Atlantic Slave Trade*; see also Morgan, *Slave Counterpoint*, 59.

166 Morgan, *Slave Counterpoint*, 63; Littlefield, *Rice and Slaves*, Table 6, 127.

167 Henry Laurens to Peter Woodhouse, 18 November 1755, in *The Papers of Henry Laurens*, vol. 2, *Nov. 1, 1755–Dec. 31, 1758*, eds. Philip M. Hamer and George C. Rogers, Jr., 16 (Columbia: University of South Carolina Press, 1970); Morgan, *Slave Counterpoint*, 71.

168 This fundamental difference is explored, albeit for a slightly later period, in Richard S. Dunn, "A Tale of Two Plantations: Slave Life at Mesopotamia in Jamaica and Mount Airy in Virginia, 1799–1828," *William and Mary Quarterly* 34, no. 1 (1977): 32–65.

169 This total does not include the population of the backcountry.

170 Morgan, *Slave Counterpoint*, 61, Table 10. These figures also include small numbers of black slaves in the backcountry. See also Wood, *Black Majority*.

171 Morgan, *Slave Counterpoint*, 95–97.

172 George Ogilvie to Margaret Ogilvie, Myrtle Grove Plantation, Santee River, South Carolina, 25 June 1774, in *Discoveries of America: Personal Accounts of British Emigrants to North America during the Revolutionary Era*, ed. Barbara DeWolfe (Cambridge: Cambridge University Press, 1997), 199.

173 Coclanis, *Shadow of a Dream*, 65–68.

174 Clowse, *Economic Beginnings in Colonial South Carolina*, Table II: "Lands Granted Annually by County: 1670–1719," 253–55.

175 Mart A. Stewart, *"What Nature Suffers to Groe": Life, Labor, and Landscape on the Georgia Coast, 1680–1920* (Athens: University of Georgia Press, 1996).

176 Merrens, *Colonial North Carolina*, 125–33.

177 Sam B. Hilliard, "Antebellum Tidewater Rice Culture in South Carolina and Georgia," in *European Settlement and Development in North America: Essays on Geographical Change in Honour and Memory of Andrew Hill Clark*, ed. James R. Gibson, 91–115 (Toronto: University of Toronto Press, 1978); David Doar, *Rice and Rice Planting in the South Carolina Low Country* (Charleston: Charleston Museum, 1936); and Joyce E. Chaplin, "Tidal Rice Cultivation and the Problem of Slavery in South Carolina and Georgia, 1760–1815," *William and Mary Quarterly* 49, no. 1 (1992): 29–61.

178 Judith A. Carney, *Black Rice: The African Origins of Rice Cultivation in the Americas* (Cambridge: Harvard University Press, 2001).

179 The areal extent of early swamp rice cultivation is not known; see Hilliard, "Antebellum Tidewater Rice Culture," 97–98.

180 "A Gentleman's Account of his Travels, 1733–34," in *Colonial South Carolina Scene*, 110–21, quotation from 118.

181 Lawrence S. Rowland, "'Alone on the River': The Rise and Fall of the Savannah River Rice Plantations of St. Peter's Parish, South Carolina," *South Carolina Historical Magazine* 88, no. 3 (1987): 121–50.

182 Ogilvie to Ogilvie, 25 June 1774, *Discoveries of America*, 199.

183 D. E. Huger Smith, "A Plantation Boyhood," in *A Carolina Rice Plantation of the Fifties*, ed. Alice Huger Smith, 59–97, especially 60–63 (New York: William Morrow and Company, 1936).

184 "Journal of Lord Adam Gordon," in *Travels in the American Colonies*, ed. Newton D. Mereness, 397 (New York: Macmillan Company, 1916).

185 Samuel Gaillard Stoney, *Plantations of the Carolina Low Country* (Charleston: Carolina Art Association, 1938); and Jonathan H. Postan, *The Buildings of Charleston: A Guide to the City's Architecture* (Columbia: University of South Carolina Press, 1997).

186 Morgan, *Slave Counterpoint*, 104–24.

187 Ogilvie to Ogilvie, 25 June 1774, *Discoveries of America*, 200.

188 Rowland, "'Alone on the River,'" 126.

189 Hilliard, "Tidewater Rice Culture," 110–12.

190 Nash, "South Carolina and the Atlantic Economy," 696.

191 Morgan, *Slave Counterpoint*, 179–87.

192 Josiah Quincy quoted in Bridenbaugh, *Myths and Realities*, 64.

193 Richard Waterhouse, "Economic Growth and Changing Patterns of Wealth Distribution in Colonial Lowcountry South Carolina," *South Carolina Historical Magazine* 89, no. 4 (1988): 203–17, especially 211–12.

194 The estimate of 2,000 made by Robert Wells in 1765 is cited in Bridenbaugh, *Myths and Realities*, 65.

195 Reverend Alexander Hewatt, quoted in Bridenbaugh, *Myths and Realities*, 113.

196 Bridenbaugh observed, "the union of merchant and planter, of countinghouse and field, was the rule rather than the exception," in *Myths and Realities*, 67.

197 Henry Laurens to Richard Shubrick, 12 October 1756, in *Papers of Henry Laurens*, vol. 2, 335.

198 Jack P. Greene, "Colonial South Carolina and the Caribbean Connection," *South Carolina Historical Magazine* 88, no. 4 (1987): 192–210, especially 208.

199 Rebecca Starr, *A School for Politics: Commercial Lobbying and Political Culture in Early South Carolina* (Baltimore: Johns Hopkins University Press, 1998).

200 Morgan, *Slave Counterpoint*, 559–80; Charles Joyner, *Down by the Riverside: A South Carolina Slave Community* (Urbana: University of Illinois Press, 1985), 196–224.

201 Morgan, *Slave Counterpoint*, 134–43, 232–35, 580–609.

202 Weir, *Colonial South Carolina*, 75–103.

203 Several scholars have stressed the region's debt to the Caribbean. Joyner claims that "the rice coast of South Carolina might be considered the northernmost of the Brit-

ish West Indies." (*Down by the Riverside*, 13). Donald Meinig has written eloquently about the "shaping influence of English West Indian experience" on Charleston (*Atlantic America*, 190); and Jack Greene has argued that "South Carolina exhibited socio-economic and cultural patterns that, in many important respects, corresponded more closely to those in the Caribbean colonies than to those in the mainland colonies to the north" ("Colonial South Carolina and the Caribbean Connection," 193).

4. **Agricultural Frontiers** (pages 126–79)

1 For overviews, see Jack P. Greene, *Pursuits of Happiness: The Social Development of Early Modern British Colonies and the Formation of American Culture* (Chapel Hill: University of North Carolina Press, 1988), 18–27, 55–80; Donald W. Meinig, *The Shaping of America: A Geographical Perspective on 500 Years of History*, vol. I: *Atlantic America, 1492–1800* (New Haven: Yale University Press, 1986), 91–109; and Virginia DeJohn Anderson, "New England in the Seventeenth Century," in *The Oxford History of the British Empire*, vol. I: *The Origins of Empire*, ed. Nicholas Canny (Oxford University Press, 1998), 193–217. An especially suggestive interpretation is in Bernard Bailyn, "Slavery and Population Growth in Colonial New England," in *Engines of Enterprise: An Economic History of New England*, ed. Peter Temin, 253–59 (Cambridge: Harvard University Press, 2000). For a review of the literature on New England agriculture, see Amy D. Schwartz, "Colonial New England Agriculture: Old Visions, New Directions," *Agricultural History* 69, no. 3 (1995): 454–81.

2 Nicholas Canny, "English Migration into and across the Atlantic during the Seventeenth and Eighteenth Centuries," in *Europeans on the Move: Studies on European Migration, 1500–1800* (Oxford: Oxford University Press, 1994). Estimates of the numbers of Puritans emigrating to New England vary considerably. Virginia DeJohn Anderson reckons 13,000 in *New England's Generation: The Great Migration and the Formation of Society and Culture in the Seventeenth Century* (Cambridge: Cambridge University Press, 1991), 15; David Cressy estimates 21,000 in *Coming Over: Migration and Communication between England and New England in the Seventeenth Century* (Cambridge: Cambridge University Press, 1987), 63. The latter figure is accepted by Canny.

3 Cressy, *Coming Over*, 68–69.

4 In addition to Anderson, *New England's Generation* and Cressy, *Coming Over*, see T. H. Breen and Stephen Foster, "Moving to the New World: The Character of Early Massachusetts Immigration," *William and Mary Quarterly* 30, no. 2 (1973): 189–222; Richard Archer, "New England Mosaic: A Demographic Analysis for the Seventeenth Century," *William and Mary Quarterly* 47, no. 4 (1990): 477–502, particularly Table IV: "Geographical Origin of Emigrants to New England, 1620–1649," and the discussion on 484; and Martyn J. Bowden, "Culture and Place: English Sub-Cultural Regions in New England in the Seventeenth Century," *Connecticut History* 35, no. 1 (1994): 68–146.

5 David Hackett Fischer, *Albion's Seed: Four British Folkways in America* (Oxford: Oxford University Press, 1989), 31–36.

6 David Grayson Allen, "*Vacuum Domicilium*: The Social and Cultural Landscape of Seventeenth-Century New England," in *New England Begins: The Seventeenth Century*, vol. I, 1–9, especially 2–3 (Boston: Museum of Fine Arts, 1982).

7 John J. Waters, "Hingham, Massachusetts, 1631–1661: An East Anglian Oligarchy in the New World," *Journal of Social History* 1 (1967): 351–70, especially note 4.

8 Bowden, "Culture and Place," 87, 89–92. See also Frank Thistlethwaite, *Dorset Pilgrims: The Story of West Country Pilgrims Who Went to New England in the 17th Century* (London: Barrie & Jenkins, 1989).

9 Archer, "New England Mosaic," 487. For support for Archer's claim, at least for migrants from London, see Alison Games, *Migration and the Origins of the English Atlantic World* (Cambridge: Harvard University Press, 1999), 49–52.

10 W. Gordon Handcock, *Soe Longe As There Comes Noe Women: Origins of English Settlement in Newfoundland* (St. John's: Breakwater, 1989), 219–32; Bowden, "Culture and Place," 86, 88.

11 Anderson, "New England in the Seventeenth Century," 193–217, especially 211–12.

12 Anderson, *New England's Generation*, 31; John J. McCusker and Russell R. Menard, *The Economy of British America 1607–1789* (Chapel Hill: University of North Carolina Press, 1985), 226–27.

13 *Historical Statistics of the United States*, Series Z 1-19, "Estimated Population of American Colonies: 1610–1780."

14 For the urban population of New England, see Jacob Price, "Economic Function and the Growth of American Port Towns in the Eighteenth Century," *Perspectives in American History* 8 (1974): 123–86, Appendix B.

15 John K. Wright, "Regions and Landscapes of New England," in *New England's Prospect* (New York: American Geographical Society, 1933), 14–49.

16 Lois Kimball Mathews, *The Expansion of New England: The Spread of New England Settlement and Institutions to the Mississippi River, 1620–1865* (Boston: Houghton Mifflin Company, 1909), 35–36, especially the maps of settlement; Marcus L. Hansen, "The Settlement of New England," in Hans Kuruth, *Handbook of the Linguistic Geography of New England*, 2nd ed. (New York: AMS Press Inc., 1973), 62–104; and Douglas R. McManis, *Colonial New England: A Historical Geography* (New York: Oxford University Press, 1975), 42–51.

17 Jean Daigle and Robert LeBlanc, "Acadian Deportation and Return," in *The Historical Atlas of Canada*, vol. I: *From the Beginning to 1800*, ed. R. Cole Harris, plate 30 (Toronto: University of Toronto Press, 1987).

18 For New Hampshire, see William H. Wallace, "A Hard Land for a Tough People: A Historical Geography of New Hampshire," *New Hampshire Profiles* 24, no. 4 (1975): 21–32, especially 21–24; and "Some Aspects of Colonial Settlement in New Hampshire," *New England–St. Lawrence Valley Association of Geographers, Proceedings* 7 (1977): 16–24. For expansion into Maine, see Jamie Eves, *Fields and Farms: A Historical Atlas of the Yankee Settlement of Maine's Penobscot River Valley 1759–1875* (unpublished ms.); and for settlement in Nova Scotia, see Graeme Wynn and Debra McNabb, "Pre-Loyalist Nova Scotia," in *Historical Atlas of Canada*, vol. I, plate 31.

19 Anderson, "New England in the Seventeenth Century," 212–15; Peter C. Mancall,

"Native Americans and Europeans in English America, 1500–1700," in *Origins of Empire*, ed. Canny, 328–50.

20 Cited in Harald E. L. Prins, "Turmoil on the Wabanaki Frontier, 1524–1678," in *Maine The Pine Tree State from Prehistory to the Present*, ed. Richard W. Judd, Edwin A. Churchill, and Joel W. Eastman, 97–119, quotation from 108 (Orono: University of Maine Press, 1995).

21 Quotation from Thistlethwaite, *Dorset Pilgrims*, 121.

22 Address of the Council and Assembly of Massachusetts to Queen Anne, 20 October 1708, cited in Emerson W. Baker and John G. Reid, "Imperialism, Colonialism, and Amerindian Coercive Power in the Early Modern Northeast: A Reappraisal," unpublished paper presented at the Omohundro Institute of Early American History and Culture 7th Annual Conference, Glasgow, 2001.

23 David L. Ghere, "Diplomacy and War on the Maine Frontier, 1678–1759," in *Pine Tree State*, 120–42.

24 John Frederick Martin, *Profits in the Wilderness: Entrepreneurship and the Founding of New England Towns in the Seventeenth Century* (Chapel Hill: University of North Carolina Press, 1991).

25 James S. Leamon, "Maine in the American Revolution, 1763–1787," in *Pine Tree State*, 143–68, especially 144–45.

26 Sumner Chilton Powell, *Puritan Village: The Formation of a New England Town* (Middletown, Conn.: Wesleyan University Press, 1963).

27 Ibid., 116–32, Figure 19.

28 Edward T. Price, *Dividing the Land: Early American Beginnings of Our Private Property Mosaic* (Chicago: University of Chicago Press, 1995), 29–64; and William H. Wallace, "Dividing the Land in New England: Summing Up," unpublished paper presented at the Eastern Historical Geography Association Annual Meeting, October 2000. See also Wallace, *Commons and Meeting Houses in New Hampshire* (Durham: University of New Hampshire Department of Geography, for the Eastern Historical Geography Association Annual Meeting, September 1991); and James L. Garvin, "The Range Township in Eighteenth-Century New Hampshire," in *New England Prospect: Maps, Place Names, and the Historical Landscape*, ed. Peter Benes, 47–68 (Boston: Boston University Press, 1980).

29 Joseph S. Wood, *The New England Village* (Baltimore: Johns Hopkins University Press, 1997), 52–70.

30 Wallace, *Commons and Meeting Houses*, 6–9.

31 Sydney V. James, *Colonial Rhode Island: A History* (New York: Charles Scribner's Sons, 1975), 255.

32 David Grayson Allen, *In English Ways: The Movement of Societies and the Transferal of English Local Law and Custom to Massachusetts Bay in the Seventeenth Century* (New York: W.W. Norton & Co., 1982).

33 McManis, *Colonial New England*, 93.

34 Howard S. Russell, *A Long, Deep Furrow: Three Centuries of Farming in New England* (Hanover, N.H.: University Press of New England, 1976), 130–31.

35 Richard Lyman Bushman, "Markets and Composite Farms in Early America," *William and Mary Quarterly* 55, no. 3 (1998): 351–74.

36 Sarah F. McMahon, "A Comfortable Subsistence: The Changing Composition of Diet in Rural New England, 1620–1840," *William and Mary Quarterly* 42, no. 1 (1985): 26–65.

37 Daniel Vickers, "Competency and Competition: Economic Culture in Early America," *William and Mary Quarterly* 47, no. 1 (1990): 3–29. See also R. Cole Harris, "The Simplification of Europe Overseas," *Annals of the Association of American Geographers* 67, no. 4 (1977): 469–83.

38 John J. McCusker and Russell R. Menard, *The Economy of British America 1607–1789* (Chapel Hill: University of North Carolina Press, 1985), 108; Richard Pares, *Yankees and Creoles: The Trade between North America and the West Indies before the American Revolution* (London: Longmans, Green and Co., 1956); and David Richardson, "Slavery, Trade, and Economic Growth in Eighteenth-Century New England," in *Slavery and the Rise of the Atlantic System*, ed. Barbara L. Solow (Cambridge: Cambridge University Press, 1991), 237–64.

39 Calculated from McCusker and Menard, *Economy of British America*, 108. These export figures do not include the coastal trade to other parts of British America. According to Head, there was a significant trade in bread and flour from Boston to Newfoundland, although much of that produce must have come originally from the Mid-Atlantic colonies. C. Grant Head, *Eighteenth-Century Newfoundland* (Ottawa: Carleton University Press, 1976), 120.

40 James F. Shepherd, "Commodity Exports from British North American Colonies to Overseas Areas, 1768–1772: Magnitudes and Patterns of Trade," *Explorations in Economic History* 8 (1970–1971): Table 4. In 1750, West India merchant James Birket visited Newport, Rhode Island, and observed that "They have . . . a good Trade to the wt India Islands with flour, Pork, Shingles, Staves, Boards, Horses, &C the Chief of which the[y] purchase from their Neighbours in Connecticut Governmt." See James Birket, *Some Cursory Remarks Made by James Birket in His Voyage to North America 1750–1751* (New Haven: Yale University Press, 1916), 30.

41 McCusker and Menard observe that "without a more precise understanding of regional patterns and interrelationships the dynamics of New England's economic and social history will remain obscure," *Economy of British America*, 107. Bushman argues that "the configuration of New England agriculture suggests a spatial, not a temporal analysis," in "Markets and Composite Farms in Early America," 361. For one important subregion, see Bruce C. Daniels, "Economic Development in Colonial and Revolutionary Connecticut: An Overview," *William and Mary Quarterly* 37, no. 3 (1980): 429–50, especially 429–34. See also Russell, *Long, Deep Furrow*, 115. As an example of what needs to be done to understand the regional variations of the rural economy, see the study of an adjacent area, albeit for a later period, in Robert MacKinnon and Graeme Wynn, "Nova Scotian Agriculture in the 'Golden Age': A New Look," in *Geographical Perspectives on the Maritime Provinces*, ed. Douglas Day, 47–60 (Halifax, N.S.: Saint Mary's University, 1988).

42 Birket, *Some Cursory Remarks*, 3, 11–12, 14, 23; Charles F. Carroll, *The Timber Economy of Puritan New England* (Providence: Brown University Press, 1973).

43 Byron Fairchild, *Messrs. William Pepperrell: Merchants at Piscataqua* (Ithaca, N.Y.: Cornell University Press, 1954), 89.

44 Robert A. Gross, *The Minutemen and Their World* (New York: Hill and Wang, 1976), particularly 80–81; Gross, "Culture and Cultivation: Agriculture and Society in

Thoreau's Concord," *Journal of American History* 69, no. 1 (1982): 42–61; and Daniel Vickers, *Farmers and Fishermen: Two Centuries of Work in Essex County, Massachusetts, 1630–1850* (Chapel Hill: University of North Carolina Press, 1994), 31–83, 205–59. For the importance of the urban market, see Karen J. Friedmann, "Victualling Colonial Boston," *Agricultural History* 47, no. 3 (1973): 189–205. Grain production in Worcester and Middlesex counties is stressed in Bettye Hobbs Pruitt, "Self-Sufficiency and the Agricultural Economy of Eighteenth-Century Massachusetts," *William and Mary Quarterly* 41, no. 3 (1984): 358–61.

45 Birket, *Some Cursory Remarks*, 18. See also Russell, *Long, Deep Furrow*, 61–62.

46 Birket, *Some Cursory Remarks*, 30.

47 William Davis Miller, "The Narragansett Planters," *Proceedings of the American Antiquarian Society* 43 (1933): 49–115; Christian McBurney, "The South Kingstown Planters: Country Gentry in Colonial Rhode Island," *Rhode Island History* 45, no. 3 (1986): 81–93.

48 Stephen Innes, *Labor in a New Land: Economy and Society in Seventeenth-Century Springfield* (Princeton: Princeton University Press, 1983).

49 Margaret E. Martin, "Merchants and Trade of the Connecticut River Valley 1750–1820," *Smith College Studies in History* 24, nos. 1–4 (1938–1939), 1–284, especially 22–23; J. Ritchie Garrison, "Farm Dynamics and Regional Exchange: The Connecticut Valley Beef Trade, 1670–1850," *Agricultural History* 61, no. 3 (1987): 1–17.

50 Daniels, "Economic Development in Colonial and Revolutionary Connecticut," 433; Russell, *Long, Deep Furrow*, 142.

51 Garrison, "Connecticut Valley Beef Trade," 3; Daniels, "Economic Development in Colonial and Revolutionary Connecticut," 432.

52 Pruitt, "Self-Sufficiency and the Agricultural Economy," 333–64.

53 Robert Blair St. George, "Artifacts of Regional Consciousness in the Connecticut River Valley, 1700–1780," in *The Great River: Art and Society of the Connecticut River Valley, 1635- 1820*, ed. Gerald W. R. Ward and William N. Hosley, Jr., 29–39 (Hartford, Conn.: Wadsworth Atheneum, 1985).

54 J. P. Brissot de Warville, *New Travels in the United States of America 1788* (Cambridge: Harvard University Press, 1964), 117.

55 Journal of Lord Adam Gordon, in *Narratives of Colonial America 1704–1765*, ed. Howard H. Peckham, 292 (Chicago: R.R. Donnelley & Sons Company, 1971).

56 Kenneth A. Lockridge, *A New England Town, The First Hundred Years: Dedham, Massachusetts, 1636–1735* (New York: W.W. Norton, 1970); and Philip J. Greven, Jr., *Four Generations: Population, Land, and Family in Colonial Andover, Massachusetts* (Ithaca, N.Y.: Cornell University Press, 1970).

57 J. M. Bumsted and J. T. Lemon, "New Approaches in Early American Studies: The Local Community in New England," *Social History* 2 (1968): 98–112, especially 110–11.

58 De Warville, *New Travels in the United States*, 119.

59 Terry L. Anderson, "Wealth Estimates for the New England Colonies, 1650–1709," *Explorations in Economic History* 12 (1975): 151–76, especially 161.

60 Innes, *Labor in a New Land*; Martin, "Merchants and Trade."

61 Kevin M. Sweeney, "River Gods in the Making: The Williamses of Western Massachusetts," in *The Bay and the River 1600–1900*, ed. Peter Benes, 101–16 (Boston: Boston University Press, 1982).

62 Vickers, "Competency and Competition." See also Rosemary E. Ommer, ed., *Merchant Credit and Labour Strategies in Historical Perspective* (Fredericton, N.B.: Acadiensis Press, 1990), 255–359, especially Stephen Innes, "Commentary," 303–308.

63 For the general argument, see Harris, "Simplification of Europe Overseas." For an overview of the New England cultural region, see Martyn J. Bowden, "The New England Yankee Homeland," in *Homelands: A Geography of Culture and Place across America*, ed. Richard L. Nostrand and Lawrence E. Estaville, 1–23 (Baltimore: Johns Hopkins University Press, 2001).

64 Bruce E. Steiner, "New England's Anglicanism: A Genteel Faith?" *William and Mary Quarterly* 27, no. 1 (1970): 122–35.

65 Fischer, *Albion's Seed*, 57–62.

66 Craig M. Carver, *American Regional Dialects: A Word Geography* (Ann Arbor: University of Michigan Press, 1987), 21–38.

67 Bowden, "Culture and Place"; "New England Yankee Homeland," 13–14.

68 Abbott Lowell Cummings, *The Framed Houses of Massachusetts Bay, 1625–1725* (Cambridge: Harvard University Press, 1979), especially 202–209.

69 There are a few exceptions, such as the stone end houses in northern Rhode Island, but overall wood was the primary building material.

70 M. W. Barley, *The English Farmhouse and Cottage* (London: Routledge and Kegan Paul, 1961), 67–70; Eric Mercer, *English Vernacular Houses* (London: Royal Commission on Historic Monuments, 1975).

71 Martyn J. Bowden, "Invented Tradition and Academic Convention in Geographical Thought about New England," *GeoJournal* 26, no. 2 (1992): 187–94, especially 191–92.

72 Michael Steinitz, "Rethinking Geographical Approaches to the Common House: The Evidence from Eighteenth-century Massachusetts," in *Perspectives in Vernacular Architecture*, vol. III, ed. Thomas Carter and Bernard L. Herman, 16–26 (Columbia: University of Missouri Press, 1989); also reprinted in Wood, *New England Village*, 71–87.

73 Kevin M. Sweeney, "Mansion People: Kinship, Class, and Architecture in Western Massachusetts in the Mid Eighteenth Century," *Winterthur Portfolio* 19, no. 4 (1984): 231–55, especially note 22. See also J. Ritchie Garrison, *Landscape and Material Life in Franklin County, Massachusetts, 1770–1860* (Knoxville: University of Tennessee Press, 1991), 150–63.

74 Some evidence is presented in Laura B. Driemeyer and Myron O. Stachiw, *The Early Architecture and Landscapes of the Narragansett Basin*, vol. III: *Bristol and the East Bay and Wickford and the West Bay* (Newport, R.I.: Vernacular Architecture Forum, 2001), 69–71, 75–77.

75 Sweeney, "Mansion People"; Garrison, *Landscape and Material Life*; William N. Hosley, Jr., "Architecture," in *The Great River*, 485–523; Amelia F. Miller, "Connecticut River Valley Doorways: An Eighteenth-Century Flowering," in *The Bay and the River*, 60–72; and Miller, *Connecticut River Valley Doorways: An Eighteenth-Century Flowering* (Boston: Boston University, 1983).

76 Kevin M. Sweeney, "Meetinghouses, Town Houses, and Churches: Changing Perceptions of Sacred and Secular Space in Southern New England, 1720–1850," *Winterthur Portfolio* 28, no. 1 (1993): 59–93.

77 William H. Pierson, Jr., *American Buildings and their Architects*, vol. I: *The Colonial and Neoclassical Styles* (Oxford: Oxford University Press, 1970), 68–69, 94–105, 131–40.

78 T. H. Breen, "Persistent Localism: English Social Change and the Shaping of New England Institutions," *William and Mary Quarterly* 32, no. 1 (1975): 3–28.

79 For overviews of the Mid-Atlantic region, see Meinig, *Atlantic America*, 119–44; and Greene, *Pursuits of Happiness*, 124–41. For summaries of the literature on the Mid-Atlantic, see Douglas Greenberg, "*Notes and Documents*: The Middle Colonies in Recent American Historiography," *William and Mary Quarterly*, 36, no. 3 (1979): 396–427; and Wayne Bodle, "Themes and Directions in Middle Colonies Historiography, 1980–1994," *William and Mary Quarterly* 51, no. 3 (1994): 355–88.

80 James T. Lemon, *The "Best Poor Man's Country": A Geographical Study of Early Southeastern Pennsylvania* (New York: W.W. Norton & Co., 1976, 1972); Wilbur Zelinsky, *The Cultural Geography of the United States* (Englewood Cliffs, N.J.: Prentice-Hall, Inc., 1973), especially 117–33; Robert D. Mitchell, "The Formation of Early American Cultural Regions: An Interpretation," in *European Settlement and Development in North America: Essays on Geographical Change in Honour and Memory of Andrew Hill Clark*, ed. James R. Gibson, 66–90 (Toronto: University of Toronto Press, 1978); John Fraser Hart, *The Land That Feeds Us* (New York: W.W. Norton & Company, 1991), 19–40; and Pierce Lewis, "American Roots in Pennsylvania Soil," in *A Geography of Pennsylvania*, ed. E. Willard Miller, 1–13 (University Park: Pennsylvania State University Press, 1995). Some qualification of the argument, at least as it relates to agriculture, is in John Hudson, *Making the Corn Belt: A Geographical History of Middle-Western Agriculture* (Bloomington and Indianapolis: Indiana University Press, 1994).

81 Jan Lucassen, "The Netherlands, the Dutch, and Long-Distance Migration, in the Late Sixteenth to Early Nineteenth Centuries," in *Europeans on the Move: Studies on European Migration, 1500–1800*, ed. Nicholas Canny, 153–91, especially 178–80 (Oxford: Oxford University Press, 1994).

82 David W. Galenson, "The Settlement and Growth of the Colonies: Population, Labor, and Economic Development," in *The Cambridge Economic History of the United States*, vol. I: *The Colonial Era*, ed. Stanley L. Engerman and Robert E. Gallman, 135–207, especially 178, Table 4.5: "Decennial Net Migration of Whites to English America, by Region (thousands)" (Cambridge: Cambridge University Press, 1996).

83 T. C Smout, N. C. Landsmen, and T. M. Devine, "Scottish Emigration in the Seventeenth and Eighteenth Centuries," in *Europeans on the Move*, 76–112, especially 90–100; see also N. C. Landsmen, *Scotland and its First American Colony, 1683–1765* (Princeton: Princeton University Press, 1985).

84 L. M. Cullen, "The Irish Diaspora of the Seventeenth and Eighteenth Centuries," in *Europeans on the Move*, 113–49, especially 128.

85 Birket, *Some Cursory Remarks*, 62.

86 Marianne S. Wokeck, *Trade in Strangers: The Beginnings of Mass Migration to North America* (University Park: Pennsylvania State University Press, 1999); Georg Fertig, "Trans-Atlantic Migration from the German-Speaking Parts of Central Europe, 1600–1800: Proportions, Structures, and Explanations," in *Europeans on the Move*, 192–235; and Aaron Spencer Fogleman, *Hopeful Journeys: German Immigration,*

Settlement, and Political Culture in Colonial America, 1717–1775 (Philadelphia: University of Pennsylvania Press, 1996). The geographical distribution of German settlement is examined in Thomas L. Purvis, "The Pennsylvania Dutch and the German-American Diaspora in 1790," *Journal of Cultural Geography* 6, no. 2 (1986): 81–99.

87 *Historical Statistics of the United States*, Series Z 1-19, "Estimated Population of American Colonies: 1610–1780"; Price, "Economic Function and the Growth of American Port Towns," Appendix B.

88 Meinig, *Atlantic America*, 127, 132, 138; Peter O. Wacker, *Land and People: A Cultural Geography of Preindustrial New Jersey: Origins and Settlement Patterns* (New Brunswick, N.J.: Rutgers University Press, 1975), 127.

89 Peter C. Mancall, "Native Americans and Europeans in English America, 1500–1700," in *Origins of Empire*, 328–50; and Daniel K. Richter, "Native Peoples of North America and the Eighteenth-Century British Empire," in *The Oxford History of the British Empire*, vol. II: *The Eighteenth Century*, ed. P. J. Marshall, 347–71 (Oxford: Oxford University Press, 1998).

90 For the central importance of the Iroquois, see Conrad E. Heidenreich, "The Great Lakes Basin, 1600–1653," plate 35; "Re-establishment of Trade, 1654–1666," plate 37; "Expansion of French Trade, 1667–1696," plate 38; and "Trade and Empire, 1697–1739," plate 39, all in *Historical Atlas of Canada*, vol. I, ed. Harris.

91 Thomas J. Sugrue, "The Peopling and Depeopling of Early Pennsylvania: Indians and Colonists, 1680–1720," *Pennsylvania Magazine of History and Biography* 66, no. 1 (1992): 3–31.

92 The following draws on useful summaries in Price, *Dividing the Land*, 207–83.

93 Robert D. Mitchell, "The Colonial Origins of Anglo-America," in *North America: The Historical Geography of a Changing Continent*, ed. Robert D. Mitchell and Paul A. Groves, 115–16. (Totowa, N.J.: Rowman & Littlefield, 1987); Sung Bok Kim, *Landlord and Tenant in Colonial New York: Manorial Society, 1664–1775* (Chapel Hill: University of North Carolina Press, 1978).

94 Lemon, *"Best Poor Man's Country,"* 55.

95 Gottlieb Mittelberger, *Journey to Pennsylvania* (Cambridge: Harvard University Press, 1960), 51.

96 De Warville, *Travels in the United States*, 208.

97 Lemon, *"Best Poor Man's Country,"* 154–57; Peter O. Wacker and Paul G. E. Clemens, *Land Use in Early New Jersey: A Historical Geography* (Newark: New Jersey Historical Society, 1995), 141–45.

98 James Logan, cited in Lemon, *"Best Poor Man's Country,"* 154.

99 Michael V. Kennedy, "'Cash for His Turnups': Agricultural Production for Local Markets in Colonial Pennsylvania, 1725–1783," *Agricultural History* 74, no. 3 (2000): 587–608.

100 Lemon, *"Best Poor Man's Country,"* 84–217.

101 Kennedy, "'Cash for His Turnups.'"

102 Birket, *Some Cursory Remarks*, 46, 64–65, 53–54.

103 McCusker and Menard, *Economy of British America*, 199, Table 9.3.

104 According to McCusker and Menard, the value of agricultural and forest products from the Mid-Atlantic was £476,989, giving an approximate per capita value, based only on the rural population, of .9578; the value of agricultural and forest products from New England was £206,077, giving a per capita value of about .4081. Ibid.

105 C. Earle and R. Hoffman, "Staple Crops and Urban Development in the Eighteenth-Century South," *Perspectives in American History* 10 (1976): 32–35; Lemon, *"Best Poor Man's Country,"* 204–207.

106 William Douglass, cited in Jerome H. Wood, Jr., *Conestoga Crossroads: Lancaster, Pennsylvania 1730–1790* (Harrisburg, Penn.: Pennsylvania Historical and Museum Commission, 1979), 109.

107 Kennedy, "'Cash for His Turnups.'"

108 Lemon, *"Best Poor Man's Country,"* 130–46, 103–105.

109 Wood, *Conestoga Crossroads.*

110 Lemon, *"Best Poor Man's Country,"* 130–37.

111 Wilbur Zelinsky, "The Pennsylvania Town: An Overdue Geographical Account," *Geographical Review* 67, no. 2 (1977): 127–47. For the connection to Ulster, see Edward T. Price, "The Central Courthouse Square in the American County Seat," *Geographical Review* 58, no. 1 (1968): 29–60; and F. H. A. Aalen et al., eds., *Atlas of the Irish Rural Landscape* (Toronto: University of Toronto Press, 1997), 185. See also Richard Pillsbury, "The Pennsylvania Culture Area: A Reappraisal," *North American Culture* 3, no. 2 (1987): 37–54.

112 Price, "Central Courthouse Square."

113 Lemon, *"Best Poor Man's Country,"* 140.

114 De Warville, *Travels in the United States*, 153.

115 Wacker, *Land and People*, 158–220; Wilbur Zelinsky, "Ethnic Geography," in *Geography of Pennsylvania*, 113–31, especially 116–18.

116 Kenneth W. Keller, "What is Distinctive about the Scotch-Irish?" in *Appalachian Frontiers: Settlement, Society, and Development in the Preindustrial Era*, ed. Robert D. Mitchell, 69–86, especially 79–83 (Lexington: University Press of Kentucky, 1991).

117 Gabrielle M. Lanier and Bernard L. Herman, *Everyday Architecture of the Mid-Atlantic: Looking at Buildings and Landscapes* (Baltimore: Johns Hopkins University Press, 1997). See also the relevant sections of Dell Upton, ed., *America's Architectural Roots: Ethnic Groups that Built America* (Washington, D.C.: Preservation Press, 1986), as well as Richard Pillsbury, "Patterns in the Folk and Vernacular House Forms of the Pennsylvania Culture Region," *Pioneer America* 9, no. 1 (1977): 12–31; and Peter O. Wacker, "Traditional House and Barn Types in New Jersey: Keys to Acculturation, Past Cultureographic Regions, and Settlement History," in *Geoscience and Man*, vol. V: *Man and Cultural Heritage: Papers in Honor of Fred B. Kniffen*, ed. H. J. Walker and W. G. Haag, 163–76 (Baton Rouge: Louisiana State University, 1974). A recent overview of the Pennsylvania culture region is in Richard Pillsbury, "The Pennsylvanian Homeland," in *Homelands*, 24–43.

118 Clifford W. Zink, "Dutch Framed House in New York and New Jersey," *Winterthur Portfolio* 22, no. 4 (1987): 265–94.

119 John Fitchen, *The New World Dutch Barn: A Study of Its Characteristics, Its Structural System, and Its Probable Erectional Procedures* (Syracuse: Syracuse University Press, 1968).

120 Lanier and Herman, *Everyday Architecture*, 90–91.

121 Henry Glassie, "Eighteenth-Century Cultural Process in Delaware Valley Folk Building," *Winterthur Portfolio* 7 (1972): 29–57, reprinted in *Common Places: Readings in American Vernacular Architecture*, ed. Del Upton and John Michael Vlach, 394–425 (Athens: University of Georgia Press, 1986).

122 Fred B. Kniffen, "Folk Housing: Key to Diffusion," *Annals of the Association of American Geographers* 55, no. 4 (1965): 549–77, reprinted in *Common Places*, 3–26.

123 Robert F. Ensminger, *The Pennsylvania Barn: Its Origins, Evolution, and Distribution in North America* (Baltimore: Johns Hopkins University Press, 1992); Charles H. Dornbusch and John K. Heyl, "Pennsylvania German Barns," *The Pennsylvania German Folklore Society*, vol. 21, 1956 (Allentown, Penn.: Schlechter's, 1958); and Lanier and Herman, *Everyday Architecture*, 181–84.

124 For surviving colonial religious buildings in the United States, see Harold Wickliffe Rose, *The Colonial Houses of Worship in America* (New York: Hastings House, Publishers, 1963).

125 Jack Greene's observation that "the Middle Colonies were becoming more like Britain as they became more settled" seems hardly supportable; see Greene, *Pursuits of Happiness*, 135.

126 A classic early statement on the southern backcountry is Frederick Jackson Turner's discussion of the "Old West" in *The Frontier in American History* (New York: Holt, Rinehart and Winston, 1962), 67–125, especially 99–106. See also Carl Bridenbaugh, *Myths and Realities: Societies of the Colonial South* (Baton Rouge: Louisiana State University Press, 1952), 119–96; and the work of historical geographer Robert D. Mitchell, *Commercialism and Frontier: Perspectives on the Early Shenandoah Valley* (Charlottesville: University Press of Virginia, 1977); "Appalachian Frontiers: The View from the East," in Mitchell and Milton B. Newton, Jr., *The Appalachian Frontier: Views from the East and the Southwest*, Historical Geography Research Series 21 (July 1988), 4–42; "Introduction: Revisionism and Regionalism," in *Appalachian Frontiers*, 1–22; and "The Southern Backcountry: A Geographical House Divided," in *The Southern Colonial Backcountry: Interdisciplinary Perspectives on Frontier Communities*, ed. David Colin Crass et al., 1–35 (Knoxville: University of Tennessee Press, 1998). Another relevant geographical account is in Harry Roy Merrens, *Colonial North Carolina in the Eighteenth Century: A Study in Historical Geography* (Chapel Hill: University of North Carolina Press, 1964), which deals, in good part, with the backcountry. For the historical literature, see Gregory H. Nobles, "Breaking into the Backcountry: New Approaches to the Early American Frontier, 1750–1800," *William and Mary Quarterly* 46, no. 4 (1989): 641–70.

127 Warren R. Hofstra and Robert D. Mitchell, "Town and Country in Backcountry Virginia: Winchester and the Shenandoah Valley, 1730–1800," *Journal of Southern History* 59, no. 4 (1993): 619–46, quotation from 622.

128 Population figures for the southern backcountry are from Mitchell, "Southern Backcountry," 21.

129 Merrens, *Colonial North Carolina*, 53–54.

130 Mitchell, "Southern Backcountry," 17; Merrens, *Colonial North Carolina*, 55–68.

131 Elizabeth A. Kessel, "Germans in the Making of Frederick County, Maryland," in *Appalachian Frontiers*, 87–104.

132 Barbara DeWolfe, *Discoveries of America: Personal Accounts of British Emigrants to North America during the Revolutionary Era* (Cambridge: Cambridge University Press, 1997), 170–86, especially 171; Duane Meyer, *The Highland Scots of North Carolina 1732–1776* (Chapel Hill: University of North Carolina Press, 1961).

133 Mitchell, *Commercialism and Frontier*, 16.

134 James H. Merrell, *The Indians' New World: Catawbas and their Neighbors from Euro-*

pean Contact through the Era of Removal (Chapel Hill: University of North Carolina Press, 1989).

135 Ibid., 195.

136 Price, *Dividing the Land*, 181–82; Mitchell, *Commercialism and Frontier*, 63–65.

137 Mitchell, *Commercialism and Frontier*, 59–92.

138 Ibid., 127–128.

139 Price, *Dividing the Land*, 141–72.

140 Mitchell, *Commercialism and Frontier*, 65, 70.

141 Hofstra and Mitchell, "Town and Country in Backcountry Virginia," 626.

142 "Diary of a Journey of Moravians from Bethlehem, Pennsylvania, to Bethabara in Wachovia, North Carolina, 1753," in *Travels in the American Colonies*, ed. Newton D. Mereness, 325–56 (New York: Macmillan Company, 1916).

143 Mitchell, "Southern Backcountry," 20.

144 Ibid.

145 Mitchell, *Commercialism and Frontier*, 172–73; Merrens, *Colonial North Carolina*, 112–15.

146 "Governor William Bull's Representation of the Colony, 1770," in *The Colonial South Carolina Scene: Contemporary Views, 1697–1774*, ed. H. Roy Merrens, 253–70, quotation from 265 (Columbia: University of South Carolina Press, 1977).

147 Mitchell, *Commercialism and Frontier*, 138–39, 178–79.

148 Richard K. MacMaster, "The Cattle Trade in Western Virginia, 1760–1830," in *Appalachian Frontiers*, 127–49, especially 128.

149 Reverend Samuel R. Houson, quoted in MacMaster, "Cattle Trade in Western Virginia," 142–43.

150 Hudson, *Making the Corn Belt*, especially 63–74.

151 Merrens, *Colonial North Carolina*, 116.

152 MacMaster, "Cattle Trade in Western Virginia," 130; Bridenbaugh, *Myths and Realities*, 138; Merrens, *Colonial North Carolina*, 134–35.

153 Bridenbaugh, *Myths and Realities*, 129; Joshua Fry and Peter Jefferson, *A Map of the Most Inhabited Part of Virginia Containing the Whole Province of Maryland with Part of Pensilvania, New Jersey and North Carolina. Drawn by Joshua Fry and Peter Jefferson in 1751* (London: Thomas Jefferys, 1751).

154 Bridenbaugh, *Myths and Realities*, 144–47; Mitchell, *Commercialism and Frontier*, 189–93.

155 Merrens, *Colonial North Carolina*, 143–45.

156 Bull, quoted in Bridenbaugh, *Myths and Realities*, 147.

157 Hofstra and Mitchell, "Town and Country in Backcountry Virginia"; Robert D. Mitchell and Warren R. Hofstra, "How Do Settlement Systems Evolve? The Virginia Backcountry during the Eighteenth Century," *Journal of Historical Geography* 21, no. 2 (1995): 123–47.

158 Mitchell, *Commercialism and Frontier*, 232.

159 "Governor William Bull's Representation of the Colony, 1770," in *Colonial South Carolina Scene*, 264.

160 Mitchell, *Commercialism and Frontier*, 104.

161 "Governor William Bull's Representation of the Colony, 1770," in *Colonial South Carolina Scene*, 256.

162 Mitchell, *Commercialism and Frontier*, 104.

163 Nobles, "Breaking into the Backcountry," 653; Richard S. Beeman, *The Evolution of the Southern Backcountry: A Case Study of Lunenburg County, Virginia* (Philadelphia: University of Pennsylvania Press, 1984), 108.

164 Edward A. Chappell, "Acculturation in the Shenandoah Valley: Rhenish Houses of the Massanutten Settlement," in *Common Places*, 27–57.

165 Paul B. Touart, "The Acculturation of German-American Building Practices of Davidson County, North Carolina," in *Perspectives in Vernacular Architecture, II*, ed. Camille Wells, 72–80 (Columbia: University of Missouri Press, 1986). See also "Diary of a Journey of Moravians," in *Travels in the American Colonies*.

166 John B. Rehder, "The Scotch-Irish and English in Appalachia," in *To Build in a New Land: Ethnic Landscapes in North America*, ed. Allen G. Noble, 95–118, especially 105–107 (Baltimore: Johns Hopkins University Press, 1992).

167 For the argument in favor of a Swedish origin, see Terry G. Jordan and Matti Kaups, *The American Backwoods Frontier: An Ethnic and Ecological Interpretation* (Baltimore: Johns Hopkins University Press, 1990); for the Germanic origin, see Fred Kniffen and Henry Glassie, "Building in Wood in the Eastern United States: A Time-Place Perspective," *Geographical Review* 56, no. 1 (1966): 40–66.

168 Ensminger, *Pennsylvania Barn*, 76–78; Rehder, "Scotch-Irish and English," 112–15.

169 Price, "Central Courthouse Square."

170 Bridenbaugh refers to the backcountry as "Greater Pennsylvania," in *Myths and Realities*, 127.

171 Cole Harris, "European Beginnings in the Northwest Atlantic: A Comparative View," in *Seventeenth-Century New England*, ed. David D. Hall and David Grayson Allen, 119–52 (Boston: Colonial Society of Massachusetts, 1984); "Simplification of Europe Overseas"; and "Making an Immigrant Society," in *The Resettlement of British Columbia: Essays on Colonialism and Geographical Change* (Vancouver: University of British Columbia Press, 1997).

172 The importance of internal consumption, rather than Atlantic exports, is stressed in Peter C. Mancall and Thomas Weiss, "Was Economic Growth Likely in Colonial British America?" *Journal of Economic History* 59, no. 1 (1999): 17–40, especially 36–37.

5. British American Towns (pages 180–203)

1 Although the fur posts on Hudson Bay were proto-urban rather than urban places, I include them in this discussion because of their singular importance in a vast area of British America.

2 Jacob M. Price, "Economic Function and the Growth of American Port Towns in the Eighteenth Century," *Perspectives in American History* 8 (1974): 123–86, especially 173. See also his "Summation: The American Panorama of Atlantic Port Cities," in *Atlantic Port Cities: Economy, Culture, and Society in the Atlantic World, 1650–1850*, ed. Franklin W. Knight and P. K. Liss, 262–76 (Knoxville: University of Tennessee Press, 1991). Also important for this entire discussion are James E. Vance, Jr., *The Merchant's World: The Geography of Wholesaling* (Englewood Cliffs, N.J.: Prentice-Hall, Inc., 1970); and the various writings of Martyn J. Bowden on the mercantile city, includ-

ing "The Internal Structure of the Colonial Replica City: San Francisco and Others," unpublished paper presented at the Association of American Geographers annual meeting, Kansas City, 1972; and "The Mercantile City in North America: Theory and Reality," unpublished paper presented to the Association of American Geographers annual meeting, San Francisco, 1994.

3 Carville V. Earle, "The First English Towns of North America," *Geographical Review* 67, no. 1 (1977): 34–50.

4 Lois Green Carr, "'The Metropolis of Maryland': A Comment on Town Development along the Tobacco Coast," *Maryland Historical Magazine* 69 (1974): 124–45.

5 Peter Borsay, *The English Urban Renaissance: Culture and Society in the Provincial Town, 1660–1770* (Oxford: Oxford University Press, 1989), 85–101; Gerald Burke, *Towns in the Making* (London: Edward Arnold, 1971), 95–116.

6 Robert Home, *Of Planting and Planning: The Making of British Colonial Cities* (London: E & FN. Spon, 1997), 8–23; James E. Vance, Jr., *The Continuing City: Urban Morphology in Western Civilization* (Baltimore: Johns Hopkins University Press, 1990), 237–81.

7 Colin G. Clarke, *Kingston, Jamaica: Urban Development and Social Change, 1692–1962* (Berkeley: University of California Press, 1975), 8–9; Gail Saunders and Donald Cartwright, *Historic Nassau* (London: Macmillan Publishers, 1979), 5–6; and J. David Wood, "Grand Designs on the Fringes of Empire: New Towns for British North America," *Canadian Geographer* 26, no. 3 (1982): 243–55.

8 Lord Adam Gordon, in Newton D. Mereness, ed., *Travels in the American Colonies* (New York: Macmillan Company, 1916), 377, 398, 410.

9 Stephen Saunders Webb, "The Strange Career of Francis Nicholson," *William and Mary Quarterly* 23, no. 4 (1966): 513–48, especially 521.

10 Edward C. Papenfuse, *"Doing Good to Posterity": The Move of the Capital of Maryland From St. Mary's City to Ann Arundell Towne, Now Called Annapolis* (Annapolis, Md.: Maryland State Archives, 1995).

11 John W. Reps, *Tidewater Towns: City Planning in Colonial Virginia and Maryland* (Williamsburg, Va.: Colonial Williamsburg Foundation, 1972).

12 Arthur Pierce Middleton, *Anglican Maryland, 1692–1792* (Virginia Beach, Va.: Donning Company Publishers, 1992).

13 Reps, *Tidewater Towns*, 132–34.

14 Anne E. Yentsch, *A Chesapeake Family and their Slaves: A Study in Historical Archaeology* (Cambridge: Cambridge University Press, 1994), 97–112.

15 Nicholson, cited in Reps, *Tidewater Towns*, 142–43.

16 Graham Hood, *The Governor's Palace in Williamsburg: A Cultural Study* (Williamsburg, Va.: Colonial Williamsburg, 1991).

17 Carl Bridenbaugh, *Seat of Empire: The Political Role of Eighteenth-Century Williamsburg* (Williamsburg, Va.: Colonial Williamsburg, 1958).

18 Vance, *Continuing City*, 207–51.

19 James E. Vance, Jr., "Land Assignment in the Precapitalist, Capitalist, and Postcapitalist City," *Economic Geography* 47, no. 2 (1971): 101–20, especially 108.

20 Martyn J. Bowden, "Growth of the Central Districts in Large Cities," in *The New Urban History: Quantitative Explorations by American Historians*, ed. Leo F. Schnore, 75–109, especially 107–108 (Princeton: Princeton University Press, 1975).

21 Ibid., 78–83.

22 Vance concludes his great work on merchant wholesaling by observing that "In North America, for the first time, the trading economy could work out its destiny unstructured by confusing inheritances"; see *The Merchant's World*, 167.

23 Bowden, "Internal Structure" and "Mercantile City in North America." For an attempt to extend the model to another region of British commercial activity, see Stephen J. Hornsby, "Discovering the Mercantile City in South Asia: The Example of Early Nineteenth-Century Calcutta," *Journal of Historical Geography* 23, no. 2 (1997): 135–50.

24 Jeanne A. Calhoun, Martha A. Zierden, and Elizabeth A. Paysinger, "The Geographic Spread of Charleston's Mercantile Community, 1732–1767," *South Carolina Historical Magazine* 86, no. 3 (1985): 182–220, especially 189–90.

25 Ibid., 190–91.

26 This is well illustrated in John Montresor, *A Plan of the City of New-York and its Environs* (London: 1775), and in Charles Blaskowitz, *A Plan of the Town of Newport in Rhode Island* (London: William Faden, 1777).

27 For an alternative argument that stresses linkage effects of particular staples, see Carville V. Earle and Ronald Hoffman, "Staple Crops and Urban Development in the Eighteenth-Century South," *Perspectives in Urban History* 10 (1976): 7–78.

28 Kenneth Morgan, "Shipping Patterns and the Atlantic Trade of Bristol, 1749–1770," *William and Mary Quarterly* 46, no. 3 (1989): 506–38, especially 518.

29 D. Wayne Moodie and Victor P. Lytwyn, "Fur-Trade Settlements," in *The Historical Atlas of Canada*, vol. I: *From the Beginning to 1800*, ed. R. Cole Harris, plate 64 (Toronto: University of Toronto Press, 1987).

30 Gordon Handcock and Alan Macpherson, "Trinity, 18th Century," in ibid., plate 26; John Mannion, "St. John's," in ibid., plate 27; Head, *Eighteenth-Century Newfoundland*, 147, 234.

31 Governor Gower, cited in C. Grant Head, *Eighteenth-Century Newfoundland* (Ottawa: Carleton University Press, 1976), 234.

32 Peter Perry, "The Newfoundland Trade: The Decline and Demise of the Port of Poole, 1815–1894," *American Neptune* 28, no. 4 (1968): 275–83.

33 William Eddis, *Letters from America* (Cambridge: Harvard University Press, 1969), 52.

34 Martyn J. Bowden, *The Mercantile Town of Bridgetown, Barbados 1630–1880: From Mini-Entrepôt to Shipping Point?* (Unpublished field guide for the Eastern Historical Geographers Annual Meeting, February 1994).

35 Malcolm H. Stern, "Portuguese Sephardim in the Americas," in *Sephardim in the Americas: Studies in Culture and History*, ed. Martin A. Cohen and Abraham Peck, 141–78 (Tuscaloosa: University of Alabama Press, 1993); Eli Faber, *A Time for Planting: The First Migration 1654–1820* (Baltimore: Johns Hopkins University Press, 1992); and Howard M. Sachar, *Farewell España: The World of the Sephardim Remembered* (New York: Knopf, 1994), 347–64. For Sephardic Jews in Barbados and Jamaica, see Eli Faber, *Jews, Slaves, and the Slave Trade: Setting the Record Straight* (New York: New York University Press, 1998).

36 Clarke, *Kingston, Jamaica*, 143, Table 1. See also B. W. Higman, "Jamaican Port Towns in the Early Nineteenth Century," in *Atlantic Port Cities*, 117–48; and Wilma R. Bailey,

"Kingston 1692–1843: A Colonial City" (Ph.D. diss., University of the West Indies, Mona, 1974). I am grateful to Barry Higman for this reference.

37 Clarke, *Kingston, Jamaica*, 14–15.

38 Bailey, "Kingston, 1692–1843," 131.

39 Mannion, "St. John's."

40 Bowden concludes his study of the *Mercantile Town of Bridgetown* by asking whether this underdevelopment was "the mark of the staple, of the dead hand of colonialism, or both?"

41 Bailey, "Kingston, 1692–1843," 120.

42 Ibid., 169.

43 Bailey describes the distribution of residences in Kingston as follows: "The pattern is very roughly that of a triangle with its apex on the Church in the Square, and vertex on Harbour Street" (ibid., 99).

44 Ibid., 145, 146, 150, 152.

45 Handcock and Macpherson, "Trinity, 18th Century"; Mannion, "St. John's."

46 David Buisseret, *Historic Architecture of the Caribbean* (London: Heinemann, 1980), 46–71.

47 Warren Alleyne and Jill Sheppard, *The Barbados Garrison and its Buildings* (London: Macmillan Caribbean, 1990); David Buisseret, *Historic Jamaica from the Air* (Barbados: Caribbean Universities Press, 1969), 16–32.

48 Bailey, "Kingston, 1692–1843," 181–83; Edward Long, *The History of Jamaica*, vol. 1 (London: 1774; new ed., Frank Cass & Co., 1970), 106.

49 Buisseret, *Architecture of the Caribbean*, 61–62; Desmond V. Nicholson, *The Story of English Harbour Antigua, West Indies* (St. John's, Antigua: Museum of Antigua and Barbuda, 1991).

50 Handcock and Macpherson, "Trinity, 18th Century"; Mannion, "St. John's."

51 Bailey, "Kingston, 1692–1843," 137–38, 204–206.

52 Ibid., 184.

53 Price, "American Port Towns," 143, 149; Darrett B. Rutman, *Winthrop's Boston: A Portrait of a Puritan Town, 1630–1649* (New York: W.W. Norton, 1972).

54 Carl Bridenbaugh, *Cities in the Wilderness: The First Century of Urban Life in America, 1625–1742* (New York: Knopf, 1955), 6; Richard S. Dunn, *Sugar and Slaves: The Rise of the Planter Class in the English West Indies, 1624–1713* (New York: W.W. Norton, 1973), 180.

55 Price, "American Port Towns," 173.

56 Ibid., Appendix B, and Sam Bass Warner, *The Private City: Philadelphia in Three Periods of Growth* (Philadelphia: University of Pennsylvania Press, 1968), 12.

57 Jay Coughtry, *The Notorious Triangle: Rhode Island and the African Slave Trade* (Philadelphia: Temple University Press, 1981).

58 Samuel Eliot Morison, cited in Vance, *The Merchant's World*, 71.

59 John J. McCusker and Russell R. Menard, *The Economy of British America 1607–1789* (Chapel Hill: University of North Carolina Prss, 1985), 192, Table 9.1.

60 Edward C. Papenfuse, *In Pursuit of Profit: The Annapolis Merchants in the Era of the American Revolution, 1763–1805* (Baltimore: Johns Hopkins University Press, 1975), especially 35–75. For growth of the grain trade and urban development, see Earle and Hoffman, "Staple Crops and Urban Development," 7–78, especially 30–32; and Peter

V. Bergstrom, *Markets and Merchants: Economic Diversification in Colonial Virginia 1700–1775* (New York: Garland Publishing, Inc., 1985), 213–19.

61 Kenneth Morgan, "The Organization of the Colonial Rice Trade," *William and Mary Quarterly* 52, no. 3 (1995): 433–52; R. C. Nash, "South Carolina and the Atlantic Economy in the Late Seventeenth and Eighteenth Centuries," *Economic History Review* 45, no. 4 (1992): 677–702.

62 Carville Earle and Ronald Hoffman, "The Urban South: The First Two Centuries," in *The City in Southern History: The Growth of Urban Civilization in the South*, ed. Blaine A. Brownell and David R. Goldfield, 23–51, especially 51 (Port Washington, N.Y.: Kennikat Press, 1977).

63 R. C. Nash, "Urbanization in the Colonial South: Charleston, South Carolina as a Case Study," *Journal of Urban History* 19, no. 1 (1992): 3–29, especially 19–21; Nash, "Trade and Business in Eighteenth-Century South Carolina: The Career of John Guerard, Merchant and Planter," *South Carolina Historical Magazine* 96, no. 1 (1995): 6–29, especially 28.

64 James F. Shepherd and Samuel H. Williamson, "The Coastal Trade of the British North American Colonies, 1768–1772," *Journal of Economic History* 32, no. 4 (1972): 783–810.

65 Price, "American Port Towns," 138.

66 David Ralph Meyer, "A Dynamic Model of the Integration of Frontier Urban Places into the United States System of Cities," *Economic Geography* 56, no. 2 (1980): 120–40.

67 Thomas M. Doerflinger, *A Vigorous Spirit of Enterprise: Merchants and Economic Development in Revolutionary Philadelphia* (Chapel Hill: University of North Carolina Press, 1986), 122–23.

68 Bridenbaugh, *Cities in the Wilderness*, 41–42.

69 M. H. Watkins, "A Staple Theory of Economic Growth," in *Approaches to Canadian Economic History*, ed. W. T. Easterbrook and M. H. Watkins, 49–73 (Toronto: McClelland and Stewart, 1967); and Albert O. Hirschman, "A Generalized Linkage Approach to Development, with Special Reference to Staples," in *Essays on Economic Development and Cultural Change in Honor of Bert F. Hoselitz*, ed. M. Nash, 67–98 (Chicago: Chicago University Press, 1977).

70 *Historical Statistics of the United States: Colonial Times to 1970* (Washington, D.C.: U.S. Bureau of the Census, 1975), 1195, Series Z 516-529, "Vessels Built in Thirteen Colonies and West Florida: 1769–1771."

71 Carl Bridenbaugh, *The Colonial Craftsman* (New York: New York University Press, 1950), 93.

72 McCusker and Menard, *Economy of British America*, 290–93.

73 Robert E. Baldwin, "Patterns of Development in Newly Settled Regions," *Manchester School of Economic and Social Studies* 24, no. 2 (1956): 161–79.

74 Bridenbaugh, *Colonial Craftsman*, 120–23.

75 Ibid., 67–69.

76 Ibid., 95.

77 Papenfuse, *In Pursuit of Profit*, 16–27, especially 17. For general context, see Richard L. Bushman, *The Refinement of America: Persons, Houses, Cities* (New York: Vintage Books, 1993), 110–22.

78 Bridenbaugh, *Colonial Craftsman*, 76–84.
79 Bridenbaugh, *Cities in the Wilderness*, 42–43.
80 For discussion of exchanges, see Fernand Braudel, *Civilization and Capitalism 15th–18th Century*, vol. II: *The Wheels of Commerce* (Berkeley: University of California Press, 1992) 97–100.
81 N. S. B. Gras, *An Introduction to Economic History* (New York: Harper and Brothers Publishers, 1922), 181–209; and Michael P. Conzen, "The Maturing Urban System in the United States, 1840–1910," *Annals of the Association of American Geographers* 67, no. 1 (1977): 88–108.
82 Bowden, "Mercantile City in North America," n.p. Vance calls this function the "information complex" in *The Merchant's World*, 148–49.
83 Bridenbaugh, *Cities in the Wilderness*, 30–34.
84 Ibid., 38, 109.
85 Ibid., 267–69, 432.
86 Robert Earle Graham, "The Taverns of Colonial Philadelphia," in *Historic Philadelphia from the Founding until the Early Nineteenth Century*, ed. Luther P. Eisenhart, 318–25, especially 320–21 (Philadelphia: Transactions of the American Philosophical Society, 1953).
87 Carl Bridenbaugh, *Cities in Revolt: Urban Life in America 1743–1776* (New York: Capricorn Books, 1962), 22; Jonathan H. Poston, *The Buildings of Charleston: A Guide to the City's Architecture* (Columbia: University of South Carolina Press, 1997), 109–11.
88 Bridenbaugh, *Cities in the Wilderness*, 360; *Cities in Revolt*, 94.
89 Calhoun et al., "Charleston's Mercantile Community"; John P. Radford, "Testing the Model of the Pre-Industrial City: The Case of Ante-bellum Charleston, South Carolina," *Transactions of the Institute of British Geographers* 4, no. 3 (1979): 392–410.
90 "Charleston at the End of the Colonial Era, 1774," in *Colonial South Carolina Scene*, 281.
91 "Governor James Glen's Valuation, 1751," in *Colonial South Carolina Scene*, 181.
92 Edmund Petrie, *Ichnography of Charleston, South Carolina* (London: 1788).
93 Calhoun et al., "Charleston's Mercantile Community," 187–89; "Charleston at the End of the Colonial Era, 1774," in *Colonial South Carolina Scene*, 281.
94 "Charleston at the End of the Colonial Era, 1774," in *Colonial South Carolina Scene*, 282. Perhaps influenced by Price's discussion of Charleston in "American Port Towns," Bowden argues in "The Mercantile City in North America" that Charleston was "a big 'shipping point'" with a stunted financial-exchange-communications sector. More recent research by Nash suggests a more significant mercantile community.
95 Poston, *Buildings of Charleston*, 155–56.
96 Ibid., 197–291, especially 228–29.
97 Calhoun et al., "Charleston's Mercantile Community," 191.
98 Borsay, *English Urban Renaissance*, especially 41–59.
99 Bushman, *Refinement of America*, 100–80.
100 For background, see Bushman, *Refinement of America*, 110–38.
101 Bernard Herman, "The Charleston Single House," in *Buildings of Charleston*, 37–41.
102 Marcia M. Miller and Orlando Ridout V, eds., *Architecture in Annapolis: A Field Guide* (Crownsville, Md.: Maryland Historical Trust, 1998), 12.

103 R. Tittler, *Architecture and Power: The Town Hall and the English Urban Community c.1500–1640* (Cambridge: Cambridge University Press, 1992).

104 Bernard Bailyn, *The New England Merchants in the Seventeenth Century* (Cambridge: Harvard University Press, 1955), 97–98.

105 William H. Pierson, Jr., *American Buildings and their Architects*, vol. I: *The Colonial and Neoclassical Styles* (New York: Oxford University Press, 1986), 94–105, 131–40.

106 "Charleston at the End of the Colonial Era, 1774," in *Colonial South Carolina Scene*, 283.

107 Cited in Webb, "Strange Career of Francis Nicholson," 528.

108 Price, "American Port Towns," Appendices C, D, and E, 177–85.

6. The Fracturing of British America (pages 204–38)

1 John Brewer, *The Sinews of Power: War, Money and the English State, 1688–1783* (New York: Knopf, 1989), especially 165–90.

2 Kenneth Donovan, "Île Royale, 18th Century," in *The Historical Atlas of Canada*, vol. I: *From the Beginning to 1800*, ed. R. Cole Harris, plate 24 (Toronto: University of Toronto Press, 1987).

3 Jean Daigle, "Acadian Marshland Settlement," in *Historical Atlas of Canada*, vol. I, plate 29. See also Andrew Hill Clark, *Acadia: The Geography of Early Nova Scotia to 1760* (Madison: University of Wisconsin Press, 1968); and John G. Reid, "1686–1720: Imperial Intrusions," in *The Atlantic Region to Confederation: A History*, ed. Phillip A. Buckner and John G. Reid, 78–103 (Toronto: University of Toronto Prss, 1994); George Rawlyk, "1720–1744: Cod, Louisbourg, and the Acadians," in ibid., 107–24; and Stephen E. Patterson, "1744–1763: Colonial Wars and Aboriginal Peoples," in ibid., 125–55.

4 Elizabeth Mancke, "Imperial Transitions," in *The "Conquest" of Acadia, 1710: Imperial, Colonial, and Aboriginal Constructions*, ed. John G. Reid et al. (Toronto: University of Toronto Press, 2004).

5 Graeme Wynn, "Pre-Loyalist Nova Scotia," in *Historical Atlas of Canada*, vol. I, plate 31; Steven G. Greiert, "The Earl of Halifax and the Settlement of Nova Scotia, 1749–1753," *Nova Scotia Historical Review* 1 (1981): 4–23.

6 William A. B. Douglas, "Nova Scotia and the Royal Navy 1713–1766" (Ph.D. diss., Queen's University, 1973), 297.

7 Julian Gwyn, *Excessive Expectations: Maritime Commerce and the Economic Development of Nova Scotia, 1740–1870* (Montreal: McGill-Queen's University Press, 1998), 28–30; for the larger context, see Gwyn, "British Government Spending and the North American Colonies 1740–1775," in *The British Atlantic Empire before the American Revolution*, edited by Peter Marshall and Glyn Williams, 74–84 (London: Frank Cass, 1980).

8 Winthrop Pickard Bell, *The "Foreign Protestants" and the Settlement of Nova Scotia: The History of a Piece of Arrested British Colonial Policy in the Eighteenth Century* (Toronto: University of Toronto Press, 1961).

9 Clark, *Acadia*, 344; Joan Dawson, *The Mapmaker's Eye: Nova Scotia through Early Maps* (Halifax, N.S.: Nimbus Publishing Ltd. and The Nova Scotia Museum, 1988), 119–21.

10 Captain John Knox, *An Historical Journal of the Campaigns in North America for the Years 1757, 1758, 1759, and 1760*, vol. I (Toronto: Champlain Society, 1914), 99.

11 Emerson W. Baker and John G. Reid, "Imperialism, Colonialism, and American Coercive Power in the Early Modern Northeast: A Reconsideration," paper presented at the Seventh Annual Conference of The Omohundro Institute of Early American History and Culture, Glasgow, 2001.

12 Conrad E. Heidenreich, "Expansion of French trade, 1667–1696," in *Historical Atlas of Canada*, vol. I, plate 38; Heidenreich and Françoise Noël, "Trade and Empire, 1697–1739," in ibid., plate 39; Heidenreich and Noël, "France Secures the Interior, 1740–1755," in ibid., plate 40; Heidenreich et al., "French Interior Settlements, 1750s," in ibid., plate 41. See also R. Cole Harris, "France in North America," in *North America: The Historical Geography of a Changing Continent*, 2nd ed., ed. Thomas F. McIlwraith and Edward K. Muller, 65–88 (Lanham, Md.: Rowman & Littlefield Publishers, 2001).

13 Winstanley Briggs, "Le Pays des Illinois," *William and Mary Quarterly* 47, no. 1 (1990): 30–56.

14 W. J. Eccles, "The Fur Trade and Eighteenth-Century Imperialism," *William and Mary Quarterly* 40, no. 3 (1983): 341–62.

15 Eric Hinderaker, *Elusive Empires: Constructing Colonialism in the Ohio Valley, 1673–1800* (Cambridge: Cambridge University Press, 1997), 31, 51.

16 Fred Anderson, *Crucible of War: The Seven Year's War and the Fate of Empire in British North America, 1754–1766* (New York: Alfred A. Knopf, 2000), 22–32; and Kenneth P. Bailey, *The Ohio Company of Virginia and the Westward Movement 1748–1792* (Glendale, Calif.: Arthur H. Clark Company, 1939).

17 Marc Egnal, *A Mighty Empire: The Origins of the American Revolution* (Ithaca, N.Y.: Cornell University Prss, 1988), 87–101.

18 Steven G. Greiert, "The Earl of Halifax and British Colonial Policy: 1748–1756" (Ph.D. diss., Duke University, 1976), especially 263–311. More generally, see Ian K. Steele, "The Anointed, the Appointed, and the Elected: Governance of the British Empire, 1689–1784," in *The Oxford History of the British Empire*, vol. II: *The Eighteenth Century*, ed. P. J. Marshall, 105–27, especially 119–21 (Oxford: Oxford University Press, 1998).

19 Hinderaker, *Elusive Empires*, 139.

20 Apart from the narrative account of the war in Anderson, *Crucible of War*, see the layout of the various campaigns in W. J. Eccles and Susan L. Laskin, "The Seven Years' War," in *Historical Atlas of Canada*, vol. I, plate 42.

21 Louis M. Waddell and Bruce D. Bomberger, *The French and Indian War in Pennsylvania 1753–1763: Fortification and Struggle during the War for Empire* (Harrisburg: Pennsylvania Historical and Museum Commission, 1996), 35–49; and Stephen Brumwell, *Redcoats: The British Soldier and War in the Americas, 1755–1763* (Cambridge: Cambridge University Press, 2002), 141–42.

22 Anderson, *Crucible of War*, 267–85.

23 Charles Morse Stotz, *The Model of Fort Pitt: A Description and Brief Account of Britain's Greatest American Stronghold* (Pittsburgh: Allegheny Conference on Community Development, 1970).

24 Daniel A. Baugh, "Great Britain's 'Blue-Water' Policy, 1689–1815," *International History Review* 10, no. 1 (1988): 33–58; and "Withdrawing from Europe: Anglo-French

Maritime Geopolitics, 1750–1800," *International History Review* 20, no. 1 (1998): 1–32.

25 Richard Middleton, "British Naval Strategy, 1755–1762," *Mariner's Mirror* 75, no. 4 (1989): 349–67; Michael Duffy, "The Establishment of the Western Squadron as the Linchpin of British Naval Strategy," in *Parameters of British Naval Power, 1650–1850* (Exeter: University of Exeter Press, 1992), 60–81.

26 Jean Daigle and Robert LeBlanc, "Acadian Deportation and Return," in *Historical Atlas of Canada*, vol. I, plate 30. Although some historians have argued that the decision to deport the Acadians was because of pressure from New England, George A. Rawlyk argues that it was solely the responsibility of the Nova Scotia military governor; see Rawlyk, *Nova Scotia's Massachusetts: A Study of Massachusetts-Nova Scotia Relations 1630 to 1784* (Montreal: McGill-Queen's University Press, 1973), 212–13.

27 Geoffrey Plank, "New England Soldiers in the St. John River Valley, 1758–1760," in *New England and the Maritime Provinces: Connections and Comparisons*, ed. Stephen J. Hornsby and John G. Reid (Montreal and Kingston: McGill-Queen's University Press, forthcoming); and M. A. MacDonald, *Rebels and Royalists: The Lives and Material Culture of New Brunswick's Early English-Speaking Settlers 1758–1783* (Fredericton, N.B.: New Ireland Press, 1990), 13–23.

28 Douglas, "Nova Scotia and the Royal Navy," 297–352.

29 W. J. Eccles and Susan L. Laskin, "The Battles for Québec, 1759 and 1760," in *Historical Atlas of Canada*, vol. I, plate 43.

30 Middleton, "British Naval Strategy, 1755–1762."

31 Anderson, *Crucible of War*, 377–84; David Lyon, *Sea-Battles in Close-Up: The Age of Nelson* (Annapolis, Md.: Naval Institute Press, 1996), 29–44.

32 Gilles Proulx, *Fighting at Restigouche: The Men and Vessels of 1760 in Chaleur Bay* (Ottawa: Parks Canada, 1999); Walter Zacharchuk and Peter J. A. Waddell, *The Excavation of the Machault: An 18th-Century French Frigate* (Ottawa: Parks Canada, 1986).

33 Pierre Tousignant, "The Integration of the Province of Quebec into the British Empire, 1763–91," in *Dictionary of Canadian Biography*, vol. IV: *1771–1800* (Toronto: University of Toronto Press, 1979), xxxii–xlix; Hilda Neatby, *Quebec: The Revolutionary Age 1760–1791* (Toronto: McClelland and Stewart Limited, 1966).

34 Chris Tabraham and Doreen Grove, *Fortress Scotland and the Jacobites* (London: B.T. Batsford/Historic Scotland, 1995), especially 87–108.

35 Yolande O'Donoghue, *William Roy, 1726–1790, Pioneer of the Ordnance Survey* (London: British Museum Publications, 1977).

36 Lester Jesse Cappon, "Geographers and Map-makers, British and American, from about 1750 to 1789," *Proceedings of the American Antiquarian Society* n.s. 81, pt. 2 (1971): 243–71; J. B. Harley, "The Contemporary Mapping of the American Revolutionary War," in *Mapping the American Revolutionary War*, ed. Harley, Barbara Bartz Petchenik, and Lawrence W. Towner (Chicago: University of Chicago Press, 1978), 1–44, especially 25–29; and Douglas W. Marshall, "The British Engineers in America: 1755–1783," *Journal of the Society for Army Historical Research* 51 (1973): 155–63. For the northern surveys, see John Clarence Webster, "Joseph Frederick Wallet Des Barres and the Atlantic Neptune," *Transactions of the Royal Society of Canada*, 3rd series, 21, no. 2 (1927): 21–40; and G. N. D. Evans, *Uncommon Obdurate: The Several Public Careers of J. F. W. DesBarres* (Salem, Mass.: Peabody Museum of Salem, 1969).

For the southern surveys, see Louis De Vorsey, Jr., ed., *De Brahm's Report of the General Survey in the Southern District of North America* (Columbia: University of South Carolina Press, 1971).

37 Nathaniel N. Shipton, "General James Murray's Map of the St. Lawrence," *The Cartographer* 4, no. 2 (1967): 93–101.

38 Michael Charlesworth, "Thomas Sandby Climbs the Hoober Stand: The Politics of Panoramic Drawing in Eighteenth-Century Britain," *Art History* 19, no. 2 (1996): 247–66; Jessica Christian, "Paul Sandby and the Military Survey of Scotland," in *Mapping the Landscape: Essays on Art and Cartography*, ed. Nicholas Alfrey and Stephen Daniels, 18–22 (Nottingham: University Art Gallery, 1990); and James Holloway and Lindsay Errington, *The Discovery of Scotland* (Edinburgh: National Gallery of Scotland, 1978): 32–46. Naval officers were also trained in topographical drawing; see John O. Sands, "The Sailor's Perspective: British Naval Topographic Artists," in *Background to Discovery: Pacific Exploration from Dampier to Cook*, ed. Derek Howse, 185–200 (Berkeley: University of California Press, 1990).

39 Charlesworth, "Thomas Sandby Climbs the Hoober Stand," 261; Bruce Robertson, "Venit, Vidit, Depinxit: The Military Artist in America," in *Views and Visions: American Landscape before 1830*, ed. Edward J. Nygren, 83–103 (Washington, D.C.: Corcoran Gallery of Art, 1986).

40 Reprinted in Anderson, *Crucible of War*, 421–50. See also F. St. George Spendlove, *The Face of Early Canada: Pictures of Canada Which Have Helped to Make History* (Toronto: Ryerson Press, 1958), 5–21.

41 R. H. Hubbard, ed., *Thomas Davies in Early Canada* (Ottawa: Oberon Press, 1972).

42 Colin M. Coates, "Like 'The Thames towards Putney': The Appropriation of Landscape in Lower Canada," *Canadian Historical Review* 74, no. 3 (1993): 317–434; P. Jasen, "Romanticism, Modernity, and the Evolution of Tourism on the Niagara Frontier, 1790–1850," *Canadian Historical Review* 72, no. 3 (1991): 283–318; and R. G. Moyles and Doug Owram, *Imperial Dreams and Colonial Realities: British Views of Canada, 1880–1914* (Toronto: University of Toronto Press, 1988), 87–113.

43 Paul Hulton, *America 1585: The Complete Drawings of John White* (London: British Museum, 1984); Amy R. W. Meyers and Margaret Beck Pritchard, *Empire's Nature: Mark Catesby's New World Vision* (Chapel Hill: University of North Carolina Press, 1998); and, more generally, Raymond Phineas Stearns, *Science in the British Colonies of America* (Urbana: University of Illinois Press, 1970).

44 A. M. Lysaght, *Joseph Banks in Newfoundland and Labrador, 1766: His Diary, Manuscripts and Collections* (Berkeley: University of California Press, 1971). See also John Gascoigne, *Science in the Service of Empire: Joseph Banks, the British State and the Uses of Science in the Age of Revolution* (Cambridge: Cambridge University Press, 1998), 34–40.

45 Lysaght, *Joseph Banks in Newfoundland*, 47.

46 For an overview, see Tony Ballantyne, "Empire, Knowledge and Culture: From Proto-Globalization to Modern Globalization," in *Globalization in World History*, ed. A. G. Hopkins, 115–40 (London: Pimlico, 2002). See also Daniel W. Clayton, *Islands of Truth: The Imperial Fashioning of Vancouver Island* (Vancouver: University of British Columbia Press, 2000).

47 Jack M. Sosin, *Whitehall and the Wilderness: The Middle West in British Colonial Policy, 1760–1775* (Lincoln: University of Nebraska Press, 1961); David Milobar, "Quebec

Reform, the British Constitution and the Atlantic Empire: 1774–1775," in *Parliament and the Atlantic Empire*, ed. Philip Lawson, 65–88, especially 70 (Edinburgh; Edinburgh University Press, 1995). See also Vincent T. Harlow, *The Founding of the Second British Empire 1763–1793*, vol. I: *Discovery and Revolution* (London: Longmans, Green and Co., 1952), 162–98.

48 Anderson, *Crucible of War*, 535–71.

49 W. J. Eccles, M. N. McConnell, and Susan L. Laskin, "Indian War and American Invasion," in *Historical Atlas of Canada*, vol. I, plate 44; Anderson, *Crucible of War*, 535–53, 617–32.

50 Mario Lalancette, "Exploitation of the Gulf of St. Lawrence," in *Historical Atlas of Canada*, vol. I, plate 54; Rosemary Ommer, *From Outpost to Outport: A Structural Analysis of the Jersey-Gaspé Cod Fishery, 1767–1886* (Montreal and Kingston: McGill-Queen's University Press, 1991), 24–26; David Lee, *The Robins in Gaspé 1766 to 1825* (Markham, Ont.: Fitzhenry & Whiteside, 1984); and Lee, *Gaspé, 1760–1867*, Canadian Historic Sites: Occasional Papers in Archaeology and History no. 23 (Ottawa: National Historic Parks and Sites, 1980); and Stephen J. Hornsby, *Nineteenth-Century Cape Breton: A Historical Geography* (Montreal and Kingston: McGill-Queen's University Press, 1992), 4–15.

51 José Igartua, "A Change in Climate: The Conquest and the *Marchands* of Montreal," in *Historical Papers 1974* (Ottawa: Canadian Historical Association, 1974), 115–134; Harry W. Duckworth, "British Capital in the Fur Trade: John Strettell and John Fraser," in *The Fur Trade Revisited: Selected Papers of the Sixth North American Fur Trade Conference Mackinac Island, Michigan, 1991*, ed. Jennifer S. Brown et al., 39–56 (East Lansing: Michigan State University Press, 1994).

52 Alexander Henry, *Travels and Adventures in Canada and the Indian Territories between the Years 1760 and 1776* (Edmonton: Hurtig, 1969), 320.

53 Harold A. Innis, *The Fur Trade in Canada* (New Haven: Yale University Press, 1930), 194.

54 Joseph Robson, *An Account of Six Years Residence in Hudson's Bay, from 1733 to 1736, and 1744 to 1747* (New York: Johnson Reprint Corp., 1965), 6.

55 Henry, *Travels and Adventures*, 41. For the routes, see Eric W. Morse, *Fur Trade Canoe Routes of Canada/Then and Now* (Toronto: University of Toronto Press, 1969).

56 Arthur J. Ray, D. Wayne Moodie, and Conrad E. Heidenreich, "Rupert's Land," in *Historical Atlas of Canada*, vol. I, plate 57; and D. Wayne Moodie, Victor P. Lytwyn, Barry Kaye, and Arthur J. Ray, "Competition and Consolidation, 1760–1825," in ibid., plate 61.

57 Ferdinand Jacobs, cited in Glyndwr Williams, ed., *Andrew Graham's Observations on Hudson's Bay, 1767–91* (London: Hudson's Bay Record Society, 1969), lxiv.

58 D. Wayne Moodie, Barry Kaye, Victor P. Lytwyn and Arthur J. Ray (Indians), "Peoples of the Boreal Forest and Parkland," in *Historical Atlas of Canada*, vol. I, plate 65; Allan Greer, *Peasant, Lord, and Merchant: Rural Society in Three Quebec Parishes, 1740–1840* (Toronto: University of Toronto Press, 1985), 178–80.

59 Neatby, *Quebec*, 75–86.

60 David Milobar, "Conservative Ideology, Metropolitan Government, and the Reform of Quebec, 1782–1791," *International History Review* 12, no. 1 (1990): 45–64; see also Neatby, *Quebec*, 144.

61 The power and influence of the metropolis is a pervasive theme in the work of Innis,

Careless, and many other Canadian scholars. See also Jack Greene, "Negotiated Authorities: *The Problem of Governance in the Extended Polities of the Early Modern World*," in *Negotiated Authorities: Essays in Colonial Political and Constitutional History* (Charlottesville: University Press of Virginia, 1994), 1–24, especially 20–21; and Elizabeth Mancke, "Another British America: A Canadian Model for the Early Modern British Empire," *Journal of Imperial and Commonwealth History* 25, no. 1 (1997): 1–36.

62 Cited in Eliga H. Gould, *The Persistence of Empire: British Political Culture in the Age of the American Revolution* (Chapel Hill: University Press of North Carolina, 2000), 106.

63 Philip J. Ayres, *Classical Culture and the Idea of Rome in Eighteenth-Century England* (Cambridge: Cambridge University Press, 1997).

64 For some rumination about including India in the British Atlantic, see David Hancock, "'An Undiscovered Ocean of Commerce Laid Open': India, Wine and the Emerging Atlantic Economy, 1703–1813," in *The Worlds of the East India Company*, ed. H. V. Bowen, Margarette Lincoln, and Nigel Rigby, 153–68 (Woodbridge: Boydell Press, 2002).

65 Historian Eric Robson observed that in 1815 "when Great Britain was the world's largest colonial power, hers was a sea and not a land Empire—composed of trading posts, islands, ports of call and settlements on sea coasts or within easy reach of the sea"; see Robson, *The American Revolution in its Political and Military Aspects 1763–1783* (New York: Oxford University Press, 1955), 234. It seems to me that these characteristics were well entrenched by the 1760s, with the anomalous exception of the thirteen colonies.

66 Ralph Davis, "English Foreign Trade, 1700–1774," *Economic History Review* 15 (1963): 300–301. See also Jacob M. Price, "The Imperial Economy, 1700–1776," in *The Eighteenth Century*, 78–104.

67 Davis, "English Foreign Trade, 1700–1774," 302–303.

68 H. V. Bowen, *Elites, Enterprise and the Making of the British Overseas Empire 1688–1775* (New York: St. Martin's Press, 1996).

69 Adams cited in Rawlyk, *Nova Scotia's Massachusetts*, 230.

70 H. V. Bowen, "British Conceptions of Global Empire, 1756–83," *Journal of Imperial and Commonwealth History* 26, no. 3 (1998): 1–27.

71 Linda Colley, *Britons: Forging the Nation 1707–1837* (New Haven: Yale University Press, 1992), 11.

72 Herman R. Friis, *A Series of Population Maps of the Colonies and the United States 1625–1790* (New York: American Geographical Society, 1968), especially "Distribution of Population" in 1770.

73 The expansionist vision of certain American elites is well treated in Egnal, *Mighty Empire*.

74 These arguments were put forward in the late 1770s by the British imperial official William Knox; see Jack P. Greene, "William Knox's Explanation for the American Revolution," *William and Mary Quarterly* 30, no. 2 (1973): 293–306; and "The Deeper Roots of Colonial Discontent: William Knox's Structural Explanation for the American Revolution," in *Understanding the American Revolution: Issues and Actors* (Charlottesville: University Press of Virginia, 1995), 10–17.

75 J. C. D. Clark, *The Language of Liberty 1660–1832: Political Discourse and Social*

Dynamics in the Anglo-American World (Cambridge: Cambridge University Press, 1994); Peter M. Doll, *Revolution, Religion, and National Identity: Imperial Anglicanism in British North America, 1745–1795* (Madison, N.J.: Fairleigh Dickinson University Press, 2000).

76 Knox, cited in Greene, "William Knox's Explanation," 303, 304, 305.

77 As Robson once observed, "The British, thinking of the Empire as a whole, as one unit, and largely ignorant of colonial conditions, never gave sufficient consideration to the existing and rapidly developing differences between the component parts"; Robson, *American Revolution*, 43.

78 Nancy F. Koehn, *The Power of Commerce: Economy and Governance in the First British Empire* (Ithaca: Cornell University Press, 1994), 5.

79 Anderson, *Crucible of War*, 562.

80 Koehn, *Power of Commerce*, 12.

81 John Shy, *Toward Lexington: The Role of the British Army in the Coming of the American Revolution* (Princeton: Princeton University Press, 1965), 241; Nicholas Tracy, *Navies, Deterrence, and American Independence: Britain and Seapower in the 1760s and 1770s* (Vancouver: University of British Columbia Press, 1988), 8–41; N. A. M. Rodger, *The Insatiable Earl: A Life of John Montagu, 4th Earl of Sandwich* (New York: W.W. Norton & Co., 1993), 135–37.

82 Koehn, *Power of Commerce*.

83 One British official noted that the 1763 peace treaty ensured that the American colonists were "accessible to us on both sides" (that is, from the Atlantic and from the interior); Ellis, cited in Shy, *Toward Lexington*, 68.

84 Woody Holton, *Forced Founders: Indians, Debtors, Slaves, and the Making of the American Revolution in Virginia* (Chapel Hill: University of North Carolina Press, 1999), 29.

85 Neil R. Stout, *The Royal Navy in America, 1760–1775: A Study of Enforcement of British Colonial Policy in the Era of the American Revolution* (Annapolis, Md.: Naval Institute Press, 1973). See also Robert Gardiner, ed., *Navies and the American Revolution 1775–1783* (Annapolis, Md.: Naval Institute Press, 1996).

86 Gage, cited in Sosin, *Whitehall in the Wilderness*, 91. Maps showing the disposition of British troops in North America in the 1760s and early 1770s are in Shy, *Toward Lexington*, 97, 112, 238, 328, 419.

87 Jesse Lemisch, "Jack Tar in the Streets: Merchant Seamen in the Politics of Revolutionary America," *William and Mary Quarterly* 25, no. 3 (1968): 371–407.

88 T. H. Breen, *Tobacco Culture: The Mentality of the Great Tidewater Plantations on the Eve of Revolution* (Princeton: Princeton University Press, 1985), 160–203; Holton, *Forced Founders*.

89 Robert W. Tucker and David C. Henderson, *The Fall of the First British Empire: Origins of the War of American Independence* (Baltimore: Johns Hopkins University Press, 1982), 319–54.

90 Tousignant, "Integration of Quebec"; Neatby, *Quebec*, 125–41.

91 Bernard Bailyn, *The Ideological Origins of the American Revolution* (Cambridge: Harvard University Press, 1967).

92 Andrew J. O'Shaughnessy, "The Stamp Act Crisis in the British Caribbean," *William and Mary Quarterly* 51, no. 2 (1994): 203–26; Wilfred B. Kerr, "The Stamp Act in Nova Scotia," *New England Quarterly* 6 (1933): 552–66.

93 For an excellent treatment of the war, see Piers Mackesy, *The War for America 1775–1783* (Lincoln: University of Nebraska Press, 1993).

94 Brendan Morrissey, *Boston 1775: The Shot Heard Around the World* (London: Osprey Publishing, 1993).

95 David Syrett, *The Royal Navy in American Waters 1775–1783* (Aldershot: Scolar Press, 1989). The failure of the British to exploit the tactical flexibility of the navy is discussed in Robson, *American Revolution*, 107.

96 Brendan Morrissey, *Saratoga 1777: Turning Point of a Revolution* (London: Osprey Publishing, 2000). The difficulties of operating away from rivers and the coast, where land forces could be supplied by sea, are discussed in Robson, *American Revolution*, 105–106.

97 David Syrett, "Home Waters or America? The Dilemma of British Naval Strategy in 1778," *Mariner's Mirror* 77, no. 4 (1991): 365–77.

98 Daniel A. Baugh, "Why Did Britain Lose Command of the Sea during the War for America?" in *The British Navy and the Use of Naval Power in the Eighteenth Century*, ed. Jeremy Black and Philip Woodfine, 149–69 (Atlantic Highlands, N.J.: Humanities Press International, Inc., 1989). See also William B. Willcox, "The British Road to Yorktown: A Study in Divided Command," *American Historical Review* 52, no. 1 (1946): 1–35.

99 David Syrett, "Count-Down to the Saints: A Strategy of Detachments and the Quest for Naval Superiority in the West Indies, 1780–2," *Mariner's Mirror* 87, no. 2 (2001): 150–62.

100 Graeme Wynn and L. D. McCann, "Maritime Canada, Late 18th Century," in *Historical Atlas of Canada Volume*, vol. I, plate 32; Hornsby, *Nineteenth-Century Cape Breton*, 19–25.

101 R. Louis Gentilcore, Don Measner, and David Doherty, "The Coming of the Loyalists," in *Historical Atlas of Canada*, vol. II: *The Land Transformed, 1800–1891*, ed. R. Louis Gentilcore, plate 7 (Toronto: University of Toronto Press, 1993).

102 Michael Craton and Gail Saunders, *Islanders in the Stream: A History of the Bahamian People*, vol. I: *From Aboriginal Times to the End of Slavery* (Athens: University of Georgia Press, 1992), 179–95.

103 Ann Gorman Condon, "1783–1800: Loyalist Arrival, Acadian Return, Imperial Reform," in *Atlantic Region to Confederation*, 184–209, especially 192–93.

104 Ibid., 195, 198.

105 Harold Kalman, *A History of Canadian Architecture*, vol. I (Toronto: Oxford University Press, 1994), 135.

106 P. J. Marshall, "Britain Without America—A Second Empire?" in *The Eighteenth Century*, 576–95, especially 588, 590.

107 Elizabeth Mancke, "Early Modern Imperial Governance and the Origins of Canadian Political Culture," *Canadian Journal of Political Science* 32, no. 1 (1999): 3–20.

108 John J. McCusker and Russell R. Menard, *The Economy of British America 1607–1789* (Chapel Hill: University of North Carolina Press, 1985), 370.

109 William H. Pierson, Jr., *American Buildings and their Architects*, vol. I: *Colonial and Neoclassical Styles* (Oxford: Oxford University Press, 1970), 297.

Index

Page numbers in *italics* refer to maps and illustrations.